Praise for

"What did it take to organize the export of natural gas from the USSR to Europe, exports that reached 100 billion cubic meters annually? As Per Högselius shows in *Red Gas*, the Europeans and Soviets surprisingly reached a mutually beneficial accommodation even during the Cold War. The lessons of *Red Gas* are important today given the recent discovery of vast deposits in the Russian A   . Policymakers, gas company officials, historians and other specialist will   y this richly researched and well-written book."

—Paul R. Josephson, Professor of History,
Colby College

"Per Hög   one of the younger historians of technology who acknowledge the   national and trans-technological impetus of modern infrastructure   ially when they serve to secure the flows of energy. *Red Gas* provides   resh perspectives on European history and beyond."

—Dirk van Laak, University of Giessen

"*Red Gas*   cise empirical study of the development of the Eurasian natural gas pi   system from the 1960s until now. In his balanced presentation of the dri   ces on both sides of the Iron Curtain Per Högselius discovers the shape   source dictatorship that no longer is built on ideological ideas but on h   ency and economic principles."

            n Rudolph, Head of Political Communication of the Evonik
Group in Brussels and Adjunct Professor for Contemporary
History at the Ruhr-University Bochum

"The firs   prehensive study of the flow of gas between the USSR and Europe, t   cellently researched book tells the fascinating story of how Western   's current dependence on Russian energy resources originated in the dy   of the Cold War. It is a novel and eye-opening work, one that is truly in   nt for students of history, environmental studies, and international rel   ."

—Helmuth Trischler, Director of the Rachel Carson
Center for Environment and Society, Munich

"*Red Gas*   oneering contribution to transnational history writing. The book illus   nicely that the 'Iron Curtain' was never as impenetrable as we might be   and it proves unequivocally the catalytic power of infrastructures in o   coming ideological and national boundaries."

—Mikael Hård, Professor of History of Technology,
Darmstadt University of Technology, Germany

"*Red Gas* opens a vital window on long-obscure dynamics of East-West cross-border problem solving, carefully and critically analyzing complex negotiations over Soviet natural gas exports amid the unwinding of Cold War tensions. Per Högselius demonstrates here both a mastery of archival sources in multiple languages and a deep appreciation for the challenges involved in Europe's 'hidden integration'—cultivating trust, reducing uncertainty, and replacing fear with interdependence. This is top-quality transnational history."
—Philip Scranton, Rutgers University

PALGRAVE MACMILLAN TRANSNATIONAL HISTORY SERIES

Akira Iriye (Harvard University) and **Rana Mitter** (University of Oxford)
Series Editors

This distinguished series seeks to: develop scholarship on the transnational connections of societies and peoples in the nineteenth and twentieth centuries; provide a forum in which work on transnational history from different periods, subjects, and regions of the world can be brought together in fruitful connection; and explore the theoretical and methodological links between transnational and other related approaches such as comparative history and world history.

Editorial board: **Thomas Bender**, university professor of the Humanities, professor of History, and director of the International Center for Advanced Studies, New York University **Jane Carruthers,** professor of History, University of South Africa **Mariano Plotkin**, professor, Universidad Nacional de Tres de Febrero, Buenos Aires, and member of the National Council of Scientific and Technological Research, Argentina **Pierre-Yves Saunier**, researcher at the Centre National de la Recherche Scientifique, France **Ian Tyrrell**, professor of History, University of New South Wales

Published by Palgrave Macmillan:

THE NATION, PSYCHOLOGY AND INTERNATIONAL POLITICS, 1870–1919
By Glenda Sluga

COMPETING VISIONS OF WORLD ORDER: GLOBAL MOMENTS AND MOVEMENTS, 1880S–1930S
Edited by Sebastian Conrad and Dominic Sachsenmaier

PAN-ASIANISM AND JAPAN'S WAR, 1931–1945
By Eri Hotta

THE CHINESE IN BRITAIN, 1800 TO THE PRESENT: ECONOMY, TRANSNATIONALISM, IDENTITY
By Gregor Benton And Terence Gomez

1968 IN EUROPE: A HISTORY OF PROTEST AND ACTIVISM, 1957–1977
Edited by Martin Klimke and Joachim Scharloth

RECONSTRUCTING PATRIARCHY AFTER THE GREAT WAR: WOMEN, CHILDREN, AND POSTWAR RECONCILIATION BETWEEN NATIONS
By Erika Kuhlman

THE IDEA OF HUMANITY IN A GLOBAL ERA
By Bruce Mazlish

TRANSNATIONAL UNCONSCIOUS
Edited by Joy Damousi and Mariano Ben Plotkin

PALGRAVE DICTIONARY OF TRANSNATIONAL HISTORY
Edited by Akira Iriye and Pierre-Yves Saunier

TRANSNATIONAL LIVES: BIOGRAPHIES OF GLOBAL MODERNITY, 1700 TO THE PRESENT
Edited by Angela Woollacott, Desley Deacon, and Penny Russell

TRANSATLANTIC ANTI-CATHOLICISM: FRANCE AND THE UNITED STATES IN THE NINETEENTH CENTURY
By Timothy Verhoeven

COSMOPOLITAN THOUGHT ZONES: INTELLECTUAL EXCHANGE BETWEEN SOUTH ASIA AND EUROPE, 1870–1945
Edited by Kris Manjapra and Sugata Bose

IRISH TERRORISM IN THE ATLANTIC COMMUNITY, 1865–1922
By Jonathan Gantt

EUROPEANIZATION IN THE TWENTIETH CENTURY
Edited by Martin Conway and Klaus Kiran Patel

NEW PERSPECTIVES ON THE TRANSNATIONAL RIGHT
Edited by Martin Durham and Margaret Power

TELEGRAPHIC IMPERIALISM: CRISIS AND PANIC IN THE INDIAN EMPIRE,
CA. 1850–1920
By D. K. Lahiri-Choudhury

THE ESTABLISHMENT RESPONDS: POWER, POLITICS, AND PROTEST SINCE 1945
Edited by Kathrin Fahlenbrach, Martin Klimke, Joachim Scharloth, and Laura Wong

EXPLORING THE DECOLONIAL IMAGINARY: FOUR TRANSNATIONAL LIVES
By Patricia A. Schechter

RED GAS: RUSSIA AND THE ORIGINS OF EUROPEAN ENERGY DEPENDENCE
By Per Högselius

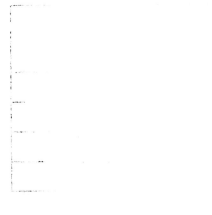

Red Gas

Russia and the Origins of European Energy Dependence

Per Högselius

SURREY LIBRARIES	
Askews & Holts	13-Feb-2013
333.7909 ECO	£20.00

RED GAS
Copyright © Per Högselius, 2013.

All rights reserved.

First published in 2013 by
PALGRAVE MACMILLAN®
in the United States—a division of St. Martin's Press LLC,
175 Fifth Avenue, New York, NY 10010.

Where this book is distributed in the UK, Europe and the rest of the world, this is by Palgrave Macmillan, a division of Macmillan Publishers Limited, registered in England, company number 785998, of Houndmills, Basingstoke, Hampshire RG21 6XS.

Palgrave Macmillan is the global academic imprint of the above companies and has companies and representatives throughout the world.

Palgrave® and Macmillan® are registered trademarks in the United States, the United Kingdom, Europe and other countries.

ISBN: 978–1–137–29371–8 (paperback)
ISBN: 978–1–137–28614–7 (hardcover)

Library of Congress Cataloging-in-Publication Data is available from the Library of Congress.

A catalogue record of the book is available from the British Library.

Design by Newgen Imaging Systems (P) Ltd., Chennai, India.

First edition: January 2013

10 9 8 7 6 5 4 3 2 1

Contents

List of Illustrations	xi
Foreword	xiii

1	Introduction	1
	Russia's Contested "Energy Weapon"	1
	Soviet Natural Gas and the Hidden Integration of Europe	2
	Dependence in the Making: A Systems Perspective	5
	The Political Nature of the East-West Gas Trade	7
	Outline of the Book	8
2	Before Siberia: The Rise of the Soviet Natural Gas Industry	13
	Soviet Power and Natural Gas for the Whole Country	13
	The Cold War Duel	15
	Soviet System-Building: Interconnecting the Republics	20
	The Rise and Stagnation of the Pipe and Equipment Industry	23
	"A Big Surplus for Export"?	26
3	Toward an Export Strategy	31
	From Central Asia to Siberia	31
	Glavgaz and the West European Natural Gas Scene	34
	Considering Exports: Opportunities and Risks	36
	Seeking Cooperation with Italy and Austria	38
	The Export Strategy Takes Shape	40
4	Austria: The Pioneer	45
	The Austrian Fuel Complex: Nazi and Soviet Legacies	45
	From SMV to ÖMV	46
	Toward Imports: ÖMV versus Austria Ferngas	48
	Rudolf Lukesch's Vision	50
	The Six-Days War as a Disturbing Event	55
	Negotiating the Gas Price	58
	The Contract	63
5	Bavaria's Quest for Energy Independence	67
	Natural Gas and the Politics of Isolation	67
	Otto Schedl's Struggle against North German Coal	69
	Toward Gas Imports: Negotiating Algeria	70
	Soviet Gas for Bavaria? The Austrian Connection	73
	Manipulated Conditions	75
	Egon Bahr and the Steel Companies as Supporters	79

	Alexei Sorokin's Charm Offensive	81
	The Soviet Option Fades Away	86
6	From Contract to Flow: The Soviet-Austrian Experience	89
	Interconnecting Austria, Czechoslovakia, and the Soviet Union	89
	Importing Soviet Gas in Practice	91
	The Galician Challenge	95
	Ukraine as a Victim	97
	Scaling Up Exports	101
	The Unseen Crisis	102
7	Willy Brandt: Natural Gas as Ostpolitik	105
	Toward a New Eastern Policy	106
	What Role for Soviet Natural Gas?	109
	From Politics to Business: Negotiating Price and Volumes	112
	Finalizing the Contract	118
	Shell and Esso: Lobbying against Unwelcome Competition	122
	Seeking Coordination with Italy and France	125
	The Significance of the Soviet-German Natural Gas Deal	129
	From European to American Imports of Soviet Natural Gas?	131
8	Constructing the Export Infrastructure	135
	Siberian Megalomania	135
	Arctic System-Building	138
	The Ukrainian Crisis and Kortunov's Death	143
	Desperation and Chaos	147
9	Trusting the Enemy: Importing Soviet Gas in Practice	151
	Enabling Transit through Czechoslovakia and Austria	151
	Doubts in Bavaria	154
	In Case of Emergency	156
	On the Verge of Breakdown	159
	Perceived Success	162
10	Scale Up or Phase Out?	167
	A Turbulent Energy Era	167
	Involving Iran	172
	Doubts in the Kremlin	177
	Envisaging the "Yamal" Pipeline	179
	Opposition from the United States	184
	The Compressor Embargo	188
	Europe's Contested Vulnerability	190
11	From Soviet to Russian Natural Gas	197
	Surging Dependence	197
	The Biggest Geopolitical Disaster of the Twentieth Century?	202
	Intentional Disruptions	204
	Managing Dependence	210
	The "Molotov-Ribbentrop" Pipeline	212

12	Conclusion	217
	Dependence in Retrospect: Four Phases	218
	Energy Weapons: Real and Imagined	220
	Understanding Europe's Enthusiasm	224
	A Gradual Learning Process	226
	The Evolution of a Transnational System	229
	The Soviet Union as a Victim	232
	A Long Duration	233

Acknowledgments	237
Notes	239
Bibliography	263
Index	269

Illustrations

Figures

1.1	Soviet/Russian natural gas exports to Western Europe, 1968–2011	4
2.1	Alexei Kortunov (1907–1973)	16
2.2	Soviet natural gas commercial reserves, 1950–1960	18
2.3	Soviet natural gas production, 1950–1965	20
2.4	Map of the Soviet pipeline system as of the early 1960s	22
3.1	Soviet natural gas commercial reserves, 1950–1966	32
4.1	Proposed international pipeline and LNG links for the supply of Austria, Italy, and Spain with natural gas	49
4.2	Thyssenrohr's pipe factory at Mülheim (Ruhr)	53
4.3	The vision of a Trans-European Pipeline for exports of Siberian natural gas to Austria, Italy, and France	56
5.1	Alexei Sorokin and Heinrich Kaun	83
6.1	Austrian minister of transportation Ludwig Weiss and Soviet gas minister Alexei Kortunov	92
6.2	ÖMV's new compressor hall at Baumgarten, built for incoming Soviet gas	93
6.3	Production of natural gas in western Ukraine, 1950–1980	96
6.4	Map of gas fields and long-distance gas pipelines in the Ukrainian SSR as of the late 1960s	98
7.1	Herbert Schelberger, Ruhrgas' chairman and main negotiator in the Soviet-German gas and pipe talks	113
7.2	Bavarian minister of economy Otto Schedl and Soviet minister of foreign trade Nikolai Patolichev in Moscow, August 1969	116
7.3	Alternative transit vision for Soviet natural gas	127
8.1	Planned Soviet gas flows from Siberia and Central Asia to the European part of the USSR	136
8.2	Soviet gas reserves, 1950–1971	137
8.3	Planned pipeline routes for the transmission of Komi and Siberian gas to Leningrad, the Baltics, and Belarus	139
8.4	Bear cub found along the Northern Lights pipeline route	141
8.5	Production of natural gas in western and eastern Ukraine, 1950–1980	145
9.1	Construction of the first transit pipeline through Czechoslovakia, July 1971	152
10.1	Exports of Iranian gas to Western Europe, with transit	

	through the Soviet Union, as envisaged in the 1975 tripartite deal	176
10.2	Gas consumption in OECD Europe, 1970–1981	186
10.3	The integrated gas system of Western Europe as of 1980	194
11.1	Italian technicians from Nuovo Pignone adjusting electronic equipment at one of the new compressor stations along the Yamal (Urengoi-Uzhgorod) pipeline	199

Table

1.1	West European dependence on Russian natural gas as of 2011, by country	3

Foreword

The history of transnationalism is about flows. Yet the term is almost always used as a metaphor relating to the movement of ideas or influence. In this book, Per Högselius examines one of the most fascinating—and literal—uses of the term: the flow of gas between the USSR and Europe. At the same time, beneath the detailed and rigorous account of a little-known story about energy supplies lies one of the most important dynamics of the Cold War. For the struggle to supply and control energy was one of the most crucial elements of the interaction between East and West. In the middle of the twentieth century, Central Western Europe, recovering from war, desperately needed reliable sources of energy to heat, build, and maintain the new, prosperous continent. Yet, intriguingly, the region had to turn to its Cold War opponent to stimulate its own growth, the consumerist effects of which were then used as an example of how *different* it was from its communist rival. These complexities are one of the most fascinating elements of this study, which breaks down simple divisions between East and West in the Cold War and yet shows how individual actors, many of them commercial rather than governmental, are engaged with each other across the blocs. The material dealing with Austria and Germany is fascinating, showing that these societies, warily seeking to re-create a place for themselves in the new Europe, could and did use energy negotiations not only to reach out to the old enemy, but also to particularize and complicate their own positions within the emerging Cold War structures. The author should also be commended for drawing on such a wide variety of primary materials, including a range of archives in several countries, giving impressive weight to his arguments.

Perhaps one of the most interesting conclusions is the dog that did not bark: the USSR could have used its control of energy to hold certain Western countries hostage, but ultimately did not do so. At a time when the post–Cold War world has witnessed repeated fear of energy hunger and indeed energy blackmail, this story is well worth remembering because such confrontations do not have to end in zero-sum results.

This fascinating study further adds to the development of transnational history in the Palgrave Macmillan Transnational History series. As it has developed, we have been able to read about a wide range of nongovernmental actors who seek ways around the boundaries imposed by the dominant political systems of the modern era, whether nation-state, empire, or indeed Cold War bloc. With his study of "red gas" and "energy transnationalism," Per Högselius has added another important element to this fast-changing historical approach.

Rana Mitter
Akira Iriye
Oxford, August 2012

1
Introduction

Russia's Contested "Energy Weapon"

How and why do countries become dependent on each other for something as vital as their energy supply? How do they build and maintain critical levels of trust across political, military, and ideological divides? And how do they cope with uncertainty and risk in these relations?

Europe's dependence on Russian natural gas has in recent years become a fiercely debated issue in European politics. The actual and potential consequences of far-reaching energy imports from the "big bear" have become a subject of growing concern not only among importing nations, but also at the level of the European Union. The gas trade has come to decisively influence EU-Russia relations and there is nowadays hardly any aspect of these that can be discussed without, directly or indirectly, taking into account natural gas. The recent "gas crises"—notably in 2006 and 2009—in which several EU member states faced acute gas shortages as a consequence of disputes between Russia and Ukraine over the extension of import and transit contracts have, in the eyes of many analysts, proved the reality of Europe's vulnerability. Moreover, some have interpreted Russia's gas disputes with Ukraine and several other ex-Soviet republics as part of a wider Russian ambition to regain political and economic influence in its "near abroad." According to this interpretation, Russian natural gas has become an "energy weapon" analogous to the OPEC's "oil weapon", and the argument is that such a weapon might be—and is possibly already being—used not only against Ukraine and other former Soviet republics, but also against Western Europe.[1]

Others, challenging this view, emphasize that Russian gas exports, to an overwhelming extent, take the form of undramatic business relations and technical cooperation from which both Russia and the EU profit, and that the frequent disputes with former Soviet republics have centered on economic rather than political issues, typically linked to the problem of nonpayment. Moreover, to the extent that the gas trade is political, it may be argued that this is not an extraordinary thing. Despite the Western ideal of an international economy based on free, depoliticized market relations, close links between politics and economics are in actual practice part and parcel of international business. Energy is one of many fields in which international trade is not a "purely economic" phenomenon. Furthermore, since natural gas emits only half as much carbon dioxide as coal (which it often replaces), Russian gas can be argued to make an important contribution to combating climate change. The main threat, according to this view, is not that Russia, for political reasons, would deliberately disrupt its gas supplies to Europe, but rather that its gas industry might fail to make the necessary investments in pipelines and gas fields and that it, as a result, will not be able to live up to and further expand its export commitments.[2]

Independent of perspective, the importance of Russian natural gas for Europe's energy supply is unlikely to decrease in coming decades. This is because of the expected depletion of North Sea and other intra-European gas resources, which are currently considered guarantors of Western Europe's security of supply and a necessary counterweight to imports from non-European sources. Gas production within the EU peaked in 1996 and has been in a phase of steady decline since around 2004. The International Energy Agency (IEA) expects gas production within the EU to decrease from 196 billion cubic meters (bcm) in 2009 to 89 bcm in 2035. The only factor that could possibly reverse this trend would be a European revolution in unconventional gas production, the probability of which is difficult to assess at the present time. Norwegian gas production will continue to increase from today's level of around 100 bcm, but not by more than 10–20 bcm, and a production peak will be reached within a decade or two. At the same time, the main scenario predicts that the EU's demand for natural gas will continue to increase, from 508 bcm in 2009 to a level of around 629 bcm in 2035.[3] This anticipated growth is closely related to European energy and climate policies, in which a gradual phase-out of coal for electricity generation plays an important role. Following the 2011 Fukushima disaster in Japan, it appears probable that natural gas, together with renewable energy sources, will replace much of Europe's nuclear power as well.

Against this background, most analysts now agree that if Europe's future energy demand is to be met, Russia's natural gas is direly needed. Other non-European gas suppliers – and, possibly, intra-European shale gas – may alleviate the situation to a certain extent, but even so any decrease in Europe's demand for Russian gas seems unrealistic. Economic recession may slow demand on the short term, but in the long run imports from the East will most probably have to increase. At the same time, growing competition from China and other countries for Siberia's gas may change the traditional logic of Russian-European interdependence in the field of natural gas. Nobody knows how this development will influence EU-Russia relations and, more generally, the overall political landscape in Europe.

Soviet Natural Gas and the Hidden Integration of Europe

How and why did Western Europe become such a massive importer of Russian natural gas? Clearly, today's dependence did not emerge overnight. The crucial formative period of the East-West gas trade can be located in the decade 1965–1975—that is, in the midst of the Cold War. Intense negotiations

Table 1.1 West European dependence on Russian natural gas as of 2011, by country (bcm, measured at 0 degrees centigrade)

	Domestic	Russian	Other	Dependence (%)
Austria	1.6	4.9	4.7	44
Belgium		7.4	21.9	25
Finland		3.8		100
France	0.7	8.6	38.3	18
Germany	10.0	30.8	53.2	33
Greece*	0.0	2.1	1.8	53
Italy	7.7	15.4	54.1	20
Netherlands	64.2	4.0	9.6	5
Switzerland*		0.3	3.3	8
Turkey	0.7	23.5	18.3	55

* Figures for 2010.
Source: BP Statistical Review of World Energy 2011.

between the Soviet Union and Italy, Austria, West Germany, Finland, and Sweden gained momentum in 1966–1967, and a number of key pioneering agreements were reached in the period from 1968 to 1970. First deliveries started to Austria as early as 1968 and to West Germany, Italy, and Finland in 1973–1974. France followed suit in 1976. Strikingly, several West European countries and regions were connected with the communist pipeline system of Eastern Europe before linking up with the grids of other EC and NATO member states.

At the time when the Berlin Wall fell and the Soviet Union collapsed, Soviet natural gas had become one of the most important sources of fuel in Western Europe. "Red" gas was taken into use on a large scale by a wide range of industrial enterprises, by power plants, by the municipal sector, and by millions of households. This was made possible through the construction of one of Europe's most critical and expensive infrastructures, which for its part formed a most remarkable case of East-West relations and of what has been labeled the "hidden integration" of Europe in the Cold War era.[4] In no other field did Western and Eastern Europe develop such close material relations during this era as in natural gas. Indeed, the gradually deepened gas trade and the construction of ever larger pipelines, generating far-reaching dependencies and vulnerabilities on both sides, ran counter to the fundamental logic of the Cold War. From the perspective of natural gas, the "Iron Curtain" takes on a new meaning and Europe looks different from what we are being told in much of the general historical literature.

Despite this peculiar and paradoxical development, and notwithstanding the central importance of Russian gas in current European and Russian affairs, little attention has been paid to their historical underpinnings. On one hand, the export of Soviet natural gas—and of Soviet oil—is often explicitly mentioned as an interesting phenomenon in the earlier literature on European postwar and Cold War history. On the other, it has, in practice,

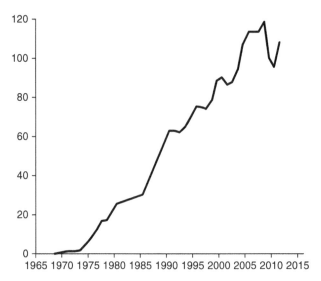

Figure 1.1 Soviet/Russian natural gas exports to Western Europe, 1968–2011 (bcm).
Sources: Stern 1980, p. 59; Stern 2005, p. 110; Oil and Gas Journal; BP Statistical Review of World Energy.

remained a "black box," discussed only in passing in connection with political or economic analyses of, for example, German chancellor Willy Brandt's New Eastern Policy, cooperation between Italy and the Soviet Union in the automotive sector, or in relation to the NATO's embargo policies concerning West European exports to the communist bloc of large-diameter steel pipes and advanced compressor technology.[5] In other words, East-West natural gas relations have never been subject to an in-depth historical inquiry in their own right. As a result, we do not know how and why Europe's dependence on Russian natural gas has actually come about.

The aim of this book is to fill this gap. *Red Gas* investigates how and why governments, businesses, engineers, and other actors sought to promote—and oppose—the establishment of an extensive East-West natural gas system at odds with Europe's formal political, military, and ideological divisions. It explains why political leaders and energy companies in several West European countries prioritized the integration of their gas supply systems with those of communist Eastern Europe, rather than first and foremost seeking integration with their Western neighbors. The book reveals how a variety of actors on either side of the Iron Curtain managed—and sometimes failed—to build and maintain sufficient levels of trust across military and ideological divides and how they used natural gas relations for a variety of purposes other than for the access to a high-quality fuel. At the center of the narrative stands the fear of unwanted consequences of energy dependence and the perceived vulnerability of actors to supply interruptions and price shocks, and the opportunities that the gas trade seemed to offer politically, economically, and environmentally—in an age obsessed with its ever-growing thirst for fuel.

Building on primary documentary sources from Russian, Ukrainian, German, and Austrian archives, the book centers empirically on the period from the mid-1960s, when the first gas export agreements were negotiated and the first East-West pipelines built, to the years around 1990, when the Berlin Wall fell, the Soviet Union collapsed, and the Cold War ended. It uncovers the complex formation of energy trade strategies from the side of governments and businesses in both the Soviet Union and the importing Western nations, and the complex process of negotiating the East-West gas contracts. The book unpacks the major conflicts between key players—both across borders and domestically—in their struggle to shape Europe's energetic future. It also tells the story of how Soviet and West European stakeholders—with mixed success—approached the task of actually creating—materially and institutionally—the new trans-European pipeline infrastructure, and of using it in practice. An underlying argument, of relevance for policymakers and analysts of today, is that we will not be able to understand the dynamic nature of Europe's current energy dependence, let alone properly deal with it, in the absence of a thorough historical understanding of how today's situation has come about.

Dependence in the Making: A Systems Perspective

How and why does a large technical system (LTS) such as the East-West gas grid come into being? Earlier studies of LTS[6] have stressed the importance of scrutinizing the activities of "system-builders" and their evolution over time. System-builders are the actors who, by definition, have the most far-reaching power to shape a system's evolution—and to kick it off in the first place. System-builders may be technically oriented innovators, but more often they are passionate business leaders or centrally placed governmental actors who have the necessary ability, mandate, and connections to bring about major infrastructural projects, turning diffuse and often controversial visions into material reality. One of their key challenges is to mobilize sufficiently strong actor networks. Having a talent in viewing the system in its totality, spotting the links between its diverse technical, political, and economic components, the successful system-builder identifies "reverse salients" in the form of weak components and links, and turns these—analytically and discursively—into "critical problems" that must be solved for the system to come about and expand along desired lines.[7]

When system-building takes place in a transnational context, however, it is an extremely demanding process to master, due to differences in standards, regulations, political traditions, and business culture in the countries involved.[8] Crucially, system-builders setting out to cooperate with "the other" have to accept that they cannot to the same degree take control over the system-building process as they may be used to in their national environments. East-West system-building in the Cold War context formed an extreme case of transnationalization, dependent as it was on what I call "system-building coalitions" that cut across the Cold War's most radical political, ideological, and military divides.

Yet transnational system-building sometimes becomes an even more dynamic process than system-building in a national context. *Red Gas* shows that effective coalitions of system-builders may turn the apparent problems of cross-border tensions and disparities into opportunities for accelerated development and growth. Natural gas system-builders in East and West spotted what I call "complementary reverse salients," or problems on either side that "fitted" each other and could be resolved precisely through increased transnational cooperation and integration. In the 1960s, for example, vast volumes of natural gas had been discovered in the Soviet Union, but the growth of the domestic Soviet gas system was retarded by the inability of the domestic steel industry to produce high-quality steel pipe. West European system-builders, for their part, knew how to build pipes, but lacked large domestic gas resources. This asymmetrical situation motivated actors in East and West to work out a countertrade arrangement in which Soviet natural gas was exported to Western Europe in return for West European deliveries of large-diameter steel pipe. Transnational coalitions of system-builders working together on resolving complementary reverse salients constituted the most fundamental driver of Europe's evolving energy dependence throughout the Cold War period.

Europe's dependence grew at a steady pace through processes of gradual learning and positive feedback. Initially, there was great suspicion on either side. In such a situation it was of a certain importance that East-West gas system-builders could point to exports of red gas across the Iron Curtain not as a totally new phenomenon, but as a logical follow-up on exports of Soviet oil. Moreover, gas system-builders effectively exploited the opportunities offered by early, inexpensive pilot projects as test cases for the future. Pilot projects and experiences of earlier cooperation helped system-builders assure themselves that they were dealing with a system with which they could communicate and cooperate in a meaningful way. To borrow a concept from social systems theory, this made it easier for "resonance" to be generated. Resonance between Soviet and West European systems in turn made it easier for system-builders to build trust.[9]

For resonance and trust to be retained, the Soviet Union also needed to show that it could provide the gas in the agreed quantity and quality, while the importers needed to demonstrate their ability to receive and pay for the gas. Failure to do either were bound to reduce the prospects for further expansion of the system. As it turned out, the Soviet Union was so obsessed with the need to ensure its Western partners of its reliability as an exporter that the country's own gas users were left to freeze when sufficient gas was not available.

Having survived its formative phase, transnational system-building became a self-reinforcing process, generating a virtuous circle of positive feedback that inspired actors on either side to gradually scale up their commitments and visions. Ultimately, through its development over nearly half a century, the system became a mature transnational infrastructure with a very high level of what students of large technical systems call "momentum." A high level of momentum made attempts to alter the system's direction of development exceedingly difficult.

In some cases, such as in connection with US-led opposition to expansion of the East-West gas trade in the early 1980s, Soviet gas exports became subject to major public and political debates, and demands for radical change—and even abandonment—of the system were voiced. By then, however, the system had grown so powerful that these demands had little chance of materializing. The robustness of the system was reconfirmed in 1989–1991, when the Berlin Wall fell, the Soviet Union collapsed, and the political map of Europe was radically redrawn. These extreme political and economic upheavals notwithstanding, the East-West gas system—and Western Europe's dependence on Russian gas—remained in place and continued to grow. The difficulty to "change direction" is clearly disturbing to actors who, in our own time, consider Europe's dependence on Russian natural gas problematic and wish to "do something" about it.[10]

The Political Nature of the East-West Gas Trade

How political have Russia's gas exports been? *Red Gas* argues that economic considerations were always more important than political ones in bringing about and sustaining the gas flow between East and West. In the absence of profit expectations, neither the Soviet Union nor Western Europe's importers would have supported the creation of the system. At the same time, the book argues that Soviet natural gas, to a certain extent, did function, and was perceived of, as an "energy weapon" and that it continues do so in an age when the gas is no longer red. The relative importance of this political dimension in relation to economic considerations has been greatly exaggerated and the true nature of the "weapon" misunderstood by many analysts, but this does not mean that it has been non-existent.

The evidence suggests that we need to broaden our view and adopt a conceptualization of "energy weapons" that reaches beyond the much-debated nightmare of politically motivated supply disruptions. An energy weapon can be so much more. This book thus widens the weapon metaphor to include issues such as dumping of red gas on Western markets, "divide and rule" strategies in which some customer countries were favored over others in Soviet attempts to splinter the Western world, rhetorical practices in which natural gas exports served to strengthen the Soviet Union's legitimacy on the international arena, and so on. While there is no evidence that the Soviet Union, up to its collapse in 1991, ever aimed to make use of the threat of supply disruptions for political blackmail, the empirical material does support the view that it sought to divide Western Europe by offering natural gas to some countries but not to others, and that national prestige was an important concern when Moscow set out to negotiate its export contracts. After the collapse of communism, politically motivated supply disruptions did occur, though usually in combination with other, less political motives.

Importantly, actors were often unaware of the real motives of their partners beyond the Iron Curtain. West Europeans were highly suspicious of Moscow's intentions, and all importers took into account politically motivated supply disruptions and aggressive price dumping as a real risk when negotiating with the Soviets and building the import infrastructure. Huge investments were made in technical facilities whose purpose was to reduce the adverse impact of unexpected Soviet moves. Whether or not the Soviet gas weapon "actually" existed, its socially constructed reality thus had a very tangible impact on the physical characteristics of the European gas system.

As it turned out, Western Europe's expensive back-up pipelines, emergency gas storage facilities, gas-quality transformation stations, and other precautionary measures did find their role in the rapidly growing East-West gas trade. The reason, however, was not that Moscow intentionally disrupted supplies, but that the export pipelines built on Soviet territory were plagued by recurring technical failures. In the construction phase of export pipelines, the everyday chaos of what was allegedly a "centrally planned economy" ensured that key equipment was often missing and that projects rarely had a chance of living up to the timetables specified in the export contracts. Seeking to enforce the deadlines, decision makers allowed pipelines and compressor stations along the international transmission routes to be built in a haste by a workforce that during the most sensitive construction phases largely consisted of probationers and conditionally released prisoners. The disastrous quality of pipelines and compressor stations built in the 1960s and 1970s inevitably gave rise to repeated technical failures and accidents later on.

Paradoxically, the real victims of the failures were not Western Europe's, but the Soviet Union's gas users. Northwestern Siberia was the world's largest gas region, but lack of pipeline capacity nevertheless made gas a scarce resource in the red empire. Soviet gas users, therefore, had to compete with West European importers for insufficient volumes of gas. Moscow, desperately seeking to ensure the West of its reliability as a partner, opted to sacrifice domestic supplies rather than cut exports. The result of this highly political choice, in terms of human suffering and industrial productivity, was devastating.

To the extent that East-West natural gas system-building was a political activity, this was true not only as far as the Soviet Union was concerned, but also in terms of West European interests. It is no coincidence that the formative phase of Soviet natural gas exports overlaps with a period of détente in East-West relations. Not only did the favorable geopolitical climate in the late 1960s and early 1970s make it easier for proponents of the East-West gas trade to mobilize support for their visions, but red gas was in itself identified as a foreign policy tool with great potential to improve the relations between the capitalist and the communist world. In some cases Western governments even subsidized the construction of pipelines across the Iron Curtain for political reasons. In the end, the perceived political opportunities were seen to far outweigh the perceived political risks.

Outline of the Book

Red Gas tells the story of East-West natural gas relations from both a Soviet and a Western perspective. It takes into account a vast body of empirical evidence from "both sides" and in original languages. The ambition has been to document Soviet natural gas exports to Western Europe from the perspective of those people and organizations who have been—or tried to be—central in envisioning, negotiating, planning, building, operating, and using the transnational gas infrastructure. The structure of the book reflects this symmetry ambition.

Chapter 2 sets the stage by outlining the historical emergence of the Soviet Union as a major natural gas producer and the rise of natural gas as a "typical communist" fuel with a special role to play in building socialism. Chapter 3 follows this up by analyzing the fierce internal debate in the Soviet Union on how to exploit the country's rapidly growing gas resources in the best way. It was in this context that the first export strategy took shape. By 1966, Moscow had made up its mind to enter the West European gas market, and negotiations were initiated with Italy, Austria, France, Finland, and Sweden.

Austria became the first capitalist country to conclude a gas agreement with the Soviet Union. Chapter 4 traces the complex negotiation process that led up to this pioneering deal. The talks took place in parallel with Austria's eager attempts to associate itself more closely with the European Economic Community (EEC), a development that was fiercely opposed by Moscow, and Soviet-Austrian natural gas relations thus became linked to a broader struggle about Austria's position in Cold War Europe. The historical contract, of great significance for the future of both Austria's and Europe's energy supply, was eventually signed in June 1968.

West Germany also bordered on the Iron Curtain and thus seemed strategically positioned to import natural gas from East European sources. In the context of the Cold War, however, a West German import of Soviet gas was bound to become much more controversial than Austria's. Germany was, in Soviet perspective, a country full of "revenge-seeking passions," still dominated politically and economically by "former Nazis and even war criminals," as Brezhnev put it at the 1966 Party Congress. The German federal government, for its part, still followed a policy of refusing to recognize East Germany as a sovereign state and the postwar borders in the east. The anti-Soviet sentiments were notable. Chapter 5 inquires how, in spite of these difficult relations, an import of Soviet gas to Germany and the construction of a transnational pipeline infrastructure for this purpose became a major topic of internal debate in the Federal Republic. The project failed to materialize, but the discussions served as a useful preparation for later negotiations.

Chapter 6 analyzes how the first Soviet gas exports worked (and how they did not work) in practice. Exports to Austria commenced in September 1968, just ten days after the Warsaw Pact's military invasion of Czechoslovakia, through which the gas was to be transited. The chapter shows how the Soviet gas ministry's system-building efforts took the form of constant crisis management in the chaos of the centralized Soviet economy. It also documents how domestic gas users—particularly in Ukraine, Belarus, Lithuania, and Latvia—faced unwanted competition from customers abroad for the same scarce gas resources.

In the aftermath of the Czechoslovak invasion, renewed negotiations were initiated between the Soviet Union and several West European countries. Moscow, now in need of rebuilding its international legitimacy and prestige, was even more eager than before to bring about natural gas exports. Most Western countries similarly judged that efforts to improve East-West relations, following the 1968 events, must not be given up, but rather intensified.

Unsuccessful earlier negotiations with Italy, France, Finland, and Sweden were revived. In addition, Germany seemed to become seriously interested in Soviet natural gas. Chapter 7 shows how German foreign minister—and later chancellor—Willy Brandt's close collaborator Egon Bahr identified natural gas as a vehicle in launching a new German Eastern policy (Ostpolitik). The chapter reconstructs the dramatic negotiations that eventually led to a first Soviet-German contract. It also traces the attempts from the side of the German, Italian, and French governments to coordinate their negotiations and thereby improve their bargaining power vis-à-vis the Soviet side, and the opposition from the Netherlands, the main competing exporter, along with several international oil companies.

Chapter 8 unveils how the Soviet Union, having signed export contracts with Germany, Italy, France, and Finland took on the immense task of bringing Ukrainian and Siberian gas in large quantities to Western Europe. The stakes were now much higher than in the initial Soviet-Austrian export arrangement. The construction of the export pipeline infrastructure was integrated into the overall Soviet system-building effort. It was a chaotic process and the Siberian pipelines were in the end delayed by many years. The export infrastructure that actually materialized looked very different from the one originally planned. In particular, Ukrainian gas came to play a more important role in meeting export obligations, an arrangement that caused severe gas shortages throughout the westernmost Soviet regions as the Kremlin, struggling to retain its reputation in Western Europe as a reliable exporter, prioritized deliveries across the Iron Curtain. Chapter 9 analyzes the same development but from a Western perspective, focusing on the practical experience of importing large quantities of Soviet gas to Germany and Italy.

The perceived functionality of the East-West gas trade stimulated further export contracts. Chapter 10 investigates how increased deliveries of Soviet gas became highly attractive following the 1973/1974 oil crisis. Moreover, a large contract was successfully negotiated according to which the Soviet Union was to play an important role as a transiteer of natural gas from far-away Iran to Europe. This was followed by a more contested West European vision of a further doubling of red gas imports. Coinciding with a period of increased East-West tension, these efforts were vehemently opposed by US president Reagan. For the first time, imports of "red" gas became subject to a vivid international and public debate, in which widely differing views of Europe's vulnerability clashed against each other. In the end, Washington was not able to prevent the Europeans from radically scaling up the East-West natural gas system.

Chapter 11, finally, analyzes the period from the late 1980s to the present. The fall of the Berlin Wall and the collapse of the Soviet Union made the future of the East-West gas trade difficult to predict. Immense difficulties to establish a stable institutional framework for trading gas among the former Soviet republics gave rise to repeated intentional supply cutoffs to Ukraine and Belarus, through which Russian natural gas was transited to Western Europe. Despite the seemingly insurmountable problems, the period saw a further steep increase in Russian gas exports.

Chapter 12, which concludes the book, discusses a number of key themes in the history of Russia's and the Soviet Union's gas exports to Western Europe. It provides additional perspectives on Russian natural gas both as a transnational infrastructure and as an "energy weapon," while also scrutinizing the overall evolutionary logic of Western Europe's growing dependence.

2
Before Siberia: The Rise of the Soviet Natural Gas Industry

Soviet Power and Natural Gas for the Whole Country

Russia's journey to become the world's largest producer and exporter of natural gas was long and bumpy. It started in earnest during World War II, at which time the Soviet energy system was almost completely based on coal. The first pipeline projects were proposed in response to Moscow's war-time energy crisis, the main cause of which was Hitler's invasion of the country's key coal regions. Cut off from vital coal supplies, Stalin's energy planners pointed to natural gas as an interesting, though yet untried alternative. The Commissariat of the Oil Industry, inspired by Soviet experiences in oil pipeline construction, argued in favor of a gas pipeline to link Moscow with newly discovered gas fields near Saratov, 800 km to the southeast. Stalin approved of the project, but lack of experience and material resources made its realization more difficult than anticipated. First proposed in 1942, the pipeline started to be built only in 1944, and it took another two years before it could be taken into operation. By then the war was over and coal deliveries from southern Russia and Ukraine had resumed. Yet the project was perceived as a success and the experience contributed decisively to Soviet postwar enthusiasm for natural gas—at least from the side of the Oil Commissariat.[1]

Having driven the Nazis out of Ukraine and eastern Poland in 1944, Oil Commissar Ivan Sedin and his deputy Nikolai Baibakov—the later Gosplan chairman—proceeded by drawing up plans for supplying Kiev, the capital of the Ukrainian SSR, with natural gas. The idea was to make effective use of the rich gas fields in Galicia, a region that had belonged to Poland in the Interwar years before being annexed by Stalin along the lines laid out in the Molotov-Ribbentrop Pact. Construction of a 500-km pipeline from Galicia to Kiev started already before the Red Army had reached Berlin. Gas from Dashava, the main Galician gas field, started flowing to Kiev in 1948, where it helped solve the city's looming fuel crisis in the chaotic aftermath of the war. The project's completion was celebrated as a great achievement, motivating

Sedin to propose an eastward extension of the pipeline. From 1951, Galician gas reached Moscow, 1,500 km away.[2]

A remarkable feature of the postwar exploitation of Galicia's gas riches was that it made the Soviet Union a "born" gas exporter. This was because a few Galician pipelines built by Polish and German engineers before and during the war continued to be in operation after Soviet annexation of Galicia. Most of them were fairly short lines serving the region's own needs. One of the lines, however, extended across what had now become the border between Soviet Ukraine and Poland. Supplying the important metallurgical town of Stalowa Wola, some 200 km to the west of the gas fields, the line had been taken into operation by Hitler's engineers in 1943. From 1944, Moscow accounted for gas transmission along this line as "exports." The volumes traded were small, but they were principally important in demonstrating that natural gas was not necessarily bound to remain within the borders of producing countries.[3]

Although natural gas, as of the early 1950s, still contributed only marginally to overall energy production in the Soviet Union, the country's leading gas men—they were almost always men—felt that they represented a new and radical branch of the overall fuel and energy complex, of vast potential importance for long-term economic development and societal prosperity. In this sense the gas industry had arguably more in common with nuclear power than with coal or oil. A variety of arguments were brought up to illustrate the superiority and advantages of natural gas as compared to other energy sources. Production of natural gas was argued to be much more economic than the production of coal, an advantage that was expected to become more pronounced with time due to technical advances in fields such as drilling and pipelaying. Natural gas was also considered superior in many industrial contexts, mainly due to the even temperature with which it burnt and the easiness with which industrial processes could be regulated when based on gas. Moreover, since natural gas was "a smoke-free fuel, which burns up completely and does not emit any polluting gases into the atmosphere," a rapid transition from wood and coal to natural gas in urban energy systems was championed for environmental reasons. Similarly, the supporters anticipated an important role for natural gas in the transport sector.[4]

Soviet gas enthusiasts cited Lenin, who was said to have pointed at natural gas as a fuel of the future that, similarly to electricity, would "make the work conditions more hygienic; it will spare millions of workers from smoke, heat and dirt." Transition to gaseous fuels, Lenin had allegedly argued, promised to save vast amounts of human work previously devoted to mining and transport of hard coal. The quote was in reality taken from Lenin's famous argument in favor of large-scale electrification, not of natural gas system-building. But by noting that natural gas had hardly been used at all before the October Revolution, the gas men could paint a picture of their fuel as a truly "socialist" form of energy. While electricity was acknowledged to constitute "the most modern type of energy for dynamos, motors, lighting, for a variety of technological processes, and electrolysis," it was argued that "for heating purposes, the most modern energy source is gas."[5]

Stalin, however, tended to prioritize the coal industry and its proven technologies. His early support to the Saratov-Moscow and Dashava-Kiev-Moscow pipeline projects had been linked to the need to respond to war-time and immediate postwar energy scarcity rather than to a long-term vision of gaseous fuels as an integral component of the overall Soviet energy economy. The extent to which the vast country actually rested on sufficient large gas resources for the new fuel to have any chance in this respect and the economic efficiency of its production and distribution over long distances was questioned by a variety of actors, especially coal industry leaders. The oil industry, in contrast, was technologically closely related to gas and the perceived affinities between the two made the Oil Ministry an outspoken supporter of the gas industry's accelerated development.

Uncertainty over the future of gas continued to prevail during the interregnum in Moscow that followed Stalin's death in March 1953. A turning point came only after Nikita Khrushchev emerged as the nation's new political leader in spring 1955. Seeking to exploit the new power balance in the Kremlin, Oil Minister Nikolai Baibakov, who had now succeeded Sedin at this post, launched a campaign emphasizing the importance of modernizing the Soviet energy system. In June 1955, *Pravda* announced that increased production of oil and gas would permit the Soviet Union "to introduce highly profitable changes in the fuel-power production balance of the country and switch a number of fuel-consuming sectors of the national economy over from solid fuel to the more economical and efficient liquid and gaseous fuel." Aware of Khrushchev's enthusiasm for economic competition with the United States, Baibakov noted that "gas supplies cover only 8 percent of the population in this country while in the United States they cover 62 percent." The Soviets needed to catch up. Khrushchev proved receptive to this line of argument and responded by approving the allocation of substantially more funds to natural gas production and transmission. Shortly afterward, he appointed Baibakov as head of the powerful State Planning Commission (Gosplan). What followed was a new emphasis on both oil and gas as key components in overall economic planning.[6]

The Kremlin's support to the gas industry was confirmed at the 20th Congress of the Communist Party, held in February 1956. A planned quadrupling of gas production from 10 bcm in 1955 to 40 bcm in 1960 was nailed down. Taking inspiration from Lenin's electrification policy, the long-term goal was from now on to make gaseous fuels available for the whole country. Red gas seemed to have taken a definite step out of the shadow of coal and oil.[7]

The Cold War Duel

The new planning targets outlined in 1955–1956 posed an extraordinary challenge to the still nascent Soviet gas industry. A radical scaling up of all activities, both physically and organizationally, was seen necessary. To deal with this task, the Council of Ministers in Moscow decided to create a separate government agency for the natural gas industry, to be spun off from the Oil Ministry. The Main Directorate of the Gas Industry (Glavnoe Upravlenie Gazovoy Promyshlennosti pri Sovete Ministrov SSSR, or for short: Glavgaz SSSR), as the new agency was called, opened up in 1956. A former colleague of Baibakov's from the Oil Ministry, Alexei Kortunov, was appointed its first

Figure 2.1 Alexei Kortunov (1907–1973)
Source: RIA Novosti.

director. Kortunov would turn out a highly capable and energetic system-builder, and his appointment would prove decisive both for the Soviet Union's and Europe's gas futures.

Born in 1907, Kortunov belonged to the same "second generation" of Soviet leaders as Nikolai Baibakov, but his professional background was much more varied. Raised in a railway worker's family in Novocherkassk in southern Russia, he had started out as a brilliant student of water administration and construction organization during the first chaotic years of Soviet rule. As a young expert he found his first major employment with the important Azovstal metallurgical plant at Mariupol in Ukraine. In 1936 he moved on to the aviation industry, which sent him to the Aerohydrodynamics Research Institute outside Moscow. When the war broke out, he was retrained as a military engineer, promoted to a commander, and eventually arrived in Germany with the Red Army.[8]

His first experience with oil and gas came three years later. Having stayed on in occupied Germany for some time, he moved several thousand kilometers east to Tuimazinsk in Bashkiria, at the heart of the Russian oil industry. There, his talent for organizing large infrastructure projects was much welcomed. Following the war, he was already a Hero of the Soviet Union, and in Bashkiria he was soon awarded the Order of Lenin for his extraordinary achievements in accomplishing complex oil production and infrastructure projects. Oil Minister Baibakov was impressed and brought Kortunov to Moscow in 1950, offering him the position as his deputy with responsibility for construction.[9] When Baibakov a few years later managed to mobilize political support for the creation of Glavgaz, he already had Kortunov in mind as the ideal director of this new organization. Loaded with experience in a variety of fields such as water management (where pipelaying was a central challenge), metallurgical engineering (the basis for the production of large-diameter steel pipes), aerohydrodynamics (of central importance for coping with gas flows), and far-reaching achievements as a construction engineer and military commander (skills that were of vital importance for large-scale construction projects), Kortunov possessed an extraordinary portfolio of competencies that would prove to be of great value when setting out to construct a Soviet-wide system for natural gas production and transmission.

Kortunov took on his new task with great enthusiasm and energy. He quickly developed a passionate relationship to natural gas as a fuel superior to oil, taking inspiration from the young and forward-looking, self-confident natural gas community that had started to take shape after the war. Under Kortunov's leadership, the Soviet gas industry proved capable of developing even more forcefully than anticipated in the already ambitious planning targets defined in 1956. At the time when Glavgaz was created, total Soviet gas production amounted to a modest 13.7 bcm, a figure that by 1957 had already reached 20.2 bcm—a growth of 47 percent from one year to the next. Kortunov self-confidently declared that it would be possible to increase natural gas production much faster than previously believed, suggesting that the country should aim for a level of 50 bcm already in 1960, as opposed to the 40 bcm mentioned in the 1956 directives. In November 1957, seemingly inspired by Moscow's success in launching the Sputnik satellite, Kortunov formulated the challenge of overtaking the United States, the undisputed world leader in natural gas production. An engineer and project manager, his interest in ideological issues was limited. But like Baibakov, he understood to use the Cold War struggle as a rhetorical tool that could be deployed to promote natural gas and secure support from the highest political level. The Soviet Union, Kortunov argued, must show that the earth's gaseous gifts could be better exploited under communism than under capitalism.[10]

Khrushchev agreed. At the May 1958 plenum of the Party's Central Committee, a new, scaled-up program was presented: "On the further development of the gas industry and the gas supply for enterprises and cities of the

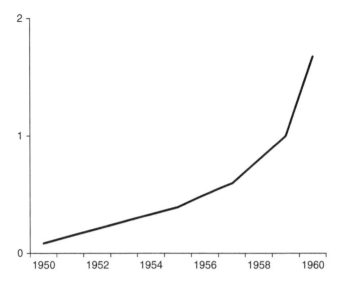

Figure 2.2 Soviet natural gas commercial reserves, 1950–1960 (tcm) (A+B categories in Soviet terminology)
Source: Based on figures in Gazovaya promyshlennost, various issues.

USSR." By 1965, it was now claimed, natural gas production must amount to no less than 150 bcm per year, compared to 30 bcm in 1958. On the long term, over a 15-year period, the program foresaw an increase to no less than 270–320 bcm. "No other branch of the national economy, not even in our country, has ever known of such growth rates," Kortunov proudly commented.[11]

A few years earlier, these dramatic growth targets would have seemed completely unrealistic, given the known gas reserves on Soviet territory. But several new, remarkable gas finds had altered the picture. Moreover, the increased funds that had started to be allocated to gas prospecting activities in 1956 allowed for a further increase in reserves. In a single year, 1959, the volume of "commercial" natural gas reserves grew by nearly 70 percent, reaching a level of 1,700 bcm. The yet-to-be-discovered resources were believed to be many times larger.

Khrushchev now thought the time ripe to make serious political use of the country's growing gas prospects. In a speech delivered at New York's Economic Club in connection with a much-publicized visit to the United States in 1959, he took up his country's rapid advances in the natural gas sector as a prime example of how the Soviet Union was allegedly about to catch up with the United States in an increasing number of fields:

> So far America occupies the first place in the world in gas production and in its known reserves, but in recent years we, too, increasingly use natural gas. Our geologists have discovered gas sources so immense that they will suffice for decades. This gives us the possibility to increase even

more the production and consumption of gas and overtake you also in this respect.[12]

Depending on the indicators chosen, it could actually be argued that the Soviet Union had already overtaken its superpower rival. Glavgaz noted that as of 1960, 40 percent of all Soviet long-distance pipelines were already larger than 720 mm in diameter, whereas in the United States only 1.1 percent of long-distance pipelines were larger than 760 mm. On average, a Soviet pipeline contained 2 million cubic meters (mcm) of gas per kilometer, whereas in the United States the corresponding figure was less than 1 mcm.[13] Kortunov's ideologically relevant conclusion was that Soviet socialist planning helped create a more large-scale, centralized, and, therefore, more efficient system than that of the United States, whose uncoordinated capitalist initiatives paved the way for suboptimal pipeline sizes and network configurations:

> It is worth noting that in the USA the competitive struggle among private capitalist companies often leads to unjustified building of small parallel pipelines and makes the establishment of a unified high-capacity system difficult, even at the level of economic regions. Therefore our socialist economy has the potential to develop the gas industry in a shorter time and with substantially lower expenditures than the USA.[14]

Kortunov and his colleagues liked to compare the Soviet gas industry not only with that of the United States, but also with competing branches of the *domestic* fuel and energy complex. The gas men observed that the share of coal in the country's overall energy balance was rapidly declining; in 1959 it still amounted to 56 percent, but four years later the share had already decreased to 46 percent. Both the oil and gas industries regarded the coal industry as belonging to yesterday's world. Increasingly, however, Glavgaz also distanced itself from the oil industry. In 1959, Kortunov observed that Soviet gas production was still seven years behind oil, but that it was rapidly catching up. Although oil production in the Soviet Union continued to increase its share in the overall energy balance, growing from 28 percent in 1959 to 35 percent in 1963, the share of gas grew even more rapidly, doubling from 6 percent to 13 percent in the same period.[15]

Kortunov's target to boost gas production to a level of 50 bcm by 1960 was in the end not met. But with actual production amounting to 47 bcm, the Kremlin was still impressed. After all, annual production had been a mere 10 bcm only five years earlier. Glavgaz was full of self-confidence, heading for the extremely ambitious and widely publicized 1965 production target of 150 bcm. And the ambitions continued to grow. At the next Party Congress, held in 1961, it was decided that national gas production, during the next 20 years, should increase fifteenfold to a breathtaking 720 bcm.[16] This extravagant ambition was primarily motivated by skyrocketing industrial demand for high-quality fuels and by the usefulness of natural gas as a raw material in the chemical industry. Apart from this, gaseous fuels were to contribute to the realization of a socialist society, facilitating convenient and hygienic housing for millions of citizens who would receive

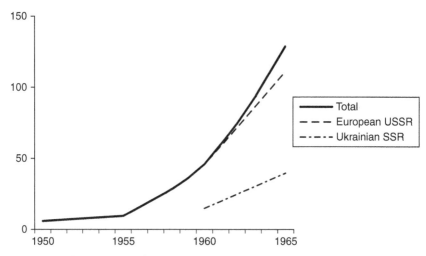

Figure 2.3 Soviet natural gas production, 1950–1965 (bcm, including associated gas)
Source: Based on figures in Slavkina 2002 and Oil and Gas Journal.

gas not only for the purpose of cooking, but also for space heating and hot water supply.[17]

Soviet System-Building: Interconnecting the Republics

The surge in Soviet gas exploration and production from around 1956 was accompanied by intense activities in long-distance pipeline construction. Since the configuration of this domestic system came to strongly influence the later export pattern, the main features of the emerging Soviet grid should be outlined here.

Throughout the 1940s and 1950s, the main purpose of long-distance gas pipeline construction was to improve fuel supply in the red empire's three largest cities: Moscow, Leningrad, and Kiev. Moscow, in particular, was prioritized. The Soviet capital had, at an early stage, become linked to two remote gas fields: Saratov on the Volga and Dashava in Galicia. Following up on these projects, Kortunov in December 1956 celebrated a first triumph as Glavgaz director as a new important pipeline from Stavropol in southern Russia to Moscow was launched. 1,300 km long and 720 mm wide, it was described as the most powerful gas pipeline ever built "in Europe" (there were more powerful ones in North America). In 1959 the line was extended to Leningrad, which up to then had relied on deliveries of manufactured gas from Estonian oil shale. Stavropol gas also became an important new energy source for the country's most important industrial region, the Donets basin (Donbass), part of which was in southwestern Russia and part in southeastern Ukraine.[18]

Other early pipelines were built to supply industrial cities located in relative proximitiy of major gas fields or oil fields that produced "associated" natural gas. These included Kuybyshev (Samara), Kazan, and Ufa in central Russia, which were supplied from nearby fields in the Volga region, and Kharkov and Dnepropetrovsk in eastern Ukraine, which were supplied from Shebelinka, a giant Ukrainian gas field that commenced production in 1956. Conveniently located in the immediate vicinity of several key industrial districts, Shebelinka was bound to become one of the most important sources of gas in the Soviet Union for the foreseeable future.

Following the creation of Glavgaz, the overall system-building strategy changed. Rather than connecting a certain gas field with a specific user region, the aim was from now on to build pipelines with an eye toward a coherent whole. Kortunov imagined the construction of an integrated system of pipelines through which major Soviet consumption centers would get access to gas from several different sources. This was a long-term vision that could not be realized overnight, but from around 1960 most transmission routes were designed in such a way as to enable their later integration into a unified, all-Soviet system. This was argued to be both efficient in an economic sense and advantageous from a security point of view.

In addition to this engineering logic, system-building was shaped by the Kremlin's political strategies. Moscow felt a need to ensure the political and economic integration of the newly annexed territories of what had been eastern Poland, the Königsberg (Kaliningrad) area in Germany's East Prussia, and the three independent Baltic states (Lithuania, Latvia, and Estonia) with the core Soviet lands, and for this purpose the vast gas reserves of Galicia in what was now western Ukraine proved highly useful. Galician gas was already being piped to Kiev and Moscow, but this was only the beginning. Encouraged by the discovery of a new large Galician field, the 1956 Party Congress decided on the construction of an additional pipeline from Dashava to Minsk, the capital of the Belarusian SSR, a distance of some 700 km, and further on to the Baltic cities of Vilnius, Klaipeda, Riga, and Liepaja. In this way three more republics—Belarus, Lithuania, and Latvia—got access to Ukrainian gas. Since all of them largely lacked own fuel resources, the arrival of this gas was welcomed by the respective republican governments. A side effect of the arrangement, however, was that dependence on Galician gas reduced the prospects to break out of the Soviet Union and orient themselves—economically and politically—toward the West, an ambition that remained very much alive particularly in the previously independent Baltic republics.[19]

The Dashava-Minsk-Riga pipeline started supplying Belarus in 1960, Lithuania in 1961, and Latvia in 1962. Meanwhile in eastern Ukraine, the vast Shebelinka field became increasingly important for supplying not only the heavily industrializing regions around Dnepropetrovsk and Kharkov, but also for strengthening gas supply in Moscow and several cities en route there, notably Kursk and Bryansk. In addition, a pipeline was built westward from Shebelinka down to Odessa on the Black Sea and from there into the tiny Moldovan SSR. Moldova in this way became the sixth Soviet republic to become reliant on Ukrainian gas.[20]

Another Soviet interconnection plan was realized in the Caucasus. From Azerbaijan's giant Karadag field on the Caspian Sea, pipelines were built westward into neighboring Armenia and Georgia. Initiated in 1958, the system was optimistically dubbed the "Friendship of the Peoples." Since the region was extremely mountainous—the pipeline reached its highest point at more than

Figure 2.4 Map of the Soviet pipeline system as of the early 1960s. An interconnected Soviet grid was at this time just about to take shape.

Source: gwf. Reproduced by permission of Oldenbourg Industrieverlag GmbH.

2,000 meters above sea level!—construction was unusually demanding from a technical and logistical point of view. Yet Georgia's capital Tbilisi received its first shipments of Azeri gas already in 1959. A year later Yerevan in Armenia, 700 km from Karadag, followed suit. The pipeline paved the way for a radical reorientation and transrepublican integration of Caucasian energy supply. By 1966, Azeri natural gas already amounted to 44 percent of Armenia's aggregate fuel supply.[21]

Further, a major challenge for Glavgaz was to improve gas supply on the eastern slopes of the Urals. This key industrial region had so far largely relied on coal that was brought in from remote places such as the Kuznetsk Basin (Kuzbass, in Siberia) and Kazakhstan. Kortunov proposed using Central Asian gas to improve and modernize the region's fuel balance. A number of large gas fields had been discovered in Uzbekistan in the 1950s, and the government's immediate response had been to make use of this gas by launching a new chemical industrial complex next to the fields. However, it soon turned out that the gas resources were so vast that neither Uzbekistan nor the other

Central Asian republics would be able to absorb them. By 1960 the commercial reserves of the largest Uzbek field, Gazli, were already estimated to be 500 bcm and some analysts in the West even believed its reserves to be the "biggest in the world."[22]

In this situation Kortunov came up with the ambitious plan to connect Gazli with the large industrial cities of Chelyabinsk and Sverdlovsk in the Urals, more than 2,000 km to the northwest. Construction of the system started in 1960. Chelyabinsk received its first Uzbek gas in 1963 and Sverdlovsk in 1965. Completion of the lines was celebrated as a great achievement both because of their unique length—Kortunov noted that they were longer than the trans-Canada pipeline (completed in 1958)—and because of the extraordinary challenge of laying the lines straight through the Central Asian desert.[23]

The next step was to connect the emerging European and non-European pipeline systems with each other. Kortunov pointed to the necessity of a paradigmatic shift in emphasis, from a focus on separate pipeline projects for the European and non-European parts of the country, respectively, to pipelines that were to link major non-European gas fields to the major consumption centers in the central Russian regions and Ukraine. The construction of a huge new pipeline connecting Central Asia with the regions around Moscow was identified as a "primary task": "on its fulfillment," Kortunov wrote in January 1965, "depends to a considerable extent the uninterrupted supply of fuel to the central regions of the country during the upcoming five-year-plan." The vision of a fully integrated Soviet gas system, in which Moscow, for example, could be supplied from both Galicia and Central Asia, was thus taking concrete shape.[24]

The Rise and Stagnation of the Pipe and Equipment Industry

The emergence of an all-Soviet pipeline grid was crucially dependent on the availability of high-quality steel pipe, compressor stations, and a variety of other materials, machinery, and equipment. Since the Soviet Union was a centrally planned economy, these goods were not always readily available for purchase on the market. Rather, everything had to be coordinated centrally with Gosplan and the responsible branch ministries. The efforts to access pipe and equipment were bound to give Alexei Kortunov and his successors much headache throughout the Cold War period. Crucially, the difficulties experienced in this context provided strong motivation for cooperation with the West—in a way that would turn out decisive for the making of the Soviet Union as a gas exporter.

In the late 1950s and early 1960s, failure to match rapidly growing gas production with scaled-up manufacture of large-diameter steel pipe was increasingly pointed at as the main reverse salient in the expanding gas supply system. The steel industry had some experience in producing pipes for the oil industry, but the manufacture of gas pipes, which would have to be strong enough to resist very high pressures, was technologically much more challenging. It was only through large, high-pressure pipelines that the transportation of natural gas over long distances became economically viable, but production of such pipes was not an easy engineering task. The steel alloy needed to be of very high quality, as would welding techniques, anticorrosion methods, and the like. The challenge increased further through Glavgaz's desire to make use of ever larger pipes. The Dashava-Kiev pipeline, taken into operation in 1948, measured 529 mm in diameter. By 1951 the Soviets had taken into operation their first 720 mm pipes. By 1955 the maximum size had grown to 820 mm and by 1959 to 1,020 mm. A 1,020 mm pipeline, according to Soviet calculations, was about four times cheaper to build and operate per cubic meter of gas than a 529 mm pipeline (under the condition that both were actually operated at their designed capacity). Similarly, 1,020 mm pipes were seen to consume less than half the amount of metal relatively to the volume of gas transported.[25]

However, the Soviets lagged far behind the West in the field of pipe manufacture, and they were painfully aware of this. American and West German steel pipes were acknowledged to be superior. Glavgaz's technical director Yuli Bokserman complained that "our metallurgical industry is slow in mastering the production of new types of steel and does not carry out work for the application of technically improved steels for the production of pipes with increased strength." Since the pipe industry sorted under the Ministry of Ferrous Metallurgy, Glavgaz could not take any direct measures for improving the situation. Instead, it opted to approach the Council of Ministers with a proposal to cover the domestic pipe deficit through imports of high-quality pipes from capitalist countries. Drawing on Khrushchev's enthusiasm for rapid development of the gas industry, Glavgaz director Kortunov managed to convince the Ministry of Foreign Trade that a significant share of hard currency earned from Soviet oil exports be reserved for this purpose.[26]

The Western pipe industry welcomed the initiative. In most parts of Europe, natural gas production and transmission had not yet reached the stage where 720 mm and 1,020 mm pipes were economically motivated, and the Soviet Union thus offered one of few markets for large-diameter pipe. Indeed, several pipe factories built in Germany and other capitalist countries would hardly have come into existence in the absence of Soviet demand. Mannesmann, the largest German pipe manufacturer, shipped around half of its total pipe output to the communist bloc. Exports gained momentum following a weakening of the US-led CoCom embargo policy in 1958. German pipe exports to the Soviet Union grew from 3,200 tons in 1958 to 255,400 tons in 1962. Italy's leading pipe manufacturer, Finsider, was not far behind: in accordance with a much-publicized countertrade deal signed in 1961, the Italians agreed to deliver no less than 240,000 tons of large steel pipe in return for massive imports of Soviet oil.[27]

The arrival of the Western pipes were of crucial importance to Soviet natural gas system-building. Without them, Khrushchev would hardly have been able to challenge the United States in his 1959 New York speech, proclaiming that the Soviet Union was about to overtake the United States in natural gas production. Western pipes played a key role not least for the construction of the large pipelines from Central Asia to the Urals, which started to be built in 1960 and which deployed 1,020 mm pipes. Had it not been for supplies from abroad, Glavgaz would hardly have been able to complete the much-celebrated lines to Chelyabinsk and Sverdlovsk on schedule.[28]

Another reverse salient was seen to lie in the field of compressor technology. Compressors were crucial for bringing long-distance pipelines to their designed transmission capacity and making their operation economical. Without compression of the gas, the capacity of a pipeline was totally dependent on the natural pressure in the underground well, which usually meant that only small volumes of gas could be transported. The need for powerful compressor stations increased with the length of the pipelines, and their importance, therefore, grew as the system expanded.

The first Soviet compressors were installed along the Saratov-Moscow and Dashava-Kiev pipelines in the late 1940s. These machines were imported from the United States, but the Soviets subsequently managed to copy the machines through reverse engineering. As a result, the first Soviet-built compressors could be taken into use from 1955, manufactured by an enterprise known as the "Engine of Revolution" in Gorky (Nizhny Novgorod). The aggregates were reported to work reliably, but following the rapid growth of pipeline length and diameter they soon became obsolete.[29]

The Gorky plant was not able to respond to the demand for larger compressors. Instead, the Leningrad-based Nevsky machine-building factory took over as the leading domestic manufacturer. This was a proud enterprise with prerevolutionary roots and a strong track record in high-quality steam engine production. During the 1917 October Revolution, workers at the plant had played an active and heroic role in liberating Russia from Tsarism. During the Soviet era, the plant continued to produce key industrial machinery and equipment, and it was an obvious candidate for taking on the challenge of producing the new, larger compressors that the gas industry was in dire need of. The new compressors were to be of the more modern centrifugal type and have gas-turbine driving gears. Their capacity was to be 4 MW, making them five times as powerful as the Gorky model.

Actual progress, however, was slow. In 1958, Glavgaz' technical director Yuli Bokserman complained that "lack of equipment" had led to a situation in which the building of compressor stations lagged far behind that of pipelines themselves. The problem was particularly disturbing for the Soviets since the United States seemed to be taking into use large compressors seemingly without notable difficulty.[30] Two years later, Kortunov complained openly in the gas industry's main branch journal, Gazovaya promyshlennost (Gas Industry), that the ambitious pipeline programs endorsed by Khrushchev were being retarded, for reasons beyond his agency's control:

> The successful solution to the task of increasing the transmission capacity of gas pipelines is seriously hampered by the Nevsky machine-building factory and the Gorky factory 'Engine of Revolution', which are much too slow in mastering the production of turbo-compressors and gas compressors with a high effect.[31]

Again, Glavgaz contemplated the possibility of imports from capitalist countries. Compressor technology, however, was a high-tech area with direct links to the military and in particular to the production of jet engines for the aerospace sector. As a result, compressors, in contrast to pipes, were from the early 1950s subject to CoCom export restrictions. The Soviets then turned to Czechoslovakia, which hosted the communist bloc's most advanced machine-building industry. The prospects for cooperation with the Czechs did not look bad, at least as far as smaller compressors were concerned. As for the larger machines, it seemed that Glavgaz would have to continue relying on domestic manufacturers.

In 1959, the first domestic 4 MW compressor, produced by Nevsky, could at last go into serial production. Glavgaz immediately started installing them along the newly built Stavropol-Moscow-Leningrad pipeline. However, the brand-new machines were found to have serious manufacture defects. Glavgaz sought to speed up Nevsky's attempts to deal with the problem by offering its assistance and cooperation. In 1960 a series of joint measures were taken for adjusting the compressors, eventually resulting in an updated version of the model. But the problems continued. None of the units reached their designed capacity and all of them turned out to be extremely sensitive to changing weather conditions.[32]

The next compressor model, with a capacity of 5 MW, proved more reliable, prompting the political leadership to award several Nevsky workers the prestigious Lenin Prize. The new machines were installed along several Soviet pipelines from the mid-1960s. In 1965, Glavgaz noted that "the construction and launch of compressor stations has significantly improved the use of long-distance gas pipelines." A full-scale model of the 5 MW machine was proudly presented at the Exhibition of Achievements of the National Economy (VDNKh) in Moscow, where it could be viewed in the gas pavillion from May 1964.[33]

In the meantime, however, the Soviet gas system continued to expand and the need for even larger compressors was quickly becoming evident. The new long-distance pipelines from Central Asia to the Urals, in particular, were seen to demand very powerful compressor units. Against this background the Soviet economic plan for 1963 foresaw the completion of pilot gas-turbine compressors with capacities of 6 MW and 9 MW. As before, however, actual progress in bringing the new technology into production was slow, and as a result, the Central Asian lines were in the end equipped with the much smaller and less economic 5 MW machines.[34]

"A Big Surplus for Export"?

The Cold War gas duel between the Soviet Union and the United States and the intense attempts from the Soviet side to access foreign pipes and compressors were not the only ways in which the Soviet gas industry interacted with the West during these early years. Kortunov and his colleagues also participated in a variety of international gas conferences and held membership in several branch organizations with a European or global scope. The most important groups were the "Ad Hoc Working Party on Gas Problems" of the United Nations Economic Committee for Europe (UNECE), and the International Gas Union (IGU), both of which the Soviets joined soon after Glavgaz' creation in 1956. Glavgaz, whose international delegations were usually led by Kortunov's deputy for international affairs, Alexei Sorokin, found participation highly rewarding, and the Soviets became very active members. Motivated by a strong perceived need to learn from the West, the Soviets, as witnessed by their West European and North American colleagues, showed an "intense curiosity and thirst for greater information" on all aspects of the gas industry in the capitalist world.[35]

Conversely, Western gas experts became increasingly interested in attending UNECE and IGU meetings as a way to keep themselves informed about Soviet developments. When Glavgaz in summer 1960 for the first time hosted an IGU Council meeting, hundreds of foreign gas experts and company representatives got a unique opportunity to acquaint themselves more closely with system-building on the other side of the Iron Curtain. In connection with the meeting, the foreign guests were invited to take part in an excursion to see, with their own eyes, the Soviet gas system in operation. Their conclusion was that Khrushchev's and Kortunov's bold statements about the Soviet system's dynamism were not taken out of the air.[36]

A year later, a large American gas industry delegation toured the Soviet Union for several weeks. The trip was sponsored by the American Gas Association (AGA) and was part of the State Department's East-West industrial exchange program. The Americans felt ambiguous about what they saw, referring to the Soviet Union as a "land of contradictions." On the one hand, they found the overall surroundings—in terms of dwellings, industrial areas, and construction sites—depressing. They judged that there was "a great excess of labor" and they were shocked to see so many women in the field, "digging bell holes or driving big D-7-type sideboom tractors." William R. Connole, a well-known former member of the US Federal Power Commission, stated that he had "never seen such incredible disregard for elementary measures of accident prevention and safety control for the thousands who were engaged in building the [pipeline] projects." The Americans were also surprised to find so much technology—notably compressor stations and a variety of construction equipment—that obviously originated in American inventiveness but had been reversely engineered and shamelessly copied. On the other hand, they were deeply impressed by the "singlemindedness of purpose" and the speed of construction that seemed to make Kortunov's directorate unique in comparison with other branches of Soviet industry. Connole thought that the extreme growth rates reported by Glavgaz were possible "only in a totally state-oriented economy," arguing that "no one who has seen the Soviet economy can underestimate its vigor or mistake its irresistible purpose once its mind is made up." The impression was that the extraordinary production goals set for 1965 would probably be met: "They will not be achieved as smoothly, or even as economically as they would be in the United States. But...the whole thing will work and work entirely well enough for Soviet purposes."[37]

The West's growing awareness of Soviet natural gas developments, and the resonance between gas system-building communities in East and West that was generated in the process, fed into discussions on the future of natural gas in Europe as a whole. As of the late 1950s, the known West European gas resources were still small, and it appeared uncertain as to whether Europe would ever be able to follow in the footsteps of North America, where natural gas had already become a fuel of great economic significance, a "public utility" available in large quantities for anyone in need of it. For most Europeans, "gas" was still synonymous to "town gas," manufactured in public gas works or coke plants, whereas natural gas remained a largely unknown energy source without a clear future. Italy, France, and Austria, where some reasonably large gas finds had been made before and during World War II, led Europe's efforts to make better use of the new fossil fuel, but even in these countries the available resources were too limited for natural gas to become anything more than a complementary energy source with dubious long-term prospects.

A turning point came in the wake of the 1956 discovery of a supergiant gas field in North Africa: Hassi R'Mel, located in the Saharan desert in what was still French Algeria. It was believed to contain up to a trillion cubic meters (tcm) of recoverable gas, making it the second-largest gas field in the world, after America's Panhandle-Hugoton. Western Europe was at the time preparing to construct several powerful oil pipelines through which crude oil in large quantities would be brought from North Sea and Mediterranean harbors to continental European locations, and it was not far-fetched to suggest an analogous arrangement for Algerian natural gas to be piped into continental Europe's industrial heartlands. The French, in particular, became very active in pointing at such a scheme as a major opportunity, and it was discussed extensively within international gas organizations such as the UNECE's gas committee. In geopolitical terms, the optimism regarding a possible pipeline from Hassi R'Mel to Europe was remarkable if seen in view of the bitter, armed conflict that at the time plagued the North African colony—a conflict that would ultimately pave way for Algeria's independence.

The reports about growing gas reserves in the Soviet Union and firsthand accounts of rapid progress in Soviet natural gas system-building stimulated West European visionaries to play with the idea of Soviet gas exports along similar lines. The research director of the French-Algerian Societé Commerciale du Methane Saharien (COMES), Gérard de Corval, suggested that "gas from the sands and gas from the steppes"—from Sahara and Central Asia, respectively—might ultimately meet in Europe.[38] Corval pointed at the possibility of market competition between the two sources, arguing that this would have decisively positive effects on the overall dynamics and competitiveness of natural gas on European energy markets. A line of price parity was likely to be formed, Corval believed, running roughly north-south through Central Europe, although it was still highly uncertain as to whether this line, to the east of which it would be more profitable to import Soviet rather than Algerian gas, would coincide with the Iron Curtain or rather be placed further east or further west. The thread was taken up by William Connole, who thought that the "volume of Soviet reserves and the stage of development of [gas] production and transmission" in the USSR would enable this country to "eventually reach western Europe with its lines." Based on firsthand impressions from his own Soviet trip, he judged that "within 5 years there'll be a big surplus for export" of Soviet gas.[39]

Kortunov responded to Corval's and Connole's prophetic speculations, which were formulated in 1959–1961, by stating that the Soviet Union did not have any plans, at least not for the time being, to sell gas to capitalist countries.[40] Shortly afterward, then, the envisioned East-West gas trade was overshadowed by the discovery of a supergiant gas field in Western Europe itself: Slochteren in the northern Netherlands. Shell and Esso, which together owned the Dutch Oil Company (NAM), had struck gas at this site already in

1959. The probable size of the field was initially kept secret by the companies and the Dutch government, but in the course of 1961 and 1962, figures started to be publicized, indicating that Western Europe rested on nearly limitless gas riches in its own ground. Longhaul imports from Algeria and/or the Soviet Union suddenly appeared less attractive. At the same time, overall East-West relations faced considerable turbulence. Construction of the Berlin Wall in summer 1961 and the Cuban missile crisis in autumn 1962 marked a new height in Cold War tensions. This seemed to make the possibility of transnational system-building across the Iron Curtain a politically unrealistic project—at least for the time being.

To sum up this chapter, the period from World War II to the early 1960s marked the breakthrough for the Soviet Union as a natural gas producer. Under Khrushchev, the gas industry enjoyed strong political support, and its growth was more rapid than that of any other branch of Soviet industry. At the end of the period, a vast union-wide pipeline grid was already emerging. Mounting problems for the domestic pipe and equipment industry to keep pace with this impressive expansion made cooperation with capitalist countries interesting. Massive volumes of steel pipe from Germany, Italy, and other countries were imported, compressors from the United States were reversely engineered, and the Soviets became highly active in key international organizations where the future of natural gas was discussed. It was in this context that the possibility of Soviet gas exports for the first time started to be discussed. A European gas system supplied from two key external sources, Algeria and the Soviet Union, started to be envisioned. As of 1961–1962, however, vast finds of natural gas in Western Europe itself made such longhaul imports less interesting, while growing geopolitical tensions also made it appear doubtful as to whether West European countries would ever dare interconnect their emerging pipeline grids with those beyond the Iron Curtain.

3
Toward an Export Strategy

From Central Asia to Siberia

Among the Soviet Union's many gas-rich regions, northwestern Siberia would eventually become by far the most important. At the time when West Europeans for the first time started considering imports of red gas, however, it was not yet known that this region rested on nearly limitless natural gas deposits. As of 1959, only two minor Siberian fields were known, the probable reserves of which amounted to a modest 42 bcm. Soviet geologists were optimistic about finding more, but it was still impossible to tell whether Siberian gas would ever become a matter of any strategic importance. Glavgaz head Alexei Kortunov's enthusiasm was still being channelled to other promising gas regions. This concerned above all Central Asia, but also Ukraine, Caucasia, and the Volga region. In 1961, Kortunov's deputy in charge of international affairs, Alexei Sorokin, told the delegates at the International Gas Congress, held in Stockholm, that "the province with the greatest perspectives regarding gas reserves is the Pricaspian Depression." The large American delegation that visited the Soviet Union shortly afterward was similarly informed that the Soviets were most optimistic concerning "the Asiatic republics." In both cases, Siberia was mentioned only in passing.[1]

In January 1962, Glavgaz elaborated on the gas industry's development prospects for the upcoming 20-year period. Although exploration in Siberia had now started to yield some promising finds, it was concluded that Siberian gas would have to play a relatively modest role throughout the period, "because of the distance of the future gas production centers from the users." By 1980, Siberia was expected to cover around 10 percent of total Soviet gas demand. Again, Central Asia was identified as a much more promising region, its gas fields expected to deliver around 25 percent of all Soviet gas. The rest would come from older gas regions in Ukraine, the Caucaus, and central Russia.[2]

At about the same time, however, the first giant Siberian gas field was discovered. It was located near the village of Punga on the left bank of the Ob, not far from an already known, but much smaller field. Shortly afterward, in April 1962, other gas finds were made a few hundred kilometers to the north, near the Taz River. A number of further discoveries followed.[3]

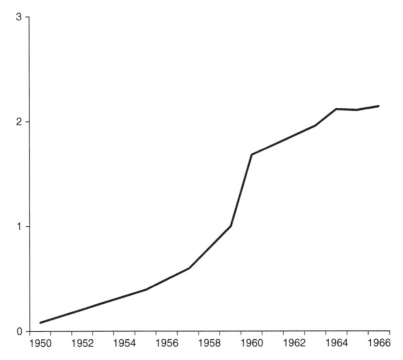

Figure 3.1 Soviet natural gas commercial reserves (as of January 1), 1950–1966 (tcm) (A+B categories in Soviet terminology)

Source: Based on figures in Gazovaya promyshlennost, various issues.

From now on, Glavgaz started to discuss Western Siberia's gas in terms of being of "immense importance for the national economy." The growing number of new Siberian fields was pointed at as a trend of "principal significance." By late 1963, the revealed reserves were already 150 bcm. The finds were welcome in view of a troublesome trend that started to become apparent from around 1962: the commercially available gas reserves of the country, which up to 1960 had increased exponentially from year to year, faced slower rates of growth. This stagnation seemed to threaten the extravagant targets set in Khrushchev's 20-year plan for the period up to 1980, when the aim was to produce no less than 720 bcm in total.[4]

The Siberian discoveries seemed to offer a new basis for sustainable growth. Glavgaz argued that "the strained fuel balance of the regions of the European part of the country and of the Urals can be alleviated *only* by bringing in fuels from far away, among which natural gas from Tyumen region [in Siberia] must take the leading place." Such an arrangement would be highly beneficial for the country as a whole, since it would allow for very large savings of coal and oil.[5]

Glavgaz, together with the Ministry of Geology and the regional party organization in Tyumen, argued that a vast investment program must be launched to quickly develop Siberia's new gas Eldorado. Yet there was no general enthusiasm over this idea. Opinions differed markedly between different actors and analysts as to how the new finds should be exploited. Glavgaz wished to start immediately integrating the newly discovered Siberian fields into the overall Soviet gas system that was about to take shape, envisaging a number of pipelines to Moscow and other European regions. Several powerful actors, however, argued in favor of a more gradual exploitation and a regional arrangement through which the gas would essentially be reserved for Siberia itself and the industrialized Urals, rather than piped westward. These actors included the state planning committee Gosplan (which meant that the relations between Baibakov and his earlier protégée Kortunov worsened) and competing branches of the Soviet fuel and energy complex. The coal and oil industries felt threatened by Kortunov's radical Siberian visions, fearing—rightly so— that the expansion of gas would take place at their expense. The Ministry of Electrification was equally skeptical because its system-builders were planning to have a large part of the gas-rich territories in northern Tyumen flooded for hydroelectric purposes—a project incompatible with large-scale gas production in the same area.[6]

The opponents, or "conservatives," believed that the share of natural gas in the overall Soviet fuel balance would not necessarily have to grow at the extreme pace envisaged in Khrushchev's 20-year plan. Central Asia and European Russia, it was argued, would be able to satisfy demand through the next couple of decades. To further strengthen their arguments, they pointed at the enormous difficulties that would likely plague gas production and pipeline construction activities in northwestern Siberia, whose geographical and climatic conditions posed completely new technological and logistical challenges. Around 70 percent of the Tyumen region was covered by nearly impenetrable swamps. In other areas the ground was permanently frozen, and no regular communication channels existed. Few people looked forward to working there. Tyumen, it was noted, was "by no means a paradise."[7]

The harsh conditions were bound to render most existing equipment useless and new solutions and methods thus needed to be developed, including new steel alloys for the pipelines. The difficulties of the steel and machine-building industries to keep pace with the rapidly growing demands of the gas industry were already well documented and it appeared far from certain that they would be able to adequately meet the even higher requirements in terms of quality and annual output with which Siberian gas production and transmission would be linked. Help from abroad might have offered a partial solution, but this possibility was hampered from 1962 by a much-debated Western embargo, instituted by the NATO in response to construction of the Berlin Wall and the Cuban missile crisis, on large-diameter pipe exports. From this perspective, it appeared highly uncertain as to whether the red empire would be able to take on the Siberian challenge.[8]

Kortunov and other "radicals" acknowledged that Siberia would demand extremely bold investments and that the realization of the vision would be very costly on the short term, but stressed that this would be compensated for in the long run by a substantially increased efficiency with regard to the country's energy supply. While Baibakov and others pointed to the enormous risks linked to the undertaking and argued that the preliminary geological estimations would first of all have to be confirmed with much greater accuracy, the radicals did not see any compelling reason to delay a massive Siberian investment program. Kortunov, who was known as a gambler prepared to take big risks in order to get extraordinary things done (a trait that had nearly killed him on several occasions during the war), thought Gosplan's careful stance unmotivated and unimaginative. In what amounted to a major power struggle, Glavgaz joined forces with other proponents, notably the Ministry of Geology, whose head Alexander Sidorenko was equally enthusiastic, and Tyumen's regional party organization, whose main representative Boris Shcherbina identified Siberian gas as a one-time opportunity to boost regional development.[9]

Another important, though somewhat unreliable supporter was Soviet leader Nikita Khrushchev, whose decisions in the 1950s had paved the way for a new dynamism in Soviet natural gas system-building. In October 1964, however, Khrushchev was unexpectedly ousted and the Kremlin came under the control of an untried leadership trio consisting of Leonid Brezhnev as first secretary, Alexei Kosygin as chairman of the Council of Ministers, and Anastas Mikoyan as head of state (the latter soon to be succeeded by Nikolai Podgorny). No firm decision had at this time yet been taken as to how and when Siberia's natural gas was to be exploited.

Glavgaz and the West European Natural Gas Scene

Kortunov and his colleagues spent enormous efforts during these years on mobilizing internal Soviet support for the gas industry's sustained growth. At the same time, they followed closely and with great interest developments abroad and particularly in Western Europe. There, the popularity of natural gas as a fuel was growing rapidly, and the prospects for trading gas internationally had started to gain momentum.

The main European focus was on the Netherlands and its giant Slochteren gas field. In September 1963 a group of West German energy companies applied for permission to build a pipeline from the Dutch border to Germany's industrial heartlands. It was widely seen as the starting point for a large-scale "Slochteren-supplied gas system that might someday extend along the Rhine Valley and into Switzerland and Austria." A few months later, it was reported that the state-owned French gas company, Gaz de France, was also negotiating with the Dutch for a contract. If realized, it would more than double France's consumption of natural gas. Belgium and the United Kingdom followed suit by initiating negotiations with the Dutch, whose gas reserves were now estimated at a breathtaking 1,100 bcm.[10]

At the same time, the end of the traumatic French-Algerian conflict and the rise of Algeria as an independent nation (1962) brought new life to the old vision of large-scale exports of Saharan gas to Western Europe. France, the former colonial power, and Britain, which had already experimented with liquefied natural gas (LNG) technology and even imported small amounts of such gas from the United States, became the first to conclude contracts with Algeria. Another possibility that was much discussed concerned possible gas imports from Libya.

By the mid-1960s, both Dutch and Saharan gas thus seemed to be on good way to reach West European markets, although it was still far from obvious which of the two would emerge as the dominant supplier. For some countries and regions, notably Spain, Portugal, and southern Italy, an import of Saharan gas appeared more suitable than imports from the Netherlands. For Belgium, Britain, and northern Germany, in contrast, Dutch gas was, both politically and logistically speaking, more attractive. France, Switzerland, Austria, southern Germany, and northern Italy were markets in between.

An important feature of the in-between markets, disfavored as they were by the considerable distance from both the Dutch and the Saharan fields, was that they took great interest in possible third sources of supply. Soviet gas was by far the most attractive of these. Two trends inspired actors to take a serious look at the prospects for imports from the East. The first was the impression that, thanks to the recent Siberian discoveries, the large Soviet "surplus for exports" that William Connole had predicted a few years earlier was indeed about to materialize. The second was the trend toward relaxation—détente—in East-West relations. In June 1963, US president Kennedy formulated his "strategy of peace" and submitted to the Soviet Union concrete proposals for putting an end to the Cold War. Shortly afterward the Partial Nuclear Test Ban Treaty was signed in Moscow. Western restrictions to trade with the communist bloc were weakened. The trend gained further momentum in the course of the next few years. By 1964, when Khrushchev was outmanoeuvred in Moscow, tensions *within* the capitalist and communist worlds, respectively, seemed to be nearly as serious as those *between* the two. "Polycentrism" was the word of the day.[11] Under such circumstances, political reservations did not necessarily dictate what could be done in terms of economic cooperation and transnational system-building.

The prospects for Western Europe to link up with the Soviet gas system were strengthened in early 1964 following an agreement-in-principle between the Soviet Union and Czechoslovakia on the construction of a gas pipeline—dubbed the "Bratstvo" (Brotherhood)—for exports of natural gas from the USSR to its socialist neighbor. This was essentially a follow-up on the much-publicized "Druzhba" (Friendship) oil-pipeline system, which had been taken into operation two years earlier and through which several Central European countries had started to import very large volumes of Soviet oil. The intention was to subsequently extend the Bratstvo to other COMECON member states. A more definite Soviet-Czech contract was signed on December 3, 1964, whereby it was revealed that the volumes to be traded were fairly small. Yet, similar to the first Dutch and Algerian gas export contracts, the deal was of principal significance in demonstrating the Soviet Union's willingness to make its natural gas available to users abroad. Moreover, the deal implied that the Soviet gas-pipeline system was about to be extended westward. As a matter of fact, Soviet gas would be available as far west as Bratislava, just on the border to Austria.[12]

When the Soviet-Czech contract was signed, several West European gas companies were in the midst of tough negotiations with the Netherlands, Algeria, and/or Libya for large-scale gas imports. The Kremlin and Glavgaz were from now on increasingly approached by West European governments and gas companies with concrete and, as it seemed, serious inquiries about possible imports of "red" gas as an alternative or complement to gas from elsewhere. This concerned in particular Austria, which bordered on Czechoslovakia, and

Italy, whose state-owned oil and gas company ENI had already gained a reputation for its massive imports of Soviet oil, and which faced a mounting gas shortage on the mid-term.

Considering Exports: Opportunities and Risks

Glavgaz director Alexei Kortunov was enthusiastic about the West European interest in Soviet, and, in particular, Siberian natural gas. In the internal Soviet discussions, he pointed at the great economic value of a possible gas export to the West as well as at "the important political significance of an influence on the fuel-energey balances of the capitalist countries." The dynamic development of the West European natural gas market, with some gas companies already having signed contracts for imports of natural gas from the Netherlands and/or Algeria, was referred to as a trend of utmost importance from a Soviet perspective. The red empire, with its vast new Siberian gas fields and its long experience of oil exports to both Central and Western Europe, was seen to have a unique opportunity to take part in a formative phase of transnational system-building.[13]

Kortunov's real goal, however, was not so much to promote foreign trade and Soviet political influence abroad as to strengthen the role of natural gas as a dominant energy source in the Soviet Union itself, and his vehement support for exports must be seen in that light. More precisely, Kortunov argued in favor of gas exports because he thought that the gas to be exported would have to come from Siberia, and that a Soviet commitment to exports would thus demand a rapid exploitation of the newly discovered fields there. In the case of exports to Czechoslovakia, the plan was to have the gas delivered from western Ukraine, but for exports of volumes large enough to motivate the construction of a pipeline all the way to countries such as Italy, the limited Galician reserves would not do.

Another factor that motivated Kortunov to eagerly support a gas export to Western Europe was that it, to judge from previous experiences of Soviet-Western energy relations, would be organized as a countertrade scheme in which the Soviet Union, in return for gas, would get access to advanced industrial items. Glavgaz had already profited considerably from imports of West European steel pipe and other equipment acquired in return for Soviet oil, and the prospects for expanding such imports were bound to grow if not only oil, but also natural gas became subject to foreign sales. Soviet attempts to import large-diameter pipes from the West had been made more difficult following the NATO's 1962 embargo policy, but the new trend toward détente in East-West relations made a loosening of this policy likely. The large-diameter pipes that Glavgaz hoped to access could conceivably be used for the construction of new long-distance pipelines from Siberia.

Kortunov was not the only one who saw advantages. The nation's political leaders also took interest in the idea, though for different reasons. The Kremlin saw both economic and political opportunities. A major agricultural crisis in 1963 had forced the Soviets to spend enormous sums on grain imports from the West, and, as a result, the Ministry of Foreign Trade needed to strengthen its hard currency earnings. Stepping up oil exports was already acknowledged as a main strategy in this connection, and natural gas could potentially become a further source of hard currency. The Ministry of Foreign Trade thought that gas sales to capitalist countries could become extremely profitable, given the prices for Dutch and Algerian gas that were up for discussion in the West.[14]

From a political point of view, the idea of gas exports seemed to offer a new way for the Kremlin to counter America's growing influence in West European economic affairs. In the previous period, the Soviets had noted with satisfaction that West European economic independence, following the end of the Marshall Plan in 1951, seemed to be strengthening at the expense of US influence. By the mid-1960s, however, Brezhnev pointed at American capital "again being invested heavily in the industries of Italy, the Federal Republic of Germany, Britain, and other countries." The United States were also believed to be looking for "some pretext to enable it, 20 years after the war, to continue keeping its troops and war bases in Europe and thereby have the means of directly influencing the economy and policy of the West European countries." Energy exports were one of few areas where the Soviets could actually provide a counterweight to American economic dominance in Western Europe. From this point of view, the idea of broadening the export base from oil to natural gas was certainly attractive. In natural gas, as a matter of fact, the Soviets would presumably be able to play an even stronger role than in oil, since competition from across the Atlantic would hardly become an issue.[15]

An opportunity that does *not* appear to have been discussed among the Soviet leadership was the possibility of using (the threat of) gas supply disruptions as a political weapon. If mentioned at all in internal Soviet strategizing, the conclusion must have been that such a weapon could not possibly have any significant effect within the foreseeable future, as natural gas at the time played only a minimal role in Western Europe's overall energy supply. This meant that the effect of a disruption would be minimal for economy and society. It also appeared improbable that the importers would dare accept more than a partial dependence on Soviet deliveries. For Soviet natural gas to possibly become an "energy weapon" in the sense that would make it subject to intense debate later on, it would first have to establish itself as a major and trustworthy component in the overall European energy system. This could not be achieved overnight.

But there were also arguments *against* large-scale gas exports to capitalist countries. One problem was that the idea did not fit neatly into the Soviet Union's overall foreign trade strategy. Soviet planners tended to be antimercantilist, viewing exports as "a necessary evil to pay for required imports." The purpose of foreign trade was to use imports to help meeting internal goals, and the role of exports was merely to cover the resulting deficit in the balance of payments. Oil had become a favorite export article in this context, since deliveries could easily be adjusted up and down so as to balance imports. Through various arrangements involving transport by pipeline, tanker, rail, and truck, Soviet oil could be shipped to customers anywhere in the world without further ado and in accordance with agreements that were not necessarily linked to any long-term commitments.[16]

Gas exports, in contrast, would demand a much more complex and less flexible arrangement in which the volumes to be exported would have to be determined years or even decades in advance. This followed from the grid-based nature of the envisaged export system. Soviet gas could be shipped economically to Central and Western Europe only by pipeline, and to finance such a system it would be necessary to agree in advance on its long-term utilization. The implication, from a planning point of view, was that gas exports could not be called upon to fill in holes in the foreign trade balance with short notice. Moreover, in contrast to oil, natural gas could be exported only to countries that were connected to the pipeline system. It could not be shipped to alternative customers in case of refusal by an importer to receive—and pay for—the gas. Exporting natural gas was thus linked to major risks for the exporter.

Most Soviet actors, though, reasoned that the opportunities linked to gas exports were sufficiently large for the risks, including the lack of flexibility, to be worth taking. The main opposition was not so much to exports as such as to Kortunov's argument that the gas would have to stem from Siberia. Gosplan, along with other "conservatives" in the internal Siberian debate, vehemently opposed this claim. Since they were strongly against a rapid exploitation of Siberia and of transmitting large volumes of Tyumen gas to regions other than Siberia itself and the Urals, the conservatives argued that exports were desirable only under the condition that they derived from non-Siberian sources. Given the highly strained Soviet fuel balance, it was far from certain as to whether such resources could be mobilized at all.

Seeking Cooperation with Italy and Austria

Kortunov argued that the Soviet Union must act quickly to capture a significant share of the emerging West European gas market. Among the possible customer countries, Italy and Austria were identified as the most promising. Both were already large-scale users of natural gas, but precisely the popularity of gaseous fuels, in combination with limited domestic resources, was seen to generate a need for import. Given the anticipated increase in gas consumption, the two countries were expected to face a gas deficit already from around 1970. Kortunov noted that the Italians and Austrians were turning not only to the Soviet Union, but to several other countries as well—including the Netherlands, Algeria, Libya, and even Iran—inquiring the prospects for gas imports. The Soviets thus faced considerable competition. Yet Glavgaz was optimistic about its chances of capturing a significant market share, a major reason being that the Soviets, in contrast to the Dutch and the Algerians, could build on a tradition of close cooperation with the respective state-owned oil and gas companies in Italy and Austria—ENI and ÖMV, respectively. The cooperation with Italy, which by virtue of its size was regarded as the most important potential market, was particularly well developed.[17]

The internal Soviet disagreements as to whether sufficient volumes of gas could be made available for export initially prevented any formal Soviet-Italian negotiations from being initiated. Instead, Glavgaz and ENI formed an informal "study group." Though not supported by any official government decree, and certainly not by Gosplan, the group started to sketch several possible export regimes. Two alternative pipeline routes were discussed: the first involved Czechoslovakia and Austria and the second Hungary and Yugoslavia. ÖMV was highly interested in the former. The Italians, however, were reluctant to let Austria take on the role of main entry point for imports of Soviet gas to Western Europe and, therefore, advocated the Hungarian-Yugoslavian option. The Soviets also tended to favor this route. Not only would it be shorter and logistically less cumbersome than an Austrian route—in particular, it would not have to cross the Alps—but it would also enable the Soviets to get a foothold in the Hungarian and Yugoslavian gas markets. Like many West European countries, these two socialist states were at the time seeking access to more gas. Hungary had already started importing small amounts of natural gas from neighboring Romania, and Yugoslavia was eagerly approaching the Algerians for a possible LNG arrangement. Regarding exports to Austria, Glavgaz thought it more advisable to arrange these independently of deliveries to Italy. More precisely, Austria might be supplied through an extension of the "Bratstvo" pipeline.[18]

The export project that Glavgaz and ENI started elaborating on in 1964–1965 was popularly referred to as the "Trans-European Pipeline." The name testified to ENI's far-reaching ambition to turn northern Italy into a major hub for the transit of Soviet gas to markets further West. While wishing to exclude Austria from the project, the Italians worked hard to persuade the main French gas company, Gaz de France, to make use of the possibility of linking up to the pipeline, and import Soviet gas by way of Italy. Switzerland, it was imagined, could also become part of the system.

The Soviets were enthusiastic about a possible inclusion of these countries, particularly France. Glavgaz had recently negotiated large-scale exports of liquid petroleum gases (LPG, that is, propane and butane, not to be confused with LNG) with the leading French company in this field, Gazocean, and was eager to broaden its sales. More importantly, France was regarded by Glavgaz as an excellent supplier of advanced equipment for the gas industry, notably anticorrosion technology and facilities for extraction of unwanted hydrogen sulfide from natural gas and the production of useful sulfur on its basis, an area where France was globally leading. Exports of natural gas to France would most probably open up for boosting this promising equipment trade.[19]

In another part of the world, Japan displayed a clear interest in gas imports from its mighty neighbor. In this case the gas fields of interest were located in eastern Siberia, whose blue gold, it was suggested, could be transported by pipeline to Sakhalin and from there to Japan in the form of LNG. In contrast to Glavgaz' European export ambitions, the idea of exports to Japan does not appear to have been opposed by Gosplan, and formal negotiations could, therefore, be launched at an early stage. In January 1966, a Soviet delegation with representatives from Glavgaz, the Ministry of Foreign Trade, and the Kremlin's Bank of Foreign Trade traveled to Tokyo for a first round of negotatiations, holding talks with several large industrial companies that represented the main intended gas users. Japan, despite its close relation with the United States, had not adhered to the NATO's 1962 pipe embargo and when German pipes had become unavailable, the role of Japan had increased in this respect, paving way for an expansion of Soviet-Japanese trade relations. Apart from pipes and equipment for the natural gas industry, the Soviet Union had bought two gas transportation vessels that were planned to be put into operation for the above-mentioned export of Soviet LPG to France. Through exports of natural gas to Japan, the Soviets hoped to sustain and further expand their access to Japanese technology.[20]

In parallel with its attempts to enable exports, Glavgaz also prepared for substantial gas *imports*. The partners eyed here were Afghanistan and Iran. In October 1963, Moscow concluded a first gas import contract with Afghanistan, paving way for the construction of a pipeline from fields in northern Afghanistan to the Uzbek SSR. The line helped strengthen the emerging Central Asian pipeline grid while also boosting deliveries to the Urals. In January 1966, then, a much larger contract was signed with Iran, foreseeing deliveries of 10 bcm annually across the Iranian-Soviet border to Azerbaijan. The main purpose of these imports was to strengthen Caucasian gas supply and thereby free large volumes of Azeri gas for transmission northward. Gosplan suggested that these volumes could become a source of gas for export to Central and Western Europe. This would then render the risky construction of export pipelines all the way from Siberia unnecessary. Glavgaz disagreed, continuing to insist that only Siberian gas would do.[21]

The Export Strategy Takes Shape

In the course of the internal Soviet debate about Siberia and the possible use of Siberian gas for export purposes, the proponents of a radical strategy for exploiting Siberia's riches—whether or not for use in the export arrangements—were strengthened by a steady stream of reports from Tyumen about new gas discoveries. By early 1965, Glavgaz estimated the probable natural gas reserves in the region at 5,000 bcm, which was "significantly more than the reserves that have been prospected so far in all other gas-bearing regions of the USSR." A year later the figure had doubled.[22] Increasingly, it became clear that Tyumen's gas riches were so immense that Siberia and the Urals would not be able to absorb this gas on their own—not even on the long term. This development tended to convince a growing number of actors that Siberian gas must be piped westward, and that an increased reliance on natural gas as a fuel would benefit the country as a whole.

Political support from Brezhnev and Kosygin seemed confirmed through the transformation of Glavgaz, in October 1965, into the Ministry of Gas Industry (Ministerstvo Gazovoi Promyshlennosti, or Mingazprom for short). Although the creation of the ministry was not an isolated or unique event, but part of a major reform package launched by Kosygin, the psychological effect was of enormous importance. For Kortunov, it signified the rise of the gas industry as an equal to the oil industry (which had had its own ministry for several decades already). Kortunov concluded that the decision was "a recognition that our gas industry has now become one of the leading branches of the national economy, taking an ever growing influence on the development of the fuel-energetic basis and the increase of the productivity of societal labor."[23]

In fall 1965, Kortunov ordered the industry's main design institute, Giprospetsgaz, to send out an expedition to the far north and come up with a concrete proposal for how Siberian gas could be brought westward. A group of five experts led by Giprospetsgaz's chief engineer Dertsakyan set out for the wilderness by boat, helicopter, and dressine. In December 1965, a radical new gas supply scheme for the whole northwestern segment of the USSR was presented, aimed at a gradual, long-term transition to Siberian gas as the main source of supply. Kortunov suggested that a massive system of pipelines be built from the northern districts of Tyumen oblast over a distance of several thousand kilometers to Leningrad, Belarus, and the Baltic republics, as well as to export markets in Central and Western Europe. If everything went well, Siberian gas might start flowing to these internal and external regions within only a few years' time.[24]

Mingazprom envisaged annual shipments of no less than 90 bcm of Siberian gas, with further capacities to be added later on. The conservatives were outraged by this extremely ambitious and, in their view, unrealistic and dangerous plan. Gosplan Chairman Baibakov and his deputy Alexander Ryabenko vehemently opposed the project, calling for "a deep reconsideration of the questions regarding the choice of the most economical variant for gas deliveries." Rather than launching a major Siberian development program, the central planners thought it necessary to first and foremost make "maximal use of the capacities of existing gas pipelines."[25]

The political leadership sided with Kortunov. At the 23rd Party Congress, held in spring 1966, Brezhnev argued that provision must be made for "further accelerated development of the oil and gas industries." Premier Kosygin stated that of these two branches, the gas industry was to be developed more rapidly, with production to grow from 128 bcm in 1965 to 225–240 bcm in 1970—an increase by more than 80 percent. Oil production, by comparison, was to grow by about 45 percent. The long-term goal, as defined by Khrushchev back in 1961, of producing 720 bcm of natural gas in 1980 had by no means been given up—and it was Siberian natural gas that would account for the main increment. The Council of Ministers decided to give Mingazprom permission to go ahead with an accelerated pace of prospecting and exploration works in northern Tyumen.[26]

Meanwhile, the international struggle among potential exporters for dominance over the emerging West European gas market had grown extremely fierce. Kortunov sought to convince Brezhnev and Kosygin that the Soviet Union must act fast if it wished to participate in this race, arguing that a Siberia-to-Europe pipeline must be built as soon as possible and that this pipeline should be mobilized for exporting 10 bcm of Siberian gas from the early 1970s. Other sources, mainly from Ukraine, should be called upon during a start-up period only. Gosplan, the government's decision in favor of Siberia notwithstanding, remained skeptical. The central planners wanted Mingazprom to investigate the feasibility of alternative export arrangements that did not include Siberian gas. In particular, the planning agency thought that exports to Western Europe had better be coordinated with imports from Iran, which, according to the 1966 agreement, were to commence in 1970.[27]

Baibakov and his colleagues doubted whether it would be economically efficient to export Siberian gas to faraway countries such as Italy. Kortunov responded with a detailed calculation that proved the opposite. The actual costs for delivering Siberian gas to the Soviet border zone in the West was expected to be 5.60 roubles per ton of reference fuel—including the costs for nearly 6,000 km of pipelines, 6.8 million tons of steel pipe, 32 compressor stations, and 4.4 billion roubles of capital expenditures. The Ministry of Foreign Trade, which supported Kortunov's plan, thought it possible to negotiate a price with the Italians of around 12–14 roubles per ton of reference fuel, which would then result in a substantial profit margin, bringing in 64–84 million roubles per year in hard currency. With costs for imported steel pipes, compressor stations, and other equipment estimated to around 400 million roubles, this meant that all costs for the export arrangements could conceivably be repaid in only 5–6 years' time. Kortunov emphasized that after this period, "the currency earnings from realizing gas exports can be used in the national economy," that is, for purposes that would not necessary be linked to the energy industry.[28]

Gosplan, however, continued to insist that Mingazprom's plans were too ambitious, expensive, and overly risky. An export agreement with Italy that centered on a huge pipeline system to be built from scratch all the way from Siberia presupposed that a variety of geological, technical, and logistic difficulties and uncertainties could be successfully resolved within a short period of time. Even if the Soviets succeeded in accessing large volumes of Western steel pipes for the longhaul transmission of Siberian gas, nobody knew whether the system would become a success. Kortunov was more than willing to take the risk, but Baibakov thought the export project might end in total failure, and that this might have negative repercussions on domestic gas supply as well. Therefore, if Soviet natural gas was to be exported, it would have to come not from Siberia, but from Ukraine, Central Asia, or Iran. Kortunov thought it ridiculous to view exports from Siberia as a risk for domestic supply security, emphasizing that the anticipated deliveries to foreign customers would be almost negligible in the country's overall supply balance, amounting to a mere 1.7 percent of total gas production in 1971 and around 4 percent of the planned level in 1975. This made it highly unlikely that exports would "damage our country's industries and cities with regard to the security of gas supply."[29]

Mingazprom did not succeed in winning over Gosplan to its side. Yet Kortunov mobilized a sufficiently strong coalition of actors for the Communist Party and the Council of Ministers to approve of the export strategy. On June 11, 1966, Kosygin formally ordered Mingazprom and the Ministry of Foreign Trade to initiate "negotiations with the Italian state-owned concern ENI for the construction of a pipeline USSR-Italy and the procurement from the Soviet Union of natural gas, and also for the sale, in this connection, by way of a long-term credit, of pipes and equipment for the gas industry."[30] In the Soviet-Italian negotiations that followed, the discussion focused explicitly on the export of Siberian gas, and on a countertrade in terms of 1.3 million tons of large-diameter steel pipe that would enable the construction of a corresponding pipeline. Kosygin's decree meant that the Soviet Union for the first time officially stated its ambition to export natural gas to the capitalist world, formulating a role for itself as a player on the West European natural gas market. It was a step into the realm of the unknown. No one knew what long-term impact the decision would have.[31]

To sum up, the period from 1964 to 1966 saw the gradual formation of a Soviet gas export strategy. It was closely linked with domestic system-building ambitions, and in particular with the possible integration of Siberia's newly discovered gas riches into the emerging Soviet pipeline grid. An internal Soviet controversy over Siberia's future spilled over into the debate over exports, leading some Soviet stakeholders to oppose exports altogether. The main point of contention, however, was not so much whether the Soviet gas system should be connected with that of Western Europe or not, but how and when exports might be realized. A series of unexpectedly large gas finds in Siberia and rapidly growing interest from the side of potential importers in the West gradually strengthened the overall Soviet interest in the possibility of sales beyond the Iron Curtain. The principal motivation were the prospects for hard currency earnings from gas sales to the West, but for Mingazprom it was also a matter of combining exports of natural gas with imports of large-diameter steel pipes from leading West European manufacturers. This would strengthen domestic system-building and, by extension, Mingazprom's position in relation to other branches of the Soviet fuel and energy complex.

Soviet archival sources clearly point to Italy as the closest cooperative partner during these early years. The tentative Hungarian-Yugoslavian transit arrangement, according to which Soviet gas would cross the Iron Curtain at Trieste, seemed to imply that other prospective West European importers would play only auxiliary roles. A few months after the formal Soviet decision to actually initiate negotiations with ENI, however, the project took on a new, unexpected turn, whereby the emphasis shifted from Italy to Austria as the main Soviet partner. The next chapter inquires how this could happen.

4
Austria: The Pioneer

The Austrian Fuel Complex: Nazi and Soviet Legacies

Austria was destined to become the first capitalist country to import red gas, and it would become more dependent on deliveries from the East than any other nation in continental Western Europe. Yet before the 1960s, Austria was better known for its own sizeable production of both oil and natural gas. In Habsburg times it had even been a major energy exporter.

The country's oil and gas activities were geographically concentrated to the province of Lower Austria in the easternmost part of the republic, where a number of promising areas had started to be explored in earnest shortly after World War I and the break-up of the Austro-Hungarian Empire. Prospecting and exploration attracted not only domestic, but also a variety of foreign actors. In 1931, one of the foreign investors, Eurogasco, which despite its name was dominated by American interests, announced Austria's first major gas find. The deposit was located at Oberlaa, only 6 km from a municipal power plant in Vienna's southeastern suburbs. Eurogasco agreed with the municipal electric utility, Wiener Elektrizitätswerke, to build a pipeline from the gas deposit to the power plant and use the gas for electricity production. Natural gas also started to be used for the production of town gas and was fed as such into Vienna's gas distribution system, thereby reducing its dependence on imported coal.

Exploration activities gained further momentum in 1938, when Nazi Germany marched into the country. Significant gas finds were made in spring 1939 at Aderklaa through a joint effort by Eurogasco, Royal Dutch Shell, and Vacuum Oil. Shortly afterward, however, the war broke out and the foreign companies lost their concessions in favor of German companies. The Germans set out to build pipelines from Aderklaa to major users in Lower Austria. A company called Südostdeutsche Ferngas AG was founded for distributing the gas.[1]

In 1945, control over Austria's hydrocarbons shifted dramatically again, as Soviet troops drove the Nazis out of Vienna and eastern Austria. The oil and gas deposits that had been under German control and which before the war had involved large American and West European investments now came under Soviet command. The Soviets initially pursued a strategy of shutting down the region's oil and gas fields and bringing home equipment and machinery as war trophies. This strategy was soon abandoned, however, as it was found more profitable to use the equipment on-site and have Austrian oil shipped east. For this purpose, an enterprise known as the Soviet Mineral Oil Administration (Sowjetische Mineralölverwaltung, SMV) was launched. The Soviets were less interested in Austria's gas, which could not as easily be brought home. Hence the gas found by SMV was mostly flared, and hardly any investments were made in developing the transmission infrastructure. As of 1955, SMV produced 766 mcm of natural gas, of which 32 percent was lost. The rest was distributed through the Nazi-built pipeline network to users in relative proximity of the gas fields.[2]

SMV was wholly owned by the Soviet Union, but it was in practice dependent upon Austrian geologists, engineers, and technicians for its everyday activities. Cooperation between Soviet and Austrian specialists provided the Austrians with something that other Western countries lacked: immediate experience of Soviet management and engineering culture. This would later on turn out to be an asset of great value.

From SMV to ÖMV

Energy played an important role in the negotiations between the Soviet Union and the Western Allies about Austria's political future. Soviet demands for war reparations in the form of oil deliveries from Austria to the Eastern bloc gave rise to controversy, and contributed to stalling the talks.[3] Only in spring 1955 could the Austrian State Treaty be signed. In the meantime, large amounts of oil had already been shipped east. The State Treaty foresaw a continuation of these shipments for another ten years. Amounting to around 40 percent of total Austrian production, they had a substantial impact on the country's overall trade balance up to the first half of the 1960s. During a short period, the forced development of the Soviet-controlled oil fields made Austria the leading oil producer in Western Europe.[4]

Organizationally, the State Treaty created a vacuum in Lower Austrian oil and gas operations. SMV ceased to exist and its Soviet personnel left the country. The government, controversially, opted for nationalization and the formation of a state-owned company, the Austrian Mineral Oil Administration (Österreichische Mineralölverwaltung, ÖMV), which seized control over most of the country's oil and gas riches. Formally based on an enterprise created in 1938, ÖMV was in practice SMV's direct successor. It was a unique agency in terms of its prehistory, which included domestic interests, West European and American investment, Nazi ownership, and Soviet state control. From now on it was in the hands of the Austrian state.

At the time of the company's creation, oil was clearly ÖMV's core business. However, as Austrian oil production started to show signs of decline and exports to the Soviet Union and other countries gradually had to give way to oil imports, the group's gas business quickly grew in relative importance. Several large investment projects were initiated. To the most important belonged the creation of a central compressor station at Auersthal, 20 km northeast of Vienna, the construction of which started in 1957, and a system of pipelines from the gas fields to the station.[5]

Another important task was to create a long-distance transmission network. For this purpose, a number of regional gas companies were set up. The first one, NIOGAS, was created in 1954 and took charge of Lower Austrian gas distribution. It was followed shortly after the ratification of the State Treaty by Oberösterreichische Ferngas (OÖ Ferngas) for Upper Austria and Steirische Ferngas for the province of Styria in the southeast. In addition, Wiener Stadtwerke (Vienna's public works) took responsibility for gas distribution in the capital region.[6]

NIOGAS and Vienna became ÖMV's most important customers. From 1960, when a pipeline was taken into operation between the networks of NIOGAS and Steirische Ferngas, Styria emerged as a third important natural gas consuming region. OÖ Ferngas also wished to get access to ÖMV's gas, but in this case the regional company failed to agree on the terms of delivery. Upper Austria was home to the country's main steel and chemical industries, both of which were highly interested in Lower Austria's gas riches, but ÖMV was pressed by the federal government to prioritize gas supply to eastern Austria, which had been under Soviet occupation and, in contrast to Upper Austria, had never received Marshall Aid. OÖ Ferngas felt discriminated. It considered building a pipeline from the gas fields to Linz at own expense, but eventually found the project too risky. As a result, Upper Austria remained without natural gas, at least for the time being, and its relations with ÖMV remained frosty.[7]

ÖMV's gas production was concentrated to three major gas fields, which up to the 1970s together accounted for around 90 percent of domestic gas production. The largest one was the Zwerndorf deposit, which had been discovered by SMV in 1952. In fact, this was not a purely Austrian gas field, as it stretched across the border into Czechoslovakia. In 1958 an intergovernmental agreement was signed foreseeing its joint exploitation, and it was subsequently renewed in accordance with updated figures on remaining reserves. The cooperative spirit that this arrangement fostered became an important asset for ÖMV in its later efforts to import natural gas from the communist bloc.[8]

Austrian natural gas production more than doubled from 0.77 bcm in 1955, when ÖMV was created, to 1.87 bcm in 1966.[9] It gained an enormous popularity as a fuel both within industry and among municipalities and households. Natural gas was promoted by ÖMV as a high-quality domestic energy source with favorable environmental characteristics—particularly in comparison to coal, which in most European countries was still the dominant fuel. Since Austria did not posssess any significant coal mines, the increased reliance on domestic natural gas, at the expense of imported coal, was also seen to strengthen national energy security.

Yet the very success of ÖMV's gas business soon turned into a problem. The rate of new domestic gas discoveries did not at all match the extraordinary growth of consumption. Anticipating further increases in aggregate demand, ÖMV concluded that domestic production would not be able to meet the country's long-term needs. As the 1960s progressed, the situation became increasingly acute. To solve the problem, ÖMV would either have to put a brake on domestic gas use or secure additional supplies from external sources. Given the fuel's growing popularity and the sizeable investments made in developing a domestic transmission and distribution system, the first alternative was not particularly attractive. ÖMV thus opted for the second path.[10]

Toward Imports: ÖMV versus Austria Ferngas

As it turned out, ÖMV was not the only Austrian gas company that from the mid-1960s started to look for gas import opportunities. The regional companies did the same. In November 1962, ÖMV's three major customers—NIOGAS, Wiener Stadtwerke, and Steirische Ferngas—formed a joint company called Austria Ferngas. Its purpose was to negotiate independent access to gas supplies from abroad and use these to counter ÖMV's de facto monopoly as a gross supplier. The main source of foreign gas eyed was the recently discovered Slochteren field in the Netherlands, whose gas, it was imagined, could be piped to Austria by way of transit through Germany.[11]

OÖ Ferngas, the Upper Austrian distributor, through whose territory a Dutch-Austrian pipeline would most probably be routed, also joined the arrangement, offering the other regional companies transit rights for Dutch gas in return for financial participation in constructing a pipeline that in a first stage would be used for supplying Upper Austria with Lower Austrian natural gas. The line was inaugurated ceremonially in early 1965, serving in particular the large nitrogen works at Linz.[12]

From the perspective of Austria Ferngas, the Upper Austrian pipeline was a first step toward access to Dutch gas. The pipeline was dimensioned so as to make it possible to use it at a later stage for transporting imported Dutch gas in the opposite direction, that is, from Upper Austria to the networks of NIOGAS, Wiener Stadtwerke, and Steirische Ferngas. The pipeline thus formed an important basis for Austria's prospective integration with the emerging West European gas system. Two West German gas companies—Ruhrgas and Thyssengas—had already signed voluminous agreements with the Netherlands for imports of Dutch gas, and the regional Austrian companies were optimistic regarding the prospects for extending this infrastructure into Austria.[13]

In reality it was still an open question as to whether it would make economic sense to transmit Dutch gas over such a long distance, and whether it would be possible to come to agreement with the German regions through which the gas would have to be transited. Moreover, Dutch gas was not ideal from an Austrian perspective, since its calorific value was much lower than the corresponding value for domestic natural gas. This would require expensive arrangements for either converting domestic gas to the quality of Dutch gas (or the other way round) or the construction of separate pipeline networks for internal distribution of Dutch and domestic gas. All in all, the future of Dutch gas in Austria remained uncertain.

Another possibility eyed by Austria Ferngas was to import Algerian gas. For this purpose, the regional companies participated in an international consortium aimed at large-scale Algerian LNG shipments to the Yugoslavian harbor of Koper on the Adriatic. The consortium elaborated on a 580-km transit system to be built through Yugoslavia and Austria to Czechoslovakia. The system would initially transport 4 bcm per year of Saharan gas, of which Yugoslavia would receive 0.5 bcm, Austria 1.5 bcm, and Czechoslovakia 1.5 bcm. In a

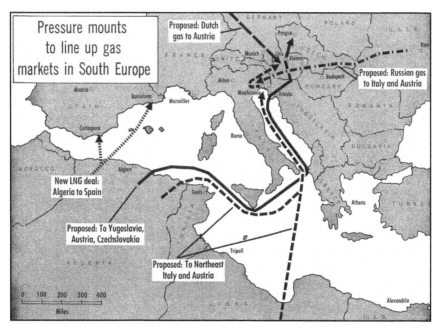

Figure 4.1 Proposed international pipeline and LNG links for the supply of Austria, Italy, and Spain with natural gas. As of 1966 it was still difficult to predict from where Austria and Italy would source their increment in gas supply.
Source: Oil and Gas Journal, February 21, 1966, p. 66. Reproduced by permission.

second phase, the total volume would be expanded to 6 bcm. Czechoslovakia was to finance the project to 50 percent, Austria to 37.5 percent, and Yugoslavia to 12.5 percent.

By winter 1966, these elaborations had reached the stage of "prenegotiations" between the Algerian, Yugoslavian, Austrian, and Czechoslovak organizations involved, but it was still unclear as to whether or not the parties would manage to agree on the gas price. The exotic combination of companies involved, none of which had any previous experience of cooperating with each other, further increased the perceived uncertainty. An important effect of the eagerness with which Austria Ferngas engaged in negotiations and formed international partnerships, however, was that it put ÖMV, the state monopolist, under pressure to defend its dominant position. Austria was such a small country that the success of one gas import project would most probably kill the prospects for others. If Austria Ferngas succeeded in bringing about imports from the Netherlands or Algeria, ÖMV would thus in effect remain without influence in the gas import business and possibly lose its firm grip on the Austrian natural gas market.[14]

ÖMV responded to this threat by devising a gas import strategy that differed considerably from that of the regional companies in terms of geographical and political orientation. With nearly all its activities concentrated to the eastern part of the country, ÖMV was reluctant to seek cooperation with the

Netherlands or Algeria. Imports of Dutch gas were bound to give OÖ Ferngas, with which ÖMV had frosty relations, a key controlling role, and in the case of Algerian imports Steirische Ferngas would most probably come to play a similar part. In both cases, ÖMV risked being bypassed.

Much more attractive, from the perspective of the state monopolist, were the prospects for cooperation with the communist bloc. Not only would imports from the East have to pass straight through the region in eastern Austria where ÖMV produced most of its own gas, but in addition such a scheme would enable the company to exploit its previous experience of cooperation with Czechoslovakia and the Soviet Union. ÖMV first approached the Czechoslovak foreign trade agency Metalimex, with which a unique agreement was reached in March 1966. It focused on "virtual" exports of Czech natural gas from the jointly operated Zwerndorf field. The contract paved the way for ÖMV to unearth some of the gas that in the bilateral production agreement had been recognized as "Czech" gas. ÖMV would increase the rate of production on the Austrian side of the border, while the Czechs, in return for economic compensation, would decrease its production. Apart from having to agree on a reasonable gas price, the partners did not face much extra work for the trade to be launched. In particular, there was no need for any cross-border pipeline. In 1967 ÖMV could in this way start importing natural gas from the East, at a volume initially amounting to 150 mcm.[15]

As for possible imports of Soviet gas, ÖMV's interest was triggered by the Soviet-Czechoslovak announcement in January 1964 that a pipeline—the "Bratstvo"—was to be constructed from Ukraine to Bratislava. ÖMV's managers showed themselves extremely excited about the possibility of linking up with this system. After the line's completion, only a few kilometers would separate the interconnected Soviet-Czechoslovak transmission system from ÖMV's main gas station at Auersthal. It almost seemed too good to be true. Rumors circulated that the Soviets and the Czechs had taken into account a possible extension of the infrastructure into Austria already in their bilateral talks. Whether or not there was any truth to this, ÖMV regarded the prospects for linking up with the planned East European grid as very promising.[16]

Vienna was largely supportive. In December 1964, the issue was formally taken up for discussion within the framework of Austria's bilateral trade consultations with the Soviet Union. The Soviets, however, at the time still unable to agree internally on its overall export strategy, did not think the time ripe for negotiations. Throughout 1965, Moscow seemed to have difficulties making up its mind as to whether—or perhaps rather in what way—gas exports to Austria should be striven for.[17] In June 1966, then, it was revealed that the Soviets did intend to enter the West European gas market, but that they viewed Italy rather than Austria as their main partner. Vienna did not receive any clear indication of whether or not Moscow considered Austria as an additional potential market. For the time being, the virtual imports from Czechoslovakia remained Austria's only source of gas from abroad.

Rudolf Lukesch's Vision

The hydrocarbons industry was not the only branch of the Austrian economy that the government decided to nationalize after the war. Other industries, particularly in natural-resource-related sectors, followed suit. The nationalized industries were managed through a state holding company, ÖIAG, which allowed the management and supervisory boards of the state-owned companies to keep each other informed about their activities in a way that would hardly have been possible in the case of distributed private ownership. This arrangement turned out to be a useful basis for identifying areas of mutual interest. The close links that were built up between ÖMV and the state-owned Austrian steel company VÖEST constituted a prime example.

The nationalization of the Austrian steel industry was relatively easy to carry out, as Austrian steel production was already concentrated to a single company. In 1946 the government created the United Austrian Iron and Steel Works (Vereinigte Österreichische Eisen- und Stahlwerke, VÖEST). Its immediate predecessor was a Nazi-era conglomerate known as Reichswerke AG für Erzbergbau und Eisenhütten "Hermann Göring." This company had been formed shortly after Hitler's annexation of Austria in 1938. On its basis, the Nazis had built up a powerful Austrian steel industry more or less from scratch. Its headquarters and main facilities were in Hitler's hometown, Linz.

VÖEST, which by the mid-1960s had around 20,000 employees, became Austria's most important state enterprise. Taking maximum advantage of postwar Austria's position as a neutral state between East and West, the company developed smooth relations with partners in both Western and Eastern Europe. Whereas the West was somewhat more important in terms of sales, the East was more important when it came to the supply of fuel and raw materials. A major headache for VÖEST, however, was the formation of the EEC, whose external customs barrier disfavored the company on the important West European steel market. In summer 1966 the EEC's six member states embarked on the third and last step in their cooperation as foreseen at the time of the creation of the community in 1957, paving the way for a total removal of intra-EEC customs barriers. At the same time, the Western steel giants increasingly sought to penetrate East European markets. VÖEST, which had traditionally enjoyed a strong position there, in a way that partly compensated for its problems on EEC markets, felt increasingly threatened. As the competitive pressure in the European steel industry increased in the 1960s, the company feared that overproduction would encourage dumping of Western products beyond the Iron Curtain, with disastrous consequences for the Austrians. The company's general director Herbert Koller described his company as being "increasingly encircled from all sides and forced into a hedgehog position."[18] In this difficult situation, VÖEST was eagerly looking for new markets and innovative business arrangements through which its position might be defended. Koller and his colleagues closely followed major trends on both Eastern and Western markets.

An important consumer of steel, the natural gas industry was of great interest to VÖEST. Although its factories did not produce natural gas pipes, they did manufacture the thick steel plates that were used as an intermediate product for pipe production. When VÖEST's board of directors in summer 1966 was informed about ENI's attempts to become an importer of Soviet natural gas, in return for large exports of large-diameter steel pipe, the company's managers were alerted. The combination of gas and pipes meant that the Soviet-Italian talks were of great interest to both ÖMV and VÖEST. Hence the project deserved being taken up for discussion within ÖIAG.

Herbert Koller and, in particular, Rudolf Lukesch, who served as VÖEST's business director, got highly interested in the more detailed information that ÖMV's chairman Ludwig Bauer was able to provide about the Soviet-Italian project. Lukesch, who was reputed for his imaginative business visions and farsighted strategizing, doubted that the Italians would have the capacity to supply the enormous amounts of pipe that the Siberian project would require and that it might, therefore, be possible for other Western steel companies to join in. If the Italians needed a partner, the most obvious candidates would be the large German steel companies Mannesmann and Thyssen. Both had a long tradition of exporting large-diameter steel pipe to the Soviet Union, and the quality of their pipes was widely recognized to be superior to that of other producers. As a result of the NATO's embargo on pipe exports to the Soviet bloc, however, their direct involvement was likely to be obstructed. Stricktly speaking, the embargo as such did hardly prevent the Germans from doing business with the East. In fact, under the influence of détente, most European NATO member states had by 1966 come to regard the embargo as obsolete and had already stopped adhering to it. Germany, however, was an exception. As a result, Mannesmann and Thyssen were unable to export pipes to the Soviet Union.[19]

For VÖEST, this state of affairs was not necessarily a bad thing. As a matter of fact, the Austrians had already profited considerably from the embargo by forming an alliance with the troubled German companies. The alliance centered on an arrangement in which German steel pipes were shipped to Austria, from where the pipes were reexported to customers in communist countries. VÖEST's condition for participating in these projects was that Austrian sheet metal was used as the main intermediate product for the pipes. Lukesch, the alliance's main architect, saw the Soviet-Italian pipeline project as an opportunity to scale up the partnership. But there was more at stake than steel: if the Austrians would actually be able to offer the Soviets German pipes, Lukesch reasoned that Moscow might also grow more interested in including Austria in the Soviet-Italian natural gas export scheme. In this way support from ÖMV for the plan was mobilized.

The German companies welcomed the proposal. Moreover, Lukesch's suspicion that the Italians felt overloaded by the task to produce the enormous volumes of steel pipe demanded by the Soviets could largely be confirmed, especially when it turned out that Mingazprom wanted pipes with a diameter of 1,220 mm, a record size not yet mastered by the Italian steel industry and its flagship Finsider. A more unexpected turn that similarly seemed to favor Austria's ambitions was that Hungary and Yugoslavia hesitated to take part in the Soviet-Italian pipeline project. The two communist countries had nothing against the project as such, but they were very reluctant to contribute to its financing. Neither of them considered the proposed pipeline to be of crucial importance. In Hungary, recent exploration activities had yielded growing domestic gas reserves, while some gas was also imported from Romania. Yugoslavia, too, possessed fairly large domestic deposits, whereas in terms of imports Tito seemed more interested in negotiating access to Algerian than Soviet gas.[20]

Against this background, if ENI insisted on the Hungary-Yugoslavia route, the Italians would most probably have to take full financial responsibility

Figure 4.2 Thyssenrohr's pipe factory at Mülheim (Ruhr).
Source: gwf. Reproduced by permission of Oldenbourg Industrieverlag GmbH.

for the construction of the transit pipeline. ÖMV and the steel companies promoted a Czechoslovak-Austrian transit as a cheaper—and more reliable—alternative. The Soviets seemed interested, particularly in view of the promised access to "Austrian" steel pipes, which, conceivably, might be paid for through gas exports to Austria. ENI was more hesitant, reluctant as it was to give up its envisioned hub position for Soviet gas in Western Europe. The issue was taken up for discussion with Mingazprom and the Ministry of Foreign Trade in Moscow in October 1966, when a major ENI delegation and a somewhat smaller Austrian group—independently of each other—visited the Soviet Union. The outcome of the consultations was made public in early November in connection with an official visit to Austria by Soviet head of state Nikolai Podgorny. The visit included a trip to VÖEST's factories at Linz, where Podgorny made a formal announcement confirming that VÖEST would participate in the Soviet-Italian pipeline project and that Austria, in return, would import Soviet natural gas. Austria would also offer its territory as a transit corridor for red gas deliveries to Italy and, possibly, France.[21]

The true motives behind this Soviet-Austrian agreement-in-principle were subject to a certain debate. West Germany, in particular, interpreted the inclusion of VÖEST and ÖMV into the Soviet-Italian plan as part of a deliberate Soviet strategy to make neutral Austria a future hub in the flow of Soviet energy to the West. The purpose, it was believed, was to disturb—and provide a counterweight to—Austria's deepening relations with the rest of Western Europe.[22] It was well known that the Kremlin disapproved of Vienna's striving, under the intense lobbying from firms such as VÖEST, for closer relations with the EEC. These attempts had taken on a new turn following parliamentary elections held in spring 1966. For the first time since the end of the war, an Austrian government could be formed that was not a grand coalition between the two largest parties—the Social Democratic Party (SPÖ) and the center-right People's Party (ÖVP)—but a majority ÖVP government. The Soviets noted with dismay how the new government, led by Chancellor Josef Klaus, immediately set out to strengthen Austria's EEC connections, seeking an "association" with (but not outright membership in) the community. Podgorny, taking up this issue for discussion while in Austria, made clear to Klaus what the Soviets thought about such an association:

> Such a treaty would quite certainly lead to a subordination of Austrian interests to those of the EEC and to a cession of a part of its economic interests to these countries, which, as is well-known, are hostile to the USSR. One must not forget that the entry of Austria into this community infringes on...the Austrian State Treaty, which states that Austria in no way may undertake a political or economic alliance with Germany. It would also circumscribe Austria as a neutral state in its obligations that follow from this neutrality.[23]

Such an aggressive rhetoric made clear that the Kremlin was deeply concerned with Austria's new political course, and it is tempting to interpret, as Bonn did, the new Soviet eagerness to include Austria in the Soviet-Italian pipeline plans as a strategy to balance or "disturb" Austria's deepening integration with the West. This interpretation makes sense in view of the fact that Podgorny's announcement about VÖEST's and ÖMV's intended cooperation with the Soviet Union was made at the same time as the sharp EEC criticism was voiced, and that the Soviets do not seem to have taken any particular interest in Austria as a gas transit country before the 1966 elections. Yet the political dimension of the emerging Soviet gas export scheme should not be exaggerated. A much more direct motive for the Soviets to support an Austrian involvement was clearly that the preferred pipeline route through Hungary and Yugoslavia no longer seemed a viable alternative, and that the Italians were unable to deliver sufficiently large pipes. If the Soviets wished to export natural gas to Italy and possibly to France as well as other Western nations, a transit through Austria increasingly appeared to be the only available option.

VÖEST director Rudolf Lukesch emphasized the one-time economic opportunities that the project were associated with from an Austrian point of view. The envisaged gas and pipe deal would effectively solve the most pressing problems of Austria's two most important state-owned enterprises: for ÖMV, the anticipated shortage of gas, and for VÖEST, the potential weakening of its position on East bloc markets at a time of crisis in the international steel industry. The Austrian government appears to have endorsed this view at an early stage. The agreement also offered a welcome opportunity for Chancellor Klaus to demonstrate his commitment to deepening Austria's relations not only with the EEC, but also with the Soviet bloc. Embarked on at a critical point in time, the gas project, it was hoped, might even soften the Kremlin's tough stance regarding Austria's EEC association.

The Six-Days War as a Disturbing Event

The Soviet-Austrian agreement-in-principle helped speed up the Trans-European Pipeline project as a whole. Thanks to Finsider's prospective cooperation with Mannesmann, Thyssen, and VÖEST, Italy's talks with the Soviet side gained new momentum. The Italian government reconfirmed its commitment to the project. Moreover, Gaz de France, which had so far not showed the same degree of eagerness as ENI and ÖMV, for the first time officially confirmed its willingness to import Soviet natural gas. Shortly afterward, the Soviets informed its West European partners that a "protocol" had been signed with Czechoslovakia, which, in contrast to Hungary and Yugoslavia, was happy to become part of the arrangement.[24]

The new dynamism with which the overall project from now on was pursued also gave rise to febrile activities in Algeria and the Netherlands, the Soviet Union's main rivals on West European markets. Algeria's Sonatrach and the Netherlands' NAM Gas Export were seen to intensify their attempts to conclude export agreements with Austria Ferngas, ENI, and Gaz de France. The Italians, who in 1965 had negotiated a highly favorable deal with Libya, to the resentment of the Algerians, skilfully played potential exporters off against each other. The French tried to do the same. In January 1967, Sonatrach was reported to have come up with "counterbids" vis-à-vis both Italy and France to the planned Soviet pipeline, and in February ENI's tentative talks with the Netherlands' NAM were allegedly resumed "at a greatly accelerated pace." When in March 1967 a Soviet delegation arrived in Rome for a first commercial round of negotiations with ENI and Finsider, the Algerian and Dutch offers were in turn used by the Italians to put the Soviets under pressure.[25]

Although the pipeline project was in practice a multilateral undertaking, the Soviets indicated that they wished to negotiate separately with each to-be-importer. Hence Austria's negotiations with the Soviets were only loosely linked to the analogous talks held with Italy and France. In early December 1966, Austrian vice-chancellor Fritz Bock, accompanied by a delegation of ÖMV and VÖEST managers, traveled to Moscow, where a detailed proposal regarding the Austrian part of the envisaged countertrade scheme was handed over to the Soviet government. At focus was natural gas, steel pipes, various equipment, and a credit arrangement. The proposal was based on the grand vision, originally developed by the Soviets and the Italians, of a 5,000 km pipeline from Siberia to Western Europe, the construction of which was estimated to require 1.5 million tons of steel pipe. The Austrians suggested that VÖEST produce 300,000 tons of high-quality thick steel plates, and that another 500,000 tons be outsourced to its German partners, who would also be responsible for turning the Austrian and German steel plates into 1,220-mm gas pipes. The remaining pipes, of a smaller diameter, would be supplied by Italy and France. In return the Soviet Union would export, on an annual basis, 10–12 bcm of natural gas to Western Europe.[26]

Although ÖMV and VÖEST were to take responsibility for the actual negotiations, the Austrian government showed itself both able and willing to play a constructive and facilitating role. Vienna's commitment to the project was thus reconfirmed in mid-March 1967 in connection with an official visit to

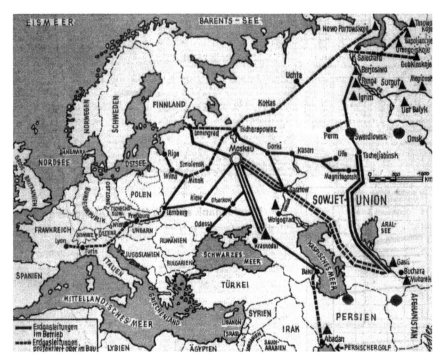

Figure 4.3 The vision of a Trans-European Pipeline for exports of Siberian natural gas to Austria, Italy, and France.

Source: Süddeutsche Zeitung, April 22, 1967. Reproduced by permission.

Moscow by Chancellor Klaus and Foreign Minister Lujo Tončić-Sorinj. Both sides emphasized the importance of the project, seeing it as "a basis for a substantial expansion of trade relations" and, therefore, also for an improvement of Soviet-Austrian relations more generally. It was agreed that "the negotiations about gas deliveries from the USSR to Austria, with the Soviet-Italian pipeline crossing Austrian territory and with the participation of Austrian companies in constructing the pipeline, should be continued."[27]

Shortly afterward, ÖMV and VÖEST convened with representatives of Mingazprom and the Soviet Ministry of Foreign Trade for a first round of concrete commercial talks. Mannesmann and Thyssen, the large German steel companies, also participated. ÖMV's delegation was led by the company's general director Ludwig Bauer and VÖEST's team by Rudolf Lukesch. Mannesmann was similarly represented by its top executive, Jos van Beveren. The Soviet delegation was headed by deputy minister of foreign affairs, Nikolai Osipov, and Alexei Kortunov's deputy from Mingazprom, Alexei Sorokin, both of whom would be frequently seen guests in Western Europe during the years to come. In their key roles in the negotiations with a variety of West European gas companies, these two men would have a substantial influence in shaping Europe's long-term gas supply.[28]

Although the first negotiation round did not generate any consensus on key issues such as the gas price, the atmosphere that characterized the talks was described as highly constructive, and there was hardly any doubt about the fact that all parties actually wanted the project to materialize. The Soviet-Italian talks also made good progress. In April 1967 a joint statement was issued by ENI and the Soviets, according to which an agreement had been reached "on the fundamental problems concerning the import by Italy of Soviet natural gas and the supply by Italy of machinery, pipes, and other equipment." On May 10, 1967, then, a three-party meeting between ÖMV, ENI—which on the occasion also represented GdF's interest—and the Soviets was organized, allowing the overall project to move further ahead. The goal was to have a contract ready by September 1967 and to start up exports in 1970.[29]

In late May, however, the talks suffered a severe setback as Rudolf Lukesch, the project's initiator and key negotiator on the Austrian side, was killed in a car accident. On June 5, then, uncertainty over the project's realization increased further as war broke out in the Middle East. The oil ministers of the Arab countries called for an oil embargo against countries friendly to Israel. Shipments of oil to the United States, Britain, and, to a lesser extent, West Germany were banned by Saudi Arabia, Kuwait, Iraq, Libya, and Algeria. The Suez canal as well as a number of oil pipelines were closed. After only a few days, the flow of Arab oil had been reduced by 60 percent. With the Middle East and North Africa supplying three-fourths of Western Europe's oil, special emergency arrangements became necessary for maintaining normal supplies.[30]

The war broke out just as the World Gas Congress, a major triannual event organized by the International Gas Union (IGU), was held in Hamburg. The Soviet Union played an active role at the conference and took the opportunity to profile itself as an important gas exporter for the future. The turmoil in the Middle East gave the Soviet delegates an opportunity to distance themselves from the hydrocarbons exporters of that region. In contrast to the Arab exporters, it was argued, the USSR had always been a reliable partner on international fuel markets. Deputy Gas Minister Sorokin stated that Moscow was "prepared to deliver natural gas immediately and in any volumes to current market prices," confirming that negotiations were already taking place with Austria, Italy, France, and Japan. Negotiations with Finland were also being prepared. At the conference, Sorokin himself was elected new IGU president.[31]

The impact of the Six-Days War on the prospects for Soviet gas exports, however, was double-edged. On the one hand, natural gas from the East could be interpreted as a welcome alternative to risky oil shipments and as a way for import-dependent nations to diversify their supply both in terms of fuel and geography. On the other hand, the war served as a reminder of the risks linked to the global energy trade, and how easily a crisis could come about as a result of geopolitical twists and turns. From this perspective it was not obvious that imports of natural gas from beyond the Iron Curtain was the optimal way to deal with Western Europe's energy problems. One analyst noted that "it must not be overlooked that gas deliveries from Algeria or the USSR are not less afflicted with political uncertainties than has been blatantly shown to be the case with Arab oil."[32]

In late June 1967, VÖEST's management board reported that the gas and steel negotiations with the Soviet Union had come to face "a certain stagnation." The board reported that the French partners, who since late 1966 had showed themselves committed to the project, appeared to have lost much of their interest. The general impression was that France, after all, had merely wanted to retain an option for gas imports from the Soviet Union as a way to put pressure on other potential exporters. This seemed to be confirmed as the French and Algerian governments in mid-June 1967 signed an agreement-in-principle foreseeing large French imports of Saharan gas to a new LNG harbor near Marseille. Deliveries would start in 1970 and reach a plateau level of 3.5 bcm in 1975. In addition, France announced that new promising gas finds had been made on its own territory, as a result of which domestic gas reserves were expected to double. In July, then, it was reported that France had "probably" withdrawn from the Soviet gas project.[33]

Italy's prospective gas imports from the East also seemed to have grown more uncertain. The Italians faced difficulties in their talks with the Soviet side, particularly concerning the financial arrangements. Just before the Six-Days War, a high-level Italian delegation had traveled to Moscow to continue the commercial talks. Upon its return, however, a conflict broke out between ENI and the Italian government. Italian minister of foreign trade, Giusto Tolloy, accused ENI of preparing the Soviet gas deal without sufficiently involving the government. Reportedly, the Soviet deal was also seen in new light after the Arab-Israeli War. The Italians, it was noted by VÖEST, feared that if "complications" of one or the other kind appeared in Soviet-Italian political relations, the Soviets might use the threat of a sudden interruption in the gas flow for political purposes.[34]

VÖEST and ÖMV were deeply worried by this course of events, fearing that the Soviet-Italian negotiations might collapse. If so, the Trans-European Pipeline would not be built and the late Rudolf Lukesch's ingenious plan for Austria's participation in the project would be jeopardized. The Austrians speculated that the Soviet Union, in case of a full Italian withdrawal from the project, might be willing to consider West Germany as an altenative customer of Siberian natural gas, so that the pipeline project could still be realized. Soviet-German relations, however, were at the time frosty, to say the least, making a German participation improbable. As a matter of fact, Rudolf Lukesch, while still alive, had at one point probed the possibilities of a German involvement with the federal government in Bonn, only to be met with suspicion.[35]

Negotiating the Gas Price

ÖMV and VÖEST had already started preparing for a possible collapse of the gas and pipe negotiations when, in mid-August 1967, the Soviets unexpectedly informed Vienna that they wished to come to Austria and continue the talks. Two weeks later, on August 30, 1967, a Soviet delegation consisting of 11 persons landed at Vienna's Schwechat airport. From there, they were brought to a romantic castle outside the Austrian capital, Schloss Hernstein-Berndorf, where a two-week stay had been prepared. Osipov, Sorokin, and the other Soviet negotiators were delighted by the luxurios feudal-aristocratic venue, where the Soviet-Austrian natural gas and pipe talks could now enter a more intensive phase.[36]

Apart from ÖMV, VÖEST, and the German steel companies, representatives of Austria's regional gas companies, who would be the actual gross receivers of the Soviet gas, had also been invited to take part in the consultations, though only at a few selected sessions. The basis for the talks was a "framework contract," which the Soviets, with Austrian approval, had worked out on beforehand, together with two more detailed documents for the gas and the pipe trade, respectively. Leaving the prospective transit of red gas to Italy and other countries aside for the moment, these documents concentrated on Austria's own imports of Soviet gas and VÖEST's corresponding exports of steel pipe.[37]

The Soviets proposed that the first gas start flowing across the Iron Curtain in 1971 at a rate of 0.3 bcm per year, a volume that would then grow stepwise by 0.3 bcm per year until 1975, when a plateau level of 1.5 bcm would be reached. The Austrians, wishing to start imports much earlier, were disappointed by this conservative offer. ÖMV proposed to start imports already in 1968 at an initial rate of 0.5 bcm, to be followed by a rapid increase so that the plateau level of 1.5 bcm was reached already in 1970. ÖMV took such an offensive stance partly out of necessity, since its domestic reserves were vaning at an accelerating pace. With regard to the timing, however, it was most probably also an expression of ÖMV's desire to bring Soviet gas onto the Austrian market before Austria Ferngas reached agreement on imports from elsewhere.[38]

The Soviet proposal foresaw exports to Austria through the Trans-European Pipeline, which was to connect the Soviet Union not only with Austria, but also with Italy and, possibly, France. Since the construction of this pipeline was expected to be completed only in 1971, the Soviets reasoned that deliveries to Austria could not start earlier. The recent uncertainties regarding Italy's imports of Soviet gas indicated that the line, if built at all, might well be pushed further into the future. ÖMV, however, suggested that in the initial phase, reserve capacity on the Soviet-Czechoslovak "Bratstvo" pipeline, which had been successfully inaugurated in June 1967, could be used. In that way exports to Austria would not have to await completion of the larger pipeline.[39]

The Austrians, well informed about developments in the Czechoslovak gas industry, knew that the Bratstvo was not planned to be used at full capacity during its first few years of operation. Moreover, talks were at the time already being conducted with Czechoslovakian agencies concerning a possible interconnection between the Austrian and Czechoslovak pipeline systems. In line with the 1966 agreement on joint exploitation of the cross-border Zwerndorf field, a "virtual" Czech-Austrian gas trade had at this time already started. Due to geological instabilities, however, it had become clear that ÖMV would not be able to continue importing 150 mcm per year from this field, as originally agreed. For 1968 only 80 mcm would be available and for 1969 120 mcm. In this situation, the Czech foreign trade organization Metalimex declared its willingness to sell the remaining annual volume from other Czechoslovak gas fields. It also offered to sell an additional 340 mcm to Austria in summer, a volume that was to be distributed over a three-year period (1968–1970). In contrast to Zwerndorf gas, however, this gas would have to be transported to Austria by pipeline. ÖMV anticipated that the interconnecting pipeline could be dimensioned for combined delivery of Czechoslovak and Soviet natural gas, and tried to persuade the Soviets that such an arrangement was advantageous for them, too. The Soviet delegation promised to look into the issue but was not willing to immediately accept the Austrian request.[40]

Another point of contention at Schloss Hernstein-Berndorf was the gas price. ÖMV argued that the contracts negotiated a few years earlier for the export of Dutch gas to Germany, Belgium, and France would have to form the point of departure in this context. The Austrians aimed for a price on par with the one charged by NAM Gas Export at the Dutch-German border, which at the time amounted to $12.50 per 1,000 cubic meters. A higher price could be accepted only to the extent that the Soviet gas was of a higher calorific value. The Soviets were surprised to hear that the Austrians thought it possible to negotiate such a low price, arguing that the point of departure must be the Dutch price plus transit costs from the gas field in the northern Netherlands to the Austrian border.[41]

This logic, however, could easily be turned upside down if Austrian gas imports were viewed from the perspective of the envisaged Soviet-Italian gas trade. ÖMV pointed to the well-known fact that ENI's much-publicized Libyan contract, signed in 1965, specified a gas price of $14.30, which, given the higher calorific value of Saharan gas, corresponded roughly to the Dutch price per unit of energy. In its negotiations with the Soviet side, ENI had regarded the Libyan price as the "absolute [upper] limit" of what it could accept. Given the shorter transit, it could then be argued the Soviets would have to accept an even lower price for deliveries to Austria. The transit costs from the Czechoslovak-Austrian to the Austrian-Italian border were estimated at $0.14 per 1,000 cubic meters. A problem, however, at least from an Austrian point of view, was that Italy had not yet agreed upon any gas price in its negotiations with the Soviet side. One reason for the stagnation of ENI's talks with the Soviets may well have been that Moscow first wanted to agree on a gas price with the Austrians.[42]

For the time being, ÖMV and the Soviets proved unable to come even to a rough price agreement. When the Soviet negotiators left Schloss Hernstein-Berndorf, the difference between the Austrian and the Soviet offers were still $4.20, which meant that the price offered by the Soviet side was about 30 percent higher than ÖMV's request.[43]

There were also problems when it came to working out a financial arrangement. The Soviet Union, with its chronic lack of hard currency, wished to obtain a large credit for the purchase of the "Austrian" steel pipes, and repay the credit through gas sales. However, the payback period would last for only six years, and a major question was how payment was to be arranged after that. The Soviets wished to get paid for its gas in cash after the credit had been repaid. The Austrian negotiators, in contrast, had been instructed by their government to seek an arrangement by which Soviet gas would be paid for through the export of Austrian industrial goods. The issue could not be resolved and would obviously have to be dealt with at a higher political level.[44]

Despite the points of disagreement, the atmosphere at the castle was reportedly very positive and constructive. The Foreign Office in Vienna, being informed that the Soviets had agreed to consider the Austrian proposal for gas imports to commence already in 1968, was pleased to receive ÖMV's and VÖEST's reports.

While a number of important issues still remained to be settled, the overall impression was that a contract would eventually come about.[45]

ÖMV felt confident enough in this respect to go ahead finalizing a contract with Czechoslovakia concerning construction of a Czechoslovak-Austrian interconnecting pipeline whose dimensions by far exceeded the anticipated gas trade between these two countries. ÖMV was to import 100 mcm of Czechoslovak gas in 1968, 170 mcm in 1969, and 160 mcm in 1970 through the line, whose diameter, however, 500 mm, allowed for a flow nearly ten times that large. The arrangement would thus enable ÖMV to import not only Czechoslovak, but also large volumes of Soviet, gas. Imports of Soviet gas, it was concluded, could now be realized in a physical sense "independently of whether or not the planned natural gas pipeline from Russia via Austria to Italy is actually built."[46]

Still, the Austrians clearly hoped that the Trans-European Pipeline, with a transmission capacity nearly ten times as large as the independent Czechoslovak-Austrian connection, would materialize. For this purpose ÖMV met with a delegation from ENI to negotiate the terms of transit through Austria. The Austrians proposed the formation of a special gas transport company, which was to be owned jointly by ÖMV, ENI, and possibly further gas recipients. ENI agreed on the fundamental principles of such an arrangement. Concerning the technical issues, a "basic agreement" was to be worked out by the two companies, addressing the exact routing of the transit. The end point in Italy was to be at Tarvisio in northeastern Italy, but the optimal routing through Austria would still have to be dealt with. The choice of route also depended on whether or not France would participate, and since this was still an open issue, two different variants of the technical study were to be worked out: one with and one without French participation.[47]

As for the bilateral Soviet-Austrian talks, a further negotiation round was held in Moscow in fall 1967. The expected progress, however, did not materialize, mainly because ÖMV and the Soviets still disagreed on the gas price. This in turn put a brake on the further negotiations regarding pipe exports, since in the absence of a gas price it was impossible to finalize any details about the countertrade.

For Moscow, the price to be agreed upon with the Austrians were of prinicipal significance, since it was likely to become a point of reference in later negotiations with other West European importers. At the same time, the Soviets were not inclined to stick to their price demands indefinitely and thereby risk losing the Austrian market to other prospective suppliers. By early December, the Soviets had grown impatient. Austria's ambassador to the Soviet Union, Walter Wodak, was pressed by the Kremlin to put pressure on ÖMV and VÖEST. In mid-January 1968, then, a new attempt was made to agree on the gas price. The outcome was now much more encouraging. Agreement could be reached on three important points:[48]

1. Gas deliveries were to commence already in 1968. The Soviet side had here given in to the Austrian request.
2. Deliveries were to start at a modest rate of 0.3 bcm in 1968 and grow

to 0.8 bcm in 1969 and 1.0 bcm in 1970. In 1971 the deliveries were to reach their plateau level of 1.5 bcm, and annual deliveries at this rate were to continue for 20 years, up to 1990. Together with the three "warm-up" years, this meant that the contract would be in force for a period of 23 years.
3. The Soviets offered to deliver the gas at a price of $15.13 per 1,000 cubic meters during the first seven years, that is, from 1968 to 1975, after which it might be renegotiated. A price float clause, regulating automatic changes in the price, would still have to be agreed upon.[49]

ÖMV's board of directors regarded the agreed price as "favorable when measured against the international level and against bids from other supplier countries," the latter referring to export offers that had actually been received from Algeria and the Netherlands.[50] Yet the price was higher than the price for Dutch gas in northern Germany and higher than the one ENI had agreed to pay for its soon-to-be-launched imports from Libya. ÖMV had wished to await further progress in the Soviet price negotiations with the Italians, since an agreed price for Soviet gas in Italy might have improved its negotiating position. But ENI did not even seem close to any agreement with the Soviets. On the contrary, the company had become even more demanding vis-à-vis the Soviet side. This was because of recent developments in Britain concerning deliveries of North Sea gas, where a price had been agreed upon that was much lower than the price ENI paid for its Libyan supplies. The British deal was not for exports, but it nonetheless changed Italy's perception of what might be achieved.[51]

Arguing that the North Sea price must be taken as a new basis for the Soviet-Italian price talks, the Italians from now on aimed for a price somewhere in the range between $10 and 11. The Soviets considered such price demands outrageous, and the actual result of the new Italian stance was that the talks collapsed. In this situation, the Austrians judged that they had no choice but to accept the Soviet Union's $15.13 bid. The alternative would have been to postpone the envisaged imports, but this was hardly an option as ÖMV judged that it urgently needed additional gas to complement its vaning domestic reserves and to forestall the conclusion of an alternative import agreement by Austria Ferngas. The future would have to tell whether the price agreed upon by ÖMV was favorable or not.[52]

Although a few issues remained to be solved, including the important one of how the gas price was to be adjusted through the 23-year period of the contract's validity, ÖMV now felt fully confident that the parties would come to agreement without much further delay. The project received further backing from the side of the Austrian and Soviet governments in connection with an official visit by Austria's new foreign minister Kurt Waldheim—the later UN secretary-general—to Moscow in March 1968. Seeking to make maximum use of the gas deal for boosting Austrian industrial exports, Waldheim managed to convince Soviet Premier Kosygin that ÖMV's imports of Soviet gas be compensated for by exports of Austrian goods to a greater extent than the Soviet side had earlier been prepared to accept. Shortly afterward, the parties also agreed on the price of the large-diameter steel pipes to be exported and on the price float clause in the gas trade. By early May 1968, all points that remained to be negotiated had been resolved and a formal ceremony for the signing of the contracts could be prepared.[53]

The Contract

The contractual arrangements that were finalized in spring 1968 were complex and consisted of four parts:[54]

1. A general framework contract
2. A detailed contract concerning gas deliveries from the Soviet Union to Austria
3. A detailed contract concerning steel pipe deliveries to the Soviet Union
4. A contract between the Foreign Trade Bank of the USSR and the Austrian Control Bank, specifying the credit arrangement

The formal contractual partner on the Soviet side, in the case of the gas trade, was Soyuznefteexport. This agency, which sorted under the Ministry of Foreign Trade, had so far been focused on organizing Soviet oil exports. Its responsibilities were now widened to include natural gas as well. Since ÖMV was already an importer of Soviet oil, its board of directors was already familiar with Soyuznefteexport and its chairman Yuri Baranovsky. This continuity in the evolving relations was appreciated by both the Soviet and the Austrian side.

The contracts were signed in Vienna on June 1, 1968. In this final arrangement, gas deliveries for 1968 had been adjusted downward to 0.13–0.20 bcm, but apart from this the preliminary agreement reached in January remained in place. For the period as a whole, ÖMV would import 32 bcm of Soviet natural gas. The figures referred to gas volumes measured at 20 degrees centigrade, which was the Soviet standard. Adjusting them to the West European standard of measuring gas volumes at 0 degrees centigrade, Austria was to receive 0.12–0.19 bcm in 1968, 0.75 bcm in 1969, 0.93 bcm in 1970, and 1.40 bcm per year from 1971. For the contractual period as a whole, Austria would receive 30 bcm.

Although the stagnation in the Soviet-Italian and Soviet-French talks made it improbable that export contracts with these countries would be signed in the near future, article 8 of the Soviet-Austrian framework contract already prepared for a possible transit to these countries. ÖMV promised to provide obstacle-free transit of gas to Italy and/or France. The contract stated that questions concerning gas transmission through Austria to these countries should be handled *without* Soviet involvement.

Further details of the gas trade were specified in the second contract, which was a much longer, 40-page document. Article 2 of this contract stated that gas deliveries were to begin on September 10, 1968. Article 3 specified the quality of the gas, which was to consist of methane to at least 92 percent, whereas for other gases and polluting substances there were maximum limits. For example, the maximum sulfur content was set at 100 mg/m^3. The lower heat value—an important indicator of the actual energy content of the gas—was to lie in the interval 8,700–9,000 kcal/m^3.[55]

In article 5 the price was set, as expected, to \$15.13/1,000 cubic meters. In principle, this price was to be valid for the entire period of the contract, 1968–1990. In case of a constant gas price, this meant that Soviet revenues for the period as a whole would amount to a staggering $450 million. However, after the first period of deliveries (1968–1975), the contract allowed for the price to be renegotiated "in case of a significant change in the comparable prices for natural gas on the European market and/or in case of an official devaluation or revaluation of the currency in which the price is expressed."

Article 7 stated that the customer should inform the provider six months in advance of the next calendar year about the volumes of gas that it wished to purchase during each quarter of that year, whereby the sum of quarterly deliveries, of course, had to be equal to the agreed annual total. In a similar manner, ÖMV was to inform the Soviet side 45 days in advance of each quarter of the year about its desired monthly deliveries during that quarter. ÖMV was thus given a certain flexibility in spreading its imports of red gas over the year. However, ÖMV had to accept certain minimum levels of daily imports. Concretely it was specified that ÖMV for each day in the winter half-year (October–March) had to accept a minimum of 3.42 mcm of gas and in the summer half-year (April–September) a minimum of 3.16 mcm. Similarly, the maximum daily delivery was set at 4.57 mcm (for both winter and summer). The average daily delivery would be 4.11 mcm, and ÖMV's flexibility hence amounted to around 20 percent.

Article 8 in the gas contract stated that in case Soyuznefteexport was unable to deliver the gas, it would have to pay a penalty amounting to 10 percent of the gas price in summer and 20 percent in winter. Article 11, however, provided for a lower penalty during the first three years, 1968–1970. Another article stated that in case of serious dispute, an arbitral court would be called upon to resolve the conflict. This court was to have its seat in Stockholm, Sweden.

The way was now paved for Austria to become physically part of the East European natural gas system. ÖMV saw this as a major opportunity rather than as a security problem. At about the same time as the Austrian-Soviet contracts were signed, ÖMV received a report from an independent international auditing company, according to which domestic natural gas production had surpassed the "acceptable limit." The Soviet contract was thus secured at a critical point in time. It would allow ÖMV to reduce domestic production rates significantly. Importantly, however, domestic gas resources were still sizeable enough for them to be drawn upon in case of any disruption in Soviet supplies, even if it would last for as much as a year. Without the domestic reserves, supply risks would clearly have been perceived as higher.[56]

The regional companies, which had worked for integration with the Netherlands and Algeria rather than with the Soviet Union, had mixed feelings about the Soviet-Austrian contract. The deal made it unlikely that imports from other sources would come about in the near future; the Austrian gas market was simply too small for that. Through their joint company Austria Ferngas, the regional actors had sought to challenge ÖMV's role as the country's main system-builder in natural gas. ÖMV's negotiators had sought to appease their regional colleagues by inviting them to take part in the talks with the Soviet side, and pressure from the federal government to accept the arrangement had been strong. In the end, three out of four regional companies agreed to cooperate with ÖMV, accepting to buy the gas that the state-owned company would receive from the Soviet Union. A distribution agreement was signed between ÖMV, on the one hand, and Wiener Stadtwerke, NIOGAS, and Steirische Ferngas—that is, the same companies that together had formed Austria Ferngas—on the other.[57]

The remaining regional actor, OÖ Ferngas, refused to join in, arguing that the Soviet gas price was unfavorable. With its proximity to Germany, the Upper Austrians argued that it would make more economic sense to import gas from the Netherlands. ÖMV sought to resolve the issue by urging Soyuznefteexport to accept a lower gas price for supplies to Upper Austria, whose large chemical industry was one of the most important gas customers in the country. In its communication with the Soviets, ÖMV argued that if such a price reduction was not granted, Upper Austria might be lost to other exporters. ÖMV and Soyuznefteexport stated their intention to do everything to solve this problem, whereby January 1969 was set as a deadline for finalizing the agreement.[58]

All in all, ÖMV concluded that the Soviet-Austrian gas negotiations—and the domestic negotiations with the regional companies—had been difficult but that the outcome was a success. The contract signed promised to open up a new era in the country's energy supply—and in its overall relations with the Soviet Union. VÖEST was also happy. The contract specified that Austria would export 520,000 tons of steel pipe worth $100 million. 270,000 tons of pipes (with a diameter of 1,220 mm and 1,020 mm) were to be delivered directly from Mannesmann and Thyssen, whereras the remaining deliveries built on VÖEST steel plates that the German companies would transform into pipes. Deliveries were to start on September 15, 1968, and end on August 31, 1970. The contract was clearly a success for VÖEST, particularly in view of the fact that it did not even possess a pipe factory! VÖEST's general director Herbert Koller pointed out that 60 percent of the pipe exports was in fact business that would go to the company's German subcontractors, Mannesmann and Thyssen. Even so, the deal was seen to give "a very welcome rear cover" for VÖEST in its struggle to survive on the international steel market.[59]

The Austrians were thus very satisfied. Yet the deal clearly did not correspond to what the involved actors had originally aimed for. In 1966, when the idea of an East-West gas pipeline had started to be discussed in earnest, Italy's ENI, together with Mingazprom, had been the driving actors. ENI, together with Gaz de France, had hoped to import up to 10–12 bcm of Soviet gas per year, but the actually contracted (Austrian) volume amounted to a mere 1.5 bcm. Austria had not been mentioned at all in the initial Soviet-Italian talks. The Trans-European Pipeline, as originally envisaged, was to have bypassed the country, taking a more southerly route through Hungary and Yugoslavia. Through timely initiatives and actions of skillful players in the Austrian steel and gas industries—notably the late Rudolf Lukesch—and with strong and outspoken support from Vienna, ÖMV and VÖEST had then unexpectedly managed to establish themselves as partners in the project. Moreover, they had managed to keep the negotiations with the Soviet side alive and bring them to a successful end despite the failure of the Soviet Union's negotiations with Italy and France. Hence what in October 1966 had been talked about as a Soviet-Italian-French project had by mid-1968 yielded a Soviet-Austrian contract. The contract was formulated in such a way as to prepare for a possible inclusion of Italy and France later on. But as of 1968, it was still an open issue as to whether the much-publicized Trans-European Pipeline would ever be built. Instead, Soviet gas exports to Western Europe were to take place through a minimal 5 km interconnection between the Soviet-Czechoslovak "Bratstvo" system and ÖMV's already existing national grid.

5
Bavaria's Quest for Energy Independence

Natural Gas and the Politics of Isolation

When the Soviet Union in the mid-1960s started considering natural gas exports to Western Europe, there was one large West European country that remained conspicuously absent from the list of potential importers: the Federal Republic of Germany. Italy, France, and Austria were consistently identified as promising to-be-importers, with Finland, Sweden, and Japan not far behind. West Germany, in contrast, was not mentioned in Mingazprom's export strategy—despite its proximity to the Eastern bloc and its vast potential market.

It is not difficult to understand why. Overall political relations between Moscow and Bonn were deeply troubled, and as of 1966 neither of the two governments seemed seriously interested in improving them. Instead, they remained openly hostile to each other. Economic exchange between the two countries, having reached a peak in 1962, was also in a phase of decline.[1]

Moscow's anti-German policies formed part of a broader Soviet strategy aimed at disturbing West European integration efforts. Not only the NATO, but also nonmilitary Western organizations such as the European Coal and Steel Community, EURATOM, and the EEC were identified by the Kremlin's foreign policy strategists as anti-Soviet forms of international cooperation. In official Soviet statements, these organizations were described as capitalist-monopolist associations that were bound to further cement the unfortunate division of Europe into political and economic blocs. The Soviet response was to promote West European disintegration by stimulating trustful cooperation with some West European countries, while isolating others. West Germany was the European country that the Soviets sought to isolate in particular.

At the 1966 Congress of the Soviet Communist Party—the same congress at which Siberia's key role in future natural gas production was officially recognized—party head Leonid Brezhnev praised the good relations that the Soviet Union enjoyed with a number of West European countries. Cooperation with Finland was said to be "characterized by trust, friendship, and cooperation," whereas "normal relations" were seen to be taking shape with the other Nordic countries. Relations with France had seen "considerable improvement," and it was optimistically suggested that "further development of Soviet-French relations may serve as an important element in strengthening European security." Relations with Italy had also "begun to improve," particularly in the economic field.² When it came to West Germany, in contrast, Brezhnev aggressively pointed at this country as the Soviet Union's main enemy in Europe:

> Today West-German imperialism is the USA's chief ally in Europe in aggravating world tension. West Germany is increasingly becoming a seat of the war danger where revenge-seeking passions are running high. West Germany already has a large army in which officers of the Nazi Wehrmacht form the backbone. Many leading posts in the Government are occupied by former Nazis and even war criminals. The policy pursued by the Federal Republic of Germany is being increasingly determined by the same monopolies that brought Hitler to power. The Rhineland politicians fancy that once they get the atomic bomb frontier posts will topple and they will be able to achieve their cherished desire of recarving the map of Europe and taking revenge for the defeat in the Second World War... Bonn is hoping to involve the USA and its other NATO partners more deeply in its revenge-seeking plans and thereby secure a revision of the results of the Second World War in its favor. It is not difficult to see that all these designs are spearheaded against the Soviet Union and other socialist countries, against peace and security in Europe and the whole world.³

Conversely, the German government did its best to criticize and isolate the Soviet Union. Foreign Minister Gerhard Schröder advocated a *Politik der Bewegung* (Policy of Movement), which could be interpreted as a German version of the Soviet strategy of isolation: Schröder made efforts to improve relations with some East European countries—notably Hungary, Romania, and Bulgaria—but not with others. Economic connections were at the heart of this policy: bilateral trade was regarded as an important vehicle in Bonn's attempt to improve relations with the favored East European countries, whereas in the case of trade with the Soviet Union it should be supported only to the extent that the Kremlin proved willing to make concessions in the Berlin issue and other key aspects of intra-German and German-Eastern relations. Aware of the Federal Republic's economic and industrial might, the German government sought to use economic levers to achieve political goals. From a German perspective, then, a natural gas import from the Soviet Union was of interest only to the extent that it could be linked to political concessions from the Soviet side. Such concessions, however, were out of the question for the Soviets. Hence a German inquiry in Moscow regarding possibilities to import Soviet natural gas appeared improbable.⁴

The same conclusion could be arrived at from an analysis of the German energy system. Natural gas still played a negligible role in German energy supply, amounting to only 2 percent of primary energy in 1965. This was a far lower figure than in countries such as Italy and Austria. Coal from the Ruhr and a few other regions still formed the backbone of Germany's fuel supply, meeting 55 percent of primary energy demand. Moreover, to the extent that natural gas was being considered as an interesting option for the future, there did not seem to be any need for imports from far away. A number of promising domestic gas deposits had been discovered in northern Germany and, above all, in the neighboring Netherlands. Since the distance between the Dutch gas fields and the main German industrial districts in the Ruhr measured no more than 200 kilometers, most analysts took it for granted that any imported gas would have to come from the Netherlands.

The first major import contract for Dutch gas was signed in 1963 by NAM Gas Export, which was controlled by Shell and Esso, and the regional German gas company Thyssengas, which was based in Duisburg. This was followed two years later by a corresponding deal with Ruhrgas, the largest German gas company with headquarters in nearby Essen. Together, the two contracts paved the way for an annual German import of 7 bcm of Dutch natural gas. Deliveries to Thyssengas were to start in September 1966 and to Ruhrgas in July 1967. The main intended users of this gas were in the Ruhr, but additional Dutch deliveries were under negotiation with gas companies in southern Germany, particularly Hessen and Baden-Württemberg. As noted in the preceding chapter, Swiss and Austrian gas companies also signaled their interest in Dutch gas.[5]

The Dutch contracts, in combination with fairly large domestic finds, meant that Germany's natural gas needs could be regarded as secured for the foreseeable future. For the country as a whole, there seemed to be no need for additional supplies, particularly not from faraway sources such as the Soviet Union. Both the strained political relations with this country and the overall German energy situation seemed to indicate that Soviet gas was bound to remain a nonissue in the Federal Republic.

Yet as we shall see in this chapter, Soviet gas did become an issue in Germany. To understand how this could happen, we will have to descend from the national and international level and, instead, turn our attention to developments in Bavaria, in the southeast of the Federal Republic.

Otto Schedl's Struggle against North German Coal

Bavaria was the largest of the West German federal states by area, making up no less than 28 percent of the Federal Republic as a whole, and the second-largest in terms of its population, which in the mid-1960s already exceeded 10 million. In the nineteenth century Bavaria had been an independent kingdom and it had always played a central role in German cultural and economic life. But it was a late industrializer, lagging behind the more dynamic north German regions in terms of economic growth and standards of living. A major reason for this backwardness was its relative lack of energy resources, in particular coal. The German coal industry had seen an enormous upswing before and during World War II as a result of military energy demand and the Nazi obsession with synthetic oil production (hydrogenation). Bavaria did not profit from this development: it did not have any notable coal resources within its borders and it faced a comparative disadvantage stemming from the need to "import" coal from northern Germany.[6]

Coal, however, was not the only energy source of value when it came to fueling industrial growth. Bavaria's southern neighbors—Austria, Switzerland, and Italy—followed a different model. Largely lacking domestic coal deposits, they relied to a much greater extent on oil, gas, and hydropower. The Bavarians, linked to their neighbors through close historical ties, felt inspired. After the appointment in 1957 of Otto Schedl as new Bavarian minister of economy and transportation, things started happening. Seeking new ways to promote Bavaria's transition from an agricultural-mercantile to a modern industrial society, Schedl quickly identified access to fuel as a critical issue and, in particular, independence of north German coal as a central challenge. The Bavarian vision that took form centered on increased reliance on imported hydrocarbons.[7]

In 1963, Schedl celebrated a first major victory as a Bavarian oil refinery center, at Ingolstadt, was inaugurated. The facilities were supplied by oil from the Middle East, which was imported by pipeline from Marseille in France into landlocked Bavaria. The investments had been substantial, but the result was a highly competitive Bavarian access to crude oil without north German involvement. Further expansion of the Ingolstadt complex was planned in relation to the completion of two other pipelines, both of which were to reach Bavaria by way of the Alps. The first was the Trans-Alpine-Pipeline (TAL), which originated in Trieste on the Adriatic and took an Austrian route. The second, known as the Central European Pipeline (CEL), had its starting point in Genova in northwestern Italy and reached Bavaria by way of Switzerland. In both projects, Bavaria cooperated with Italy's state-owned oil and gas company ENI and, in the TAL case, with ÖMV, ENI's Austrian counterpart.[8]

Not surprisingly, Schedl's vision of a Bavarian energy system based almost completely on hydrocarbons and a trustful cooperation with France, Italy, Austria, and Switzerland was opposed by north German coal interests. The federal government, for which protection of the coal industry was an important task, was similarly displeased with Schedl's Bavarian ambitions. Bonn made an attempt to use regulatory instruments to delimit Bavaria's energy imports, but the results were meager. The government proved unable to slow more than marginally a development that, under Schedl's guidance, appeared increasingly unstoppable.

Toward Gas Imports: Negotiating Algeria

Apart from importing petroleum from far away, Bavaria possessed a small oil industry of its own. It had its center to the east of Munich, just below the Alps. Production was mainly in the hands of the German subsidiary of Mobil Oil, the US-based international company.[9] The available volumes of Bavarian oil were not at all large enough to satisfy regional demand, but an unexpected side effect of oil exploration was that some promising natural gas deposits were discovered. Otto Schedl put high hopes on these gas finds as an additional contribution to Bavaria's emancipation from north German coal. Production started in 1957 on a pilot scale and grew to around 100 mcm in 1960 and 239 mcm in 1964, at which time it already corresponded to 16 percent of Germany's total natural gas production.[10] Whereas manufactured gas still dominated the overall German gas industry, Bavaria saw a rapid increase in the use of natural gas. By 1965, 29 percent of all gas consumed in Bavaria was natural gas, as opposed to only 13 percent in Germany as a whole.[11]

The gas finds provided impetus for the creation of a regional pipeline network. For this purpose, several cities and towns located in reasonable proximity of the gas fields came together, setting up a regional distribution company. Bayerische Ferngas AG (Bayerngas), as it was named, had the city of Munich as its largest shareholder. By 1965, a network measuring 350 km had taken shape, enabling the gas to be piped to Munich as well as to a number of mid-sized Bavarian cities such as Augsburg and Landshut.[12]

The new fuel was seen to have a "modern image" and grew popular with both industries and households. Rapidly growing consumption, however, soon led to the recognition that local reserves would not last for long. By 1965 the available Bavarian reserves had grown to 5–6 bcm, but annual production was already 0.5 bcm and continued to grow from year to year. Hence the very popularity of natural gas forced Bavaria to look for alternative supplies.[13]

Dutch gas was the first source that was considered in this context. Initially the expectation was that imports from the Netherlands would mainly be used to supply the German industrial regions near the Dutch border. In 1965, however, a preliminary agreement was reached between NAM Gas Export and Gasversorgung Süddeutschland (GVS, with headquarters in Stuttgart) for the delivery of Dutch gas to Baden-Württemberg. Other regional gas companies with an interest in Dutch gas were Gas-Union and Saar Ferngas, headquartered in Frankfurt and Saarbrücken, respectively. Bavaria, which bordered on Baden-Württemberg, was another region of interest to the Dutch. It was of special interest not only as a promising market in its own right, but also as a prospective transit corridor for Dutch gas destined for Austria and perhaps across the Iron Curtain to Czechoslovakia. Esso and Shell, together with Ruhrgas, became very active in lobbying the Bavarian government and Bayerngas for support to such an arrangement.[14]

Otto Schedl, however, did not welcome the Dutch and north German ideas. To him, the prospects for a gas supply from the northwest were too reminiscent of the logic of "importing" coal from the Ruhr. Bavaria's greater distance from the gas fields were bound to translate into higher energy prices and thus unfavorable conditions for Bavarian industry vis-à-vis northern Germany. This was precisely what Schedl had devoted so much effort to avoiding. In addition, Dutch and north German gas had a much lower calorific value than Bavarian natural gas. This meant that the Bavarians would either have to build a separate network for Dutch and north German gas or invest in expensive equipment to make the two gas types compatible with each other.[15]

Schedl, therefore, looked for alternative arrangements. With Ingolstadt and its refinery complex as a symbol of Bavarian independence and competitiveness, and with two major oil pipeline projects under way from Italian harbors, he felt inspired to seek similar opportunities with respect to imports of natural gas. Of particular interest were imports from North Africa. Minor shipments of Algerian LNG had started to Britain in October 1964 and to France in 1965. Several other countries were negotiating potential imports from Algeria, whereby not only LNG, but also piped gas was being discussed as a visionary option, favored by ENI in Italy and Franco in Spain. Pipeline routes were sketched across the Mediterranean to the Italian and Iberian peninsulas, from where some of the gas, it was imagined, could be piped on to France, Switzerland, Belgium, and Germany.[16] Another interesting project was the emerging cooperation between Czechoslovakia, Yugoslavia, and Austria Ferngas for the import of Algerian LNG, as discussed in the previous chapter. Czechoslovakia was to be the most important recipient according to this plan, which was based on the construction of a transit pipeline from Koper on the Adriatic through Slovenia and Austria to Czechoslovakia.

For Bavaria, the intensifying scramble for North African gas was highly interesting. An advantage was that an import of Algerian gas to Bavaria most probably could be arranged without north German involvement. At the same time, an import of Saharan gas would make future Bavarian imports of Dutch gas more acceptable, since the Dutch would then have to compete with the Algerians. On the other hand, an import of Algerian LNG to Bavaria would certainly not be straightforward, as it involved expensive and technically risky liquefaction, sea transport, and pipeline transit through foreign territory. The alternative of a pipeline all the way from the Sahara to Bavaria, by way of the Mediterranean and southern Europe, also involved a number of risks. Schedl was aware of the risks, but had positive experience of analogous arrangements for oil imports from far away.

The Bavarians got an excellent opportunity to approach the Algerians as Czechoslovakia's negotiations with the state-owned Algerian oil and gas company Sonatrach unexpectedly collapsed in February 1966. This was a consequence of the Algerian side's refusal to accept the Czechoslovak demand for a countertrade arrangement according to which Algeria would import Czechoslovak industrial goods. Schedl suggested that natural gas exports to Czechoslovakia could be replaced by exports to Bavaria. Both the Algerians and the other partners in the project were positive to the idea.[17]

However, the argument that Algerian gas could be brought all the way to Bavaria at competitive cost was subject to dispute. Many analysts were pessimistic regarding the prospects for Algeria to actually profit from exporting natural gas to Bavaria. Schedl identified the small size of the Bavarian natural gas market as the main obstacle that had to be removed in this context. His way of doing this was to actively expand the future market by approaching the main regional and municipal energy companies as well as other large potential customers in southern Germany, seeking to convince them that a transition from traditional fuels (notably coal and manufactured gas) to imported natural gas would be highly profitable on the long term. Schedl envisaged the formation of a consortium of large customers who together would have substantial bargaining power vis-à-vis Sonatrach. The consortium was to be led by Bayerngas, the only Bavarian actor with any experience in natural gas transmission.

Schedl thought that if only the gas price was sufficiently low, industrial gas users and particularly southern Germany's heat and power plants would have strong incentives to switch from oil to gas as their fuel basis. The replaced oil was in turn expected to put coal from the Ruhr under pressure. It was estimated that Bavaria together with the neighboring state of Baden-Württemberg would be able to absorb up to 5 bcm of natural gas annually from 1970, a large share of which could potentially be supplied from Algeria. The market could then conceivably grow by an additional 5–10 bcm in the course of the 1970s.[18]

Schedl and his advisors judged that the price of Algerian gas in Bavaria would have to be 0.7 German Pfennig per million calories (Pf/Mcal) or lower for such a rapid growth to be feasible. Anticipating transit costs of around 0.1 Pf/Mcal from Yugoslavia, the price of the gas at the receiving terminal on the Adriatic would have to be 0.6 Pf/Mcal or lower. When the Bavarians presented their analysis to the Algerian side for the first time in spring 1966, however, it turned out that Sonatrach had a very different price conception. Its stance was that a price below 0.8–0.9 Pf/Mcal at the Adriatic terminal would make the project unprofitable. Schedl thought that the Algerians would be more willing to accept a lower price once the larger consortium of south German gas customers had been formed. Unfortunately, this was a troublesome and time-consuming process.[19]

In September 1966 the Bavarians and Algerians, together with the Austrian and Yugoslavian parties, met in Munich. Sonatrach and the Algerian government were now eager to come to quick agreement with the Europeans and had largely given in to the Bavarian demands, offering 0.65 Pf/Mcal. The Yugoslavian and Austrian partners were more than satisfied with this price, but Schedl still regarded it as too high, continuing to insist that the price must be 0.6 Pf/Mcal or lower.[20]

In the next negotiation round, held in Paris in October, the Bavarians remained absent. This strongly upset Algeria's young minister of industry and energy, Belaid Abdessalam, who had hoped to finalize at least a preliminary contract. Schedl went to Algiers in an attempt to rescue the talks, explaining that the German consortium had now been officially formed, and that a report about its stance to the project could be expected within six weeks. Abdesselam, however, judged that the Bavarians had no serious interest in Algerian gas after all, and that they were merely trying to use the negotiations as a way of putting alternative suppliers—mainly Shell and Esso—under pressure.[21]

By autumn 1966, the overall atmosphere in the Bavarian-Algerian talks was thus worsening. As of late November, Schedl's close advisor Hans Heitzer thought it "extremely doubtful that the project will be realized."[22]

Soviet Gas for Bavaria? The Austrian Connection

While still negotiating with Sonatrach and Abdessalam, Schedl was informed about the attempts from the side of Austria to join the Italian-French scheme for Soviet natural gas imports. The prospects for Soviet gas deliveries to Austria, as announced by Soviet head of state Nikolai Podgorny in early November 1966, aroused both fear and enthusiasm at the Bavarian Ministry of Economy. On the one hand, given the limited size of the Austrian gas market, a Soviet-Austrian gas deal would most probably kill any short- and mid-term prospects for Austria to come to terms with Algeria. Since Bavaria, as envisaged in the four-country elaborations, was to import Algerian gas by way of Austria, this would most probably put an end to Bavaria's Algerian visions as well. On the other hand, however, the vision of Austrian imports of Soviet natural gas opened up a completely new supply opportunity. Linz, whose large chemical industry had been spotted as one of the most important potential customers of "red" gas, was located just 50 km beyond the Bavarian border, and it was not far-fetched to suggest that a Soviet-Austrian pipeline terminating at Linz could be extended across the border into Germany.[23]

In order to probe this possibility, Schedl contacted VÖEST, the Austrian steel company. VÖEST's business director Rudolf Lukesch, the key architect behind the Soviet-Austrian gas and pipe deal, welcomed the Bavarian interest. A Soviet gas export to Bavaria promised to enlarge VÖEST's business opportunities, since greater Soviet gas exports would have to be compensated for by additional pipe sales. Lukesch, therefore, promised to push for an inclusion of Bavaria in the envisaged Soviet-Austrian deal, which at this time was just starting to be negotiated.[24]

The idea of linking up with Eastern Europe's natural gas infrastructure certainly fitted well in Schedl's wider strategy of "liberating" Bavaria from its dependence on north German coal. Yet the project was bound to become controversial. After all, the Cold War was raging and the Soviet Union explicitly identified West Germany as its main European enemy. As pointed out above, Germany was in Soviet eyes a country full of "revenge-seeking passions," dominated politically and economically by "former Nazis and even war criminals." The German federal government, for its part, still refused to recognize Europe's postwar borders and East Germany as a sovereign state. The NATO's embargo on pipe exports to the East was still adhered to by the German government and the overall anti-Soviet sentiments in the country were notable.

But there were signs of change. The federal government in Bonn had the impression that the Soviets, despite the political rhetoric, were highly interested in expanding their economic exchange with Germany. This was certainly of interest to the Germans, too, where lower rates of economic growth tended to make East European markets more attractive. In October 1966, then, the anti-Soviet, center-right coalition government in Bonn, led by Chancellor Ludwig Erhard, collapsed. As a result, the Social Democrats, who were much more positive to cooperation with the East, got an unexpected opportunity to enter Bonn's corridors of power. The Erhard government was replaced by a "grand coalition" formed by the two largest parties, the Christian Democratic Union (and its sister party in Bavaria, the Christian Social Union) and the Social Democrats. Kurt Georg Kiesinger of the Christian Democrats became new chancellor, whereas Willy Brandt, until then mayor of West Berlin and chairman of the Social Democratic Party, was appointed vice chancellor and minister of foreign affairs. Karl Schiller, a highly trusted social democratic colleague of Brandt's from the Senate of Berlin, was appointed minister of economy.

The Kiesinger-Brandt government took office on December 1, 1966. Issues regarding peace and détente figured prominently in the government declaration. In terms of foreign policy, the declaration was strongly influenced by Brandt's and his closest advisor Egon Bahr's concept of *Wandel durch Annäherung* (change through rapproachment) as a new guiding priniciple in Germany's relations with the communist countries in Eastern Europe. A vision of a European peace order was painted, without which German reunification—the major long-term goal—was seen unattainable. It was suggested that Germany and the Soviet Union conclude a renunciation of force treaty. Moreover, diplomatic relations with Germany's eastern neighbors were said to be desirable, which in effect meant that the new government denounced the Hallstein Doctrine.[25]

Although Otto Schedl was a Christian Democrat, the change of government in Bonn must have appeared favorable to him. Schedl's vision in terms of international relations was to turn Bavaria into a "gate to the southeast," building on the "traditional influence" of Bavaria in that European region. His overall view on Europe appears to have been one of a united, tightly interconnected continent as the natural order of things, and of the Iron Curtain as a painful, artificial divide between East and West. According to Schedl, Central and East European countries, including the Soviet Union, should not be seen as enemies, but as allies. The real enemy, Schedl argued, was found much farther away: in China, where Mao Zedong had just proclaimed Cultural Revolution:

> We experience in our time a fanatically pursued strengthening of force in the interior of Asia, a formidable exertion and in recent history surprising, yes, terrifying deployment of power by the largest Asian people, the Chinese...We are thereby confronted with a historical process of immense reach, which inevitably also forces us to quickly and fundamentally rethink the habits of our political work with regard to indispensible necessities, the traditional foundations of our political cohabitation. The barely imaginable threat from Asia, in its recent relentlessness and consequence, threatens both Russia and the Central European countries, just as it threatens Germany and Western Europe and to the same degree the whole free world...Only together do we have a chance to prevent such an aggression, which challenges the very survival of humanity—provided that we have the power to overcome the divide among ourselves.[26]

The prospects for natural gas imports from the Soviet Union fitted well into this overall view.

While Schedl had reason to welcome the change of government in Bonn, another encouraging development was the November 9, 1966, decision by the NATO Council to annul the embargo on exports of large-diameter steel pipe to the communist bloc. This meant that the door was now formally open again not only for complex indirect export arrangements in the form of cooperative ventures with Austria—as envisaged in the Soviet-Austrian gas negotiations that had just been initiated—but also for direct pipe exports from Germany. All in all, the idea of importing Soviet natural gas to Bavaria in return for German pipe exports no longer seemed totally unrealistic.

Manipulated Conditions

Schedl saw no reason to consult the federal government in Bonn before initiating his contact with the Austrians. Bonn, however, had its own information channels and was alerted at an early stage about the Bavarian plans. In mid-December 1966 a report was filed by Willy Schlieker, a prominent retired steel industrialist who was able to mobilize his personal network to access inside information about the Soviet-Austrian gas plans. Reporting that the Austrians and Bavarians together hoped to annually import around 5 bcm of Soviet gas, of which at least 3 bcm would be for Bavaria, Schlieker told Bonn that he considered it probable that Bavaria access Soviet gas at a more favorable price than Dutch or Algerian gas.[27]

Schlieker's report was one of the first things that landed on newly appointed foreign minister Brandt's desk. The Foreign Office found it difficult, however, to assess what the report actually meant. It was, therefore, forwarded to the Ministry of Economy, which was responsible for the government's energy policy and where Bonn's main expertise in energy issues resided. There, Karl Schiller had just taken over as new minister. As a former colleague of Brandt's he had been taking active part in the attempts led by Egon Bahr, Brandt's advisor, to work out a new foreign policy strategy first for Berlin and then for Germany as a whole. Most of Schiller's advisors at the Ministry of Economy, however, had made their careers under Adenauer and Erhard as chancellors. In their spontaneous initial reaction, they were, predictably, highly suspicious of the idea of gas imports from the Soviet Union. Viewing the issue from an all-German rather than from a Bavarian perspective, the only real advantage they saw was the possible combination of gas imports and steel exports. The advisors also acknowledged that the entry of additional gas suppliers on the German gas market, generally speaking, was to be welcomed, since this would reduce the risk of the whole market being "caught up in the hands of a single monopolistic group," that is, Ruhrgas on the transmission side and Shell and Esso on the producing side. However, these advantages by far did not outweigh the risks.[28]

The prime reason for the ministry to take a skeptical stance to Schedl's initiative was that it seemed to threaten the German coal industry, whose situation at the time was already precarious. Despite active support and subsidies from the previous government, domestic coal production had in the early 1960s entered a phase of decline. By 1966 total production amounted to 118 million tons, down from 148 million tons only three years earlier. The main reason was seen to lie in the toughening competition from imported oil.[29] For the advisors in Bonn, a possible import of Soviet natural gas was bound to further worsen this trend:

> Given the critical situation in the German coal mines, which will most certainly persist for many years and which has an enduring effect on the whole heat energy market, a gas supply over very long distances at politically, from the supplier side, manipulated conditions, can from our side at the moment not be advocated to any notable extent.[30]

A worsening of the competitive situation for north German coal on the Bavarian market was, of course, precisely what Otto Schedl aimed for. The Federal Ministry of Economy, however, viewed the issue from a more conspiratory angle, judging that the Soviets might keep the gas price artificially low in an attempt to deliberately weaken the German coal industry. The ministry's leading gas expert, Norbert Plesser, thought it improbable that the Soviets would be able to deliver natural gas all the way to Bavaria at a competitive price without politically motivated state subsidies:

> Mr. Schlieker's cost estimates hardly appear realistic. However, it may be expected that the Soviet price bid, for political reasons, will be manipulated to be sufficiently low, if there is a serious intention to deliver natural gas to the FRG.[31]

And then there was the security issue, which was particularly worrisome if Soviet gas was to be imported in such large quantities as envisaged by Schedl:

> In the case of a far-reaching dependence of the Federal Republic's gas supply, or parts of it, on Soviet deliveries, it must be feared that different political considerations from the Soviet side could lead to an increase in the price or to a curbing or suspension of deliveries.[32]

In stark contrast to the positive stance taken by the Austrian and the regional Bavarian governments, Bonn thus viewed the prospects for imports of Soviet gas with great suspicion. The government's stance to Soviet gas differed markedly from an assessment made only two months earlier regarding the expected effects of gas imports from Algeria. In the Algerian case, the consequences for the German coal industry had been regarded as more or less negligible, since "the Bavarian market for coal has been lost anyway." Moreover, the planned gas imports corresponded to only one-tenth of the oil that was flowing into Bavaria through the new South European oil pipelines. Algerian gas was regarded as a "desired alternative to Dutch gas and a diversification of the gas basis in the Federal Republic." Referring to successful British and French imports of Algerian LNG, the Algerian alternative was considered acceptable "from a security point of view."[33]

Otto Schedl continued to refrain from any direct discussions with the Federal Ministry about his intentions to import Soviet natural gas. Nor did he take any direct contact with Soviet authorities, using, instead, his Austrian partners as mediators. A proposal was being worked out by the Austrians and Bavarians, which, if realized, would make Bayerngas, the regional distributor, an important actor not only for the supply of Bavaria itself, but also for the transit of Soviet gas to Italy. The plan envisaged a pipeline route through the Danube valley to Linz and onward to Rosenheim in Bavaria, near the Austrian border. From there, it would turn south, traversing the Austrian region of Tyrol on its way to the Italian market. The pipeline would thus enter and exit Austria twice. This was certainly not the simplest or most direct way to transit Soviet gas to Italy, but both the Austrians and the Bavarians regarded it as advantageous. The arrangement would allow Tyrol and above all Linz, with its large chemical industry, to be supplied with Soviet gas, and make southern Germany an integral part of the Trans-European Pipeline. It would also allow for the gas pipeline to be built alongside the TAL oil pipeline. This, it was believed, would substantially reduce planning and construction costs.[34]

The Soviets, however, were skeptical to the proposal. On the one hand, they were not fond of the complicated pipeline route. On the other, it rightly seemed to them that the Bavarians acted without support from Bonn. In the absence of such support, an actual commitment from the German side did not appear probable.

To deal with the latter problem, VÖEST director Lukesch in early March 1967 paid a visit to Bonn, where he was received by Minister Schiller's top energy advisor Gerhard Woratz. Informing Woratz about the envisaged Soviet export scheme, in which Austria, Italy, and France were so far involved, Lukesch lobbied for support for Bavaria's participation as a fourth importer. He stressed that this would pave the way for a very large countertrade arrangement in terms of pipe exports to the Soviet Union, from which both the Austrian and German steel industry would profit. Lukesch estimated that Austria and Germany would together be able to export as much as 800,000 tons of pipe. In the absence of a Bavarian participation, the volume would be much lower.[35]

Woratz, who was not primarily an expert on natural gas but in charge of the government's overall energy policy, in which coal from the Ruhr and the growing competition between coal and oil was at the center, did not openly oppose a Bavarian gas import. He noted, however, that the federal government had received no indication from the Soviet side that Moscow actually had an interest in exporting natural gas to Germany. Woratz considered it improbable that the Soviets would come up with such an interest. He also argued that the issue was very complex and that it would have to be investigated much more thoroughly, particularly with respect to how the current ideas could be brought into unity with Bonn's overall energy policy conceptions. Lukesch was thus forced to return to Linz without having managed to persuade the federal German stakeholders.[36]

Meanwhile, Foreign Minister Brandt's initially optimistic efforts to launch a new Eastern Policy faced problems. In a speech to the European Council in January 1967, Brandt pointed at Germany's historical role as a "bridge between Western and Eastern Europe" and optimistically explained that his government intended to rebuild this bridge, which had been destroyed by the Cold War. A few days later, the federal government attempted a first step toward realizing this vision by establishing diplomatic relations with Romania, hoping that this would set in motion a process in which similar ties would be established with other communist countries too. In reality the move only served to worsen Bonn's relations with the Soviet Union, and in particular with the GDR and Poland. East German leader Walter Ulbricht and his Polish counterpart, Władysław Gomułka, were furious about Brandt's diplomatic strategy. At a Warsaw Pact meeting in February the two countries obtained a pledge from the Soviet Union and the other pact members to adhere to a "reverse Hallstein doctrine," making Bonn's official recognition of the GDR and Poland's borders a precondition for the establishment of diplomatic relations with West Germany. This policy—known as the Ulbricht doctrine—efficiently blocked much of Brandt's ambitions in the East. The overall evolution of Soviet-German relations did thus not give reason for optimism with regard to a possible Bavarian participation in the Soviet-Austrian natural gas project.[37]

Egon Bahr and the Steel Companies as Supporters

In early April 1967, the Federal Ministry of Economy's energy advisors finalized their updated report on the Bavarian-Soviet gas issue and submitted it to Minister Karl Schiller. The report repeated that an import from the East would be subject to "politically, from the supplier side, manipulable conditions," but the opposition was no longer as categoric as it had been a few months earlier. It was argued that an import of Soviet gas to southern Germany might possibly be considered, though only "to the extent that the south German gas market does not get into a dependence on Russian natural gas deliveries."[38]

"Dependence," however, was a vague concept that needed further specification. The energy experts sought to quantify the degree of dependence that could possibly be accepted. According to the ministry's forecast, the Bavarian gas market could be expected to grow to around 3 bcm by 1975. This was the same volume that Schedl hoped to import from the Soviet Union. If the Bavarian plan was realized, it would hence make Bavaria totally dependent on Soviet deliveries. This was not considered acceptable. However, Schiller's advisors reasoned that a volume of around 1 bcm could possibly be taken into consideration—under the condition that reserve capacities could be made available:[39]

> There must be a guarantee that an interruption of the Russian deliveries can be compensated—without supply disturbances in the market—by supplies from elsewhere (German, Dutch, or other natural gas as well as refinery gas, etc.).[40]

A volume of 1 bcm of Soviet gas was also seen as acceptable from the perspective of Germany's troubled coal industry:

> The tenuous situation of coal vis-à-vis oil, natural gas, and nuclear power in the south German market would no longer be *decisively* influenced by an import of Russian natural gas within the above mentioned limits.[41]

The Ministry of Economy thus seemed to take a small step toward embracing the possibility of red gas imports. Schedl and Lukesch, however, wanted more. Supported by Mannesmann and Thyssen, the German steel companies, they argued that an import of only 1 bcm of Soviet gas would ruin the cost-effectiveness of the envisaged deliveries. Bavaria would have to import a larger volume of Soviet gas or none at all. The steel companies also repeated Lukesch's main point that a failure to involve Bavaria in the Soviet-Austrian arrangement would lead to a corresponding reduction in the volume of steel pipes to be exported, from 800,000 tons to only 500,000 tons.[42]

Attempting to "rescue" the tentative plan for Soviet gas exports to Bavaria, the German pipe manufacturers approached both the Bavarian government and the Federal Ministry of Economy. Ernst Wolf Mommsen, Thyssen's highly respected president, pushed Schedl and his advisors to officially state—ideally through a joint initiative with the federal government—their interest in importing Soviet gas, so as to make clear to Moscow that the Bavarian interest was real and serious. Mommsen recommended Bonn and Munich to initiate a direct contact with the Soviet side.[43]

This idea was vehemently opposed by several key persons at the Federal Ministry of Economy. Briefing Minister Schiller about the latest developments, State Secretary Fritz Neef argued that Bonn must "not engage itself with a direct initiative vis-à-vis the Russians." According to Neef, who belonged to the advisors who had been at their posts already under the previous, center-right government, a *passively encouraging* stance from the side of Bonn to a Bavarian gas import could possibly be accepted. It would be quite a different thing, however, for the government to make an *active effort* to support the project. Neef was against an active effort because of the "significance from a foreign policy perspective and the uncertainty concerning the outcome" of such an initiative.[44]

Mannesmann, the other large steel company, similarly took action to rescue Bavaria's gas imports and thus secure larger pipe sales. Apart from following Mommsen's example, recommending the Ministry of Economy to establish a contact with the Soviet side, the company's managers also lobbied the Foreign Office. Mannesmann's head of public relations, Reinhart von Eichborn, met on several occasions with Egon Bahr, the main architect behind Brandt's new foreign policy, to discuss a possible German participation in the Soviet-Austrian deal. By now, German media were already speculating wildly about the Trans-European Pipeline. Rumours flourished about "bids" from Moscow and about "cheap natural gas." In reality there was no clear indication that the Soviet Union was willing to sell any gas at all to Germany, and even less clear was the extent to which it would be cheap. Even so, the press coverage and the conversations with von Eichborn appear to have stimulated Bahr to take a more active interest in the gas issue as a potential instrument of foreign policy that might help improve the strained German-Soviet relations.[45]

In late April the development took on a new turn following an unexpected move on the Russian side. The Soviet Union's trade representation in Germany, which so far had been ordered not to take any initiatives, received orders to investigate under what conditions natural gas exports to the Federal Republic could be organized. Based in Cologne, the representation was an important link between Bonn and Moscow at a time when a legal foundation for the economic exchange between the two countries, in the form of a trade treaty, was lacking, and it was a logical point of departure for inquiries concerning the possibilities of gas sales to Germany.

The representatives' director, Samsonov, set about thoroughly investigating the issue. His first step was to try and get an idea of how the Netherlands, with which the Soviet Union would have to compete on the German gas market, had managed to arrange its gas exports to the Federal Republic. Together with Rudolf Kröning, a German industrialist in Soviet service, Samsonov contacted NAM Gas Export, whose board of directors were somewhat perplexed by the Soviet approach but eventually did agree to meet for a general discussion. Samsonov and Kröning also met with NAM's largest German customer, Thyssengas, which had signed the first major German contract with the Dutch back in 1963. Thyssengas had successfully started up imports from the Netherlands on a minor scale in September 1966, and as of spring 1967 it was still the only German gas company with any experience of actually importing Slochteren gas. The Soviet contacts with Thyssengas were facilitated by the fact that Thyssen's president Mommsen enjoyed a good reputation in Moscow.[46]

Thyssengas appears to have been happy sharing its experiences of importing Dutch natural gas with Samsonov and Kröning. The company's managers advised the Soviet representatives to proceed with their inquiries by discussing the possibility of red gas imports with the German gas industry's main branch organization, the Association of German Water and Gas Works (Verband der deutschen Gas- und Wasserwerke, VGW). The members of this organization were mainly municipal gas distributors, many of whom turned out to be enthusiastic about the prospects for Soviet gas to strengthen competition among gross suppliers. A meeting with the Soviet representatives was organized at VGW's Frankfurt headquarters. It became an important occasion for the Soviets to get firsthand information concerning the nature of Germany's interest in Soviet natural gas, and to learn more about the latest trends in the dynamically evolving German gas industry. At the meeting, Samsonov explicitly stated that the Soviet Union was seriously interested in exporting natural gas to Germany.[47]

At the Ministry of Economy, the new Soviet initiative gave rise to confusion. The ministry had so far had the impression that the Soviet Union would not take any serious interest in gas exports to Germany, and that Otto Schedl's wild ideas were, therefore, more or less hypothetical. Now it seemed that the issue "could become acute" in short notice. Minister Schiller ordered his energy advisors to closely monitor Moscow's activities on the international arena, so that the government might respond quickly to any new unexpected moves. At the same time, Schiller emphasized that no direct initiative should be taken from the ministry's side. The strategy was to wait and see.[48]

Alexei Sorokin's Charm Offensive

The Soviet charm offensive continued in connection with the Congress of the International Gas Union (IGU), which was held in Hamburg in June 1967. As we saw in the previous chapter, possible Soviet gas exports to Western Europe was a much discussed topic at this conference. The Soviet Union was represented by a large delegation counting 32 gas experts, headed by Gas Minister Kortunov's deputy for international affairs, Alexei Sorokin. Immediately after the congress, the whole group embarked on a major tour through Germany, arranged in advance by Samsonov. The objective was to visit potential business partners, government agencies, and other organizations of interest.[49]

One of the key visits was to Mannesmann's headquarters in Düsseldorf. The Soviets were received by the company's chief executive, Jos van Beveren, who enthusiastically showed the guests around at the modern pipe factories. Sorokin and his colleagues, familiar as they were with analogous plants in the Soviet Union, were impressed. Van Beveren indicated that both Mannesmann and Thyssen would be willing to deliver pipes to the Soviet Union not only by way of VÖEST, as envisaged in the Soviet-Austrian countertrade scheme, but also as exporters in their own right. Sorokin inquired about the volumes of pipes Mannesmann would be able to deliver. Van Beveren explained that Mannesmann and Thyssen would probably be able to export up to 200,000 tons each on an annual basis, starting already in 1968 if necessary. It was noted that such exports would most probably have to be part of a larger countertrade scheme in which Germany, in return for the pipes, would import Soviet natural gas. The Soviet representatives confirmed, however, that they had not yet taken any direct contact with the federal government concerning the prospects for gas exports to the Federal Republic.[50]

From Düsseldorf, the Soviet delegation traveled to Munich. There, they were received by Bayerngas' general director, Presuhn, who took the guests on a tour around the regional and municipal gas infrastructure facilities in Bavaria. In the evening, the Soviets were invited to an informal dinner, where in a series of toasts the prospects for further cooperation were held high. Bavarian minister of economy Otto Schedl, while not formally in charge of arranging the visit, now got an excellent opportunity to meet directly with key Soviet officials. The Soviets repeatedly stressed their preparedness to deliver natural gas to Bavaria, regretting that they had not yet received any "application" from the Bavarian side in this respect.[51]

An analogous visit was paid to Baden-Württemberg, where the Soviets were welcomed by the Technical Works of Stuttgart and its general director Heinrich Kaun. Kaun also functioned as chairman of VGW and Alexei Sorokin already knew him from previous IGU meetings. Kaun was instrumental in organizing a meeting between the Soviet delegation and Otto Schedl's counterpart in Baden-Württemberg, Minister of Economy Hans-Otto Schwarz, who like Schedl had started to take serious interest in the possibility of importing red gas.[52]

Sorokin's visit to Germany generated increased interest in the issue from the side of the Foreign Office. The earlier elaborations undertaken by Brandt's advisor Egon Bahr and Mannesmann's Reinhart von Eichborn were intensified. Jos van Beveren now joined the discussions. The three men sat down with a European map, sketching alternative ways in which Soviet gas pipelines could potentially be extended into West Germany. They also suggested that not only Bavaria and Baden-Württemberg might be of interest as markets for Soviet gas, but other parts of Germany as well. This was a radical departure from previous conceptions, all of which had focused on southern Germany.[53]

Two alternative East-West pipelines were discussed. The first, roughly corresponding to the original Bavarian vision, was based on transit through Czechoslovakia and, possibly, Austria. In the other, Soviet gas would reach West Germany by way of Poland, the GDR, and West Berlin. Bahr thought that the second variant must be prioritized "because of the shorter distance." The pipeline should be of the same size as the planned Soviet-Austrian-Italian pipeline, with a maximum transmission capacity of 10–11 bcm. Of these, 6–7 bcm should be delivered to the Federal Republic whereas some 4 bcm might be reserved for customers in the transit countries. Von Eichborn and van Beveren, viewing the project from the perspective of the steel industry, thought either pipeline scheme would pave the way for a profitable countertrade in pipes.[54]

Bahr increasingly portrayed an import of Soviet natural gas—particularly if transited through Poland and the GDR—as an interesting potential

Figure 5.1 Alexei Sorokin (left) and Heinrich Kaun. In his position as deputy gas minister with responsibility for international cooperation, Sorokin was the most important Soviet gas industry representative on the international arena for more than 20 years. Kaun was a key person on the German side in bringing people together and forming a transnational coalition of system-builders.

Source: gwf. Reproduced by permission of Oldenbourg Industrieverlag GmbH.

contribution to the new government's emerging Eastern policy. In an internal report to Brandt, submitted shortly after Sorokin's visit to Bavaria and Baden-Württemberg, Bahr identified several advantages of a Soviet-German gas deal:[55]

- the project, if realized, would serve as proof of Germany's desire to reach a more relaxed relation with the Soviet Union;
- it was in line with the Soviet interest in decreased German dependence on the large international (primarily American) oil companies;
- it would open up for cooperation with the GDR and thus allow the federal government to demonstrate that it, in contrast to the previous government, did not seek to isolate the East Germans. Due to the Soviet Union's direct involvement, Bahr thought it would be difficult for Ulbricht to oppose the project;
- it was well in line with Brandt's elaborations regarding the future role of Berlin as a hub of East-West détente;
- it could be expected to stimulate not only the steel industry, but also the troubled economy of West Berlin; and
- it would cost the state nothing. The government would, at most, have to issue a loan guarantee as part of the financial arrangement for pipe exports.

Bahr did not see any serious risks linked to imports of red gas:

> The objections...would mainly concern the issue of whether a dependence on Soviet deliveries or disturbances in the Soviet zone [i.e., the GDR] might thereby arise, but taking into account the very small volumes involved, compared to overall demand, these dependencies would not be serious. The negative stance of the Western oil companies to such a project would be certain, but this should not constrain the federal government's scope for manoeuvring with respect to the preponderant political advantages.[56]

Bahr also thought it improbable that West Germany's closest ally, the United States, would oppose a large-scale import of Soviet natural gas. This was because Brandt and Bahr had at an early stage discussed their "change through rapprochement" policy with the Johnson administration. US secretary of state Dean Rusk, predicting—wrongly, as it would turn out—that Brandt advance to become chancellor in the 1965 elections to the Bundestag, had informally approved of the proposed Eastern policy already in August 1964. Since then, there had been no signs that the American stance had changed.[57]

Moreover, Bahr saw the natural gas project as a promising way to bring Christian Democratic interests in line with those of the Social Democrats. Notwithstanding the strong emphasis on détente in Kiesinger's government declaration, the Christian Democrats had been hesitant to Brandt's and Bahr's attempts to bring about a decisive shift away from the former government's "policy of movement" to a more flexible and proactive policy vis-à-vis the Soviet Union. The natural gas project, however, was seen to have good chances of being accepted by both coalition parties, since it was closely linked to very large volumes of steel exports. Bahr had already received informal support from Minister of Finance Franz-Josef Strauss—who headed Bavaria's CSU—prompting Bahr to conclude that the project could "strongly contribute to removing the barriers to our Eastern Policy."[58]

Yet Bahr does not appear to have been informed about the detailed elaborations regarding the project's links to key energy policy and security issues that the Ministry of Economy's advisors had undertaken. The volumes of Soviet gas that had been up for discussion in the informal talks between Bahr and the Mannesmann managers, 6–7 bcm, were dramatically much larger than the 1 bcm for Bavaria that Schiller's advisors had thought acceptable from a security point of view. Moreover, the tentative northern route of the East-West pipeline, which was strongly favored by Bahr, gave rise to a set of completely new questions. No analysis had been carried out regarding the extent to which vast new volumes of gas from the East were really necessary or desirable from an economic or energy policy perspective, nor whether they could actually be accepted, as Bahr thought, from an energy security perspective. The Foreign Office clearly did not have the competence to assess the likely impact of a Soviet gas import on the dynamics of German energy markets—a sensitive issue that was further complicated by the fact that the West German natural gas industry was in a formative phase of development.

At the time, key decisions were pending regarding the structure, control, and division of labor in overall German gas transmission and distribution.[59]

The energy experts at the Ministry of Economy were surprised to hear about what they regarded as naive elaborations by the Foreign Office. Bahr's ideas were regarded as "completely hypothetical." An import of 6–7 bcm of Soviet gas to northern Germany was "out of the question," because the supply of northern and central Germany was already arranged for in the form of domestic and Dutch gas supplies. Contracts for the period up to the late 1970s had already been signed and no additional gas was needed for the foreseeable future. If additional contracts were signed at this stage, major disturbances of the German coal and oil markets would result, running counter to the government's efforts to integrate natural gas in a "harmonious" way into the overall German energy system. The only market in northern Germany where Soviet gas could possibly make a positive contribution was West Berlin.[60]

The security issue constituted another problem:

> A Soviet gas supply to the German market could only be accounted for to the extent that, in case of a supply disruption, a switch to other supply sources is feasible. Soviet deliveries could thus, a priori, only account for a relatively small supply share. A different assessment would only be conceivable if the danger of supply disruptions from the East could be prevented by an arrangement where, in the framework of a true interconnected network, the dependence on natural gas imports would not be unilateral. Such a situation could result, for example, if the Soviet zone [i.e., East Germany] for its part would wish to be supplied, through a pipeline across the territory of the Federal Republic, with Dutch or north German natural gas.[61]

All in all, the Ministry of Economy thought it would do more harm than good to continue the elaborations undertaken by the Foreign Office concerning the prospects for a large-scale Soviet gas supply, that is, not only to Bavaria, but to northern Germany as well. The energy advisors, continuing to view red gas more as a problem than an opportunity, were quite happy to note that the government had still not received any formal indication from Moscow that the Soviets were interested in any exports at all.

Bahr's "naive" ideas also met with opposition from within the Foreign Office. A report from the Political Division stressed the importance, from a tactical point of view, of letting the Soviets take the initiative. The Division was particularly critical of Bahr's envisaged pipeline route through the GDR, since this would materially strengthen the Soviet Union's presence in East Germany. More generally, the idea of importing natural gas from the East had better be avoided altogether, since "the Soviets, through the building of pipelines, are given an additional argument for their propaganda" about the "Soviet Union as a power of peace." The tone in such statements testified to the lingering presence of politicians and advisors from the Adenauer and Erhard era.[62]

On the other hand, Bahr's anticipation that the United States would not object to a German import of Soviet natural gas was confirmed. In an inquiry to the Ministry of Economy in September 1967, First Secretary Dux at the US Embassy in Bonn asked for information about the current stand on the gas project. Plesser at the ministry responded that there were currently no plans for Soviet gas imports to Germany. Plesser explicitly asked Dux whether Washington would object to a delivery of Soviet natural gas to German gas users, should such a possibility appear. Dux indicated that the Johnson administration—whose main foreign policy focus at the time was clearly on developments elsewhere in the world—would not raise obstacles. However, the United States would not regard it as advisable if "entire areas were to be connected to the Soviet natural gas pipeline, without the existence of a corresponding substitutability through other gas deliveries."[63]

The Soviet Option Fades Away

Meanwhile in Bavaria, Minister of Economy Otto Schedl, whose initiatives had forced the federal government to investigate the Soviet gas issue in the first place, continued to work for a Bavarian participation in the Soviet-Austrian project. France's waning interest sparked new hopes that the Soviet Union would become more interested than before in supplying Germany, and that Bavaria might take France's intended place in the envisaged Trans-European scheme. Schedl's state secretary Franz Sackmann thought "the fact that France, obviously, no longer has any interest in importing Russian natural gas keeps the possibility open that the south German area may after all be made accessible, by way of Austria, for Russian gas."[64]

But the optimism did not prevail. Neither in the Soviet-Austrian nor in the Soviet-Italian negotiations was a Bavarian participation any longer mentioned as a possibility to be considered. The Soviets still judged that the federal German government would not support a Bavarian involvement. Alexei Sorokin's charm offensive was not followed up by any further Soviet moves. Thyssen and Mannesmann, who saw their pipe export opportunities shrink, were disappointed. Again, they turned to Bonn, lobbying—in vain—for a direct government initiative.[65]

Egon Bahr, arguing that the natural gas project had better be taken up for discussion at the highest possible level, proposed that a meeting be organized between Kiesinger, Brandt, Schiller, Strauss, and Minister of Inner-German Affairs Herbert Wehner. The Ministry of Economy's energy experts, however, argued that such a meeting would be both risky and unnecessary, since "a treatment of the Soviet natural gas project at a high political level would possibly become known and that would probably not facilitate a Soviet preparedness to involve the German [steel] industry." The same conclusion was eventually drawn by a majority of advisors at the Foreign Office. The divisions at the Foreign Office that had been involved in the discussions convened in late September 1967 for a thorough discussion, only to conclude that the whole thing was not "acute" and that no top-level consultation was necessary, at least not for the time being.[66]

In October, Bonn's overall relations with Moscow worsened again following the reception of a Soviet memorandum in which the "unconditional recognition of the GDR" was defined as a precondition for any further negotiations on concrete cooperative projects. Bonn regarded this as an "extreme position which could not be agreed to." Chancellor Kiesinger responded by publicly stating that he currently saw "no possibility for successful talks" with Moscow. This made the Kremlin even more furious, since the chancellor had broken a German promise to keep the emerging foreign policy talks between the two countries strictly confidential. In a stern Soviet declaration issued in December, Brezhnev threatened to completely cut off the dialogue with the Kiesinger-Brandt government. Foreign Minister Brandt, for his part, considered resigning from his post.[67]

All in all, as winter fell over Europe in 1967 it seemed highly improbable that Bavaria or any other part of Germany would become an importer of red gas. Schedl, Bahr, and the steel industry had failed to mobilize internal support for the project, and overall Soviet-German relations had reached a low-water mark that did not seem to permit any far-reaching cooperation. Meeting informally with representatives of the German steel industry and with regional politicians and gas companies, Soviet representatives had showed themselves highly interested in exporting natural gas to Germany, but Moscow was not prepared to take any formal initiative without Bonn's explicit support. Bonn, for its part, was equally reluctant to take any initiative, preferring to "wait and see." Instead, a supply of southern Germany with gas from the Netherlands started to be seen as the most promising option for Bavaria. Rather than continuing to dwell on the Soviet alternative, the internal German debate increasingly came to focus on how Dutch exports to southern Germany might be organized and by whom the necessary pipelines were to be owned and controlled.[68]

Apart from Dutch gas, imports from Algeria were once again pointed at as a possibility, for both Bavaria and Baden-Württemberg. With the Soviet-Austrian agreement taking clearer shape, the earlier vision of imports from Algeria by way of Yugoslavia and Austria seemed to have reached a dead-end. In July 1967, however, France had contracted large-scale imports of Algerian natural gas, opening up the possibility of German imports of Algerian natural gas by way of transit through France. Sonatrach was very eager to exploit this opportunity, announcing its intention to set up, for this purpose, a branch office in Munich.[69]

Otto Schedl, though, was not as enthusiastic about the Algerian option as he had once been. Since June 1967, there were no more direct contacts between Munich and Algiers. Like the federal energy advisors in Bonn, Schedl noted that the French had agreed to pay quite a high price for its Algerian gas, making it improbable that Bavaria, which in addition to the French border price would have to pay for transit through France, would be able to conclude an economically viable deal with the Algerians. Instead, Schedl concentrated his efforts to make sure that Bavaria received Dutch gas at the lowest possible price, and that a Dutch contract would not prevent Bavaria from signing contracts with other suppliers later on.[70]

To sum up this chapter, red gas became an issue in West Germany as a result of Otto Schedl's ambition to strengthen Bavarian energy independence. Schedl initially eyed Algeria as the ideal partner and far-reaching talks were held with the government of this country and its state-owned oil and gas company

Sonatrach. Following the tentative agreement in November 1966 regarding red gas exports to Austria, the Bavarian efforts shifted eastward. A year later, however, the impression was that Bavaria would be left without Soviet gas. This was so despite the fact that Schedl had managed to gain informal support for his Soviet vision from a variety of stakeholders in Bonn, the German steel industry, and the Soviet Union itself. Foreign Minister Willy Brandt and his strategist Egon Bahr not only appropriated Schedl's ideas, but also expanded them, seeking to maximize their utility in a foreign policy context. Yet the timing was slightly wrong. Soviet-German relations showed signs of improvement, but "change through rapprochement" was still a contested Eastern policy. A culture of suspicion regarding all things Soviet lingered on among the key advisors at the Foreign Office and the Federal Ministry of Economy. On the Soviet side, this was mirrored by a refusal to actively approach Bonn with a concrete export offer—despite a clear willingness, as expressed in numerous informal conversations with the Bavarians and the German gas and steel industries, to cooperate.

6
From Contract to Flow: The Soviet-Austrian Experience

As of June 1968, when the pioneering Soviet-Austrian gas export contract was signed, the future of red gas in Europe was still impossible to predict. Gas companies and governments in a whole range of Western countries—plus Japan—had shown a clear interest in imports from the East, but the actual outcome of the talks that had been held so far was meager. Italy's negotiations with Moscow seemed to have collapsed, France had more or less withdrawn from the talks, and Bavaria, as we have seen, appeared to be heading toward a gas supply from the Netherlands rather than from the Soviet Union. The Soviets had made even less progress in negotiating with Finland, Sweden, and Japan.

One reason for this slow progress was that West European actors felt uncertain as to whether the Soviet Union could be trusted. As we have seen, the critics questioned Moscow's true intentions regarding its efforts to enter the West European gas market, fearing "manipulation" of the envisaged trade for political or other purposes. There were also worries that the Soviet Union might turn out technically or organizationally incapable of meeting the expectations. Against this background, the Austrian deal was of critical importance. Any problems in ÖMV's cooperation with Mingazprom would discourage other West European gas companies from following the Austrian example. If successful, however, exports to Austria would likely be interpreted as a confirmation of the East-West gas trade's overall feasibility. The Austrian reference might in that case help Moscow secure additional export contracts, potentially forming the point of departure for a much more large-scale gaseous integration across the Iron Curtain. The Soviets were well aware of what was at stake, and hence took whatever measures it could to ensure that the Austrian project became a success. As we will see in this chapter, this required huge sacrifices.

Interconnecting Austria, Czechoslovakia, and the Soviet Union

The Soviet-Austrian gas export contract had stated that natural gas deliveries from the Soviet Union through Czechoslovakia to Austria must commence only three months after its signing, on September 10, 1968. This was possible thanks to the already existing pipeline infrastructure. In Czechoslovakia, the eastern section of the "Bratstvo" pipeline from the Soviet Union had been completed a year earlier, allowing Soviet natural gas to reach Bratislava, on the Austrian border, in June 1967. Through a short extension of the same pipeline across the Iron Curtain it would be possible to supply the capitalists beyond the Danube.

The Bratstvo originated in western Ukraine, whose historically important gas fields had been charged with the task of supplying Czechoslovakia until Siberian gas became available. It exited the Soviet Union outside the town of Uzhgorod in the scenic trans-Carpathian region of what since 1945 was the westernmost corner of the Ukrainian SSR, and entered Czechoslovakia at the village of Ruská. Measuring 820 mm in diameter on Soviet and 720 mm on Czechoslovak territory, the pipeline was 539 km long, of which 189 km was in Ukraine and 350 km in Czechoslovakia. The project was the centerpiece of an arrangement through which the Czechs, struggling to respond to the rapidly growing popularity of gas as a fuel, aimed to substitute imported natural gas for domestic, mostly manufactured, gas. According to the first Soviet-Czechoslovak contract, signed in December 1964, gas imports from the Soviet Union were to increase from 270 mcm in 1967 to 500 mcm in 1968 and 800 mcm in 1969, before a plateau level of 1 bcm was reached in 1970.[1]

The Czechs had originally hoped to balance this import from the East with sizeable supplies from Algeria, but in February 1966 the talks with Sonatrach and the Algerian government had, disappointingly, broken down as a result of the failure to agree on a countertrade scheme. To compensate for the nonarrival of Algerian gas, Prague set out to negotiate additional imports from the Soviet Union. The result was a second Soviet-Czechoslovak contract, signed in 1967. Approved in February 1968 by the Soviet Council of Ministers, it paved the way for an additional 1.5 bcm of Soviet gas per year, starting in 1970. Hence Moscow's long-term export obligations to Czechoslovakia amounted to 2.5 bcm per year. Actual deliveries amounted to 265 mcm in 1967 and 587 mcm in 1968, which meant that they were so far nicely in line with Soviet contractual obligations.[2]

At the same time, and independently of the Soviet-Czechoslovak agreement, Austria and Czechoslovakia agreed to trade natural gas with each other. As we saw in Chapter 4, the cooperation initially centered on "virtual" exports of Czechoslovak gas to its western neighbor, but the arrangement was subsequently expanded to include piped shipments as well. The traded volumes were much smaller than the ones Czechoslovakia received from the Soviet Union, but they were of principal significance since they paved the way for construction of the first European natural gas pipeline that crossed the Iron Curtain. Though built before the Soviet-Austrian contract was signed, the pipeline was dimensioned in such a way as to allow for both Czechoslovak and Soviet gas to be pumped through it. Yet it was a fairly small and above all a very short pipeline. Since the distance between the Austrian and the Czechoslovak gas systems was only 5 km, its completion can hardly be described as a major engineering feat.[3]

Apart from the interconnecting line itself, several new facilities for receiving, measuring, and compressing the imported gas were planned, most of which were to be erected near the village of Baumgarten on the Austrian-Czechoslovak border. Importantly, the compressors in the new import system were to be selected in such a way that they could be used rationally even if the bigger project concerning a possible transit of Soviet gas to Italy and France was not realized. ÖMV applied for permission from its supervisory board to order the necessary equipment three months before its contract with the Soviet Union was signed. The order was placed in haste, since the gas was to start flowing half a year later. ÖMV argued that "since the working out of project and bid normally takes at least 3–4 months, a competitive process has to be refrained from." Accordingly, only one company was asked to come up with a bid for the construction of the Baumgarten station: the American firm Pritchard. The reason for choosing this firm was that it had been responsible for constructing ÖMV's gas stations in the past and that it was hence very familiar with the Austrian gas system. The compressors themselves were ordered from Clark, another US-based company. ÖMV would thus rely heavily on American technology for its import of Soviet natural gas. Baumgarten was also to host, on a permanent basis, seven Soviet specialists in charge of measuring volume, temperature, pressure, and chemical composition of the arriving gas. Official measurements were to be carried out with equipment belonging to the Soviet Union, but ÖMV also installed its own control apparatus so as to verify the Soviet figures.[4]

Another essential component of the emerging import system was gas storage. ÖMV planned to turn several partly depleted gas fields into storage facilities. Experiments had started at one of these fields, Matzen, well before negotiations with the Soviet Union were initiated, but the activities grew markedly in importance as a result of the Soviet deal. The need for storage capacities was formally motivated by the fact that deliveries from beyond the Iron Curtain would take place at a more or less constant rate throughout the year. The Soviet-Austrian contract only allowed for minor daily and seasonal variations in the flow (up to around 20 percent). This flexibility was not sufficient to match the shifting pattern of use. For this reason it was necessary to store some excess gas that arrived during periods of low use and make it available at a later point when demand was high. Nothing was openly said about the possibility of making use of gas storage facilities for a different purpose: as a reserve capacity to be called upon in case of unexpected—intended or unintended—disruptions in the gas flow.[5]

Importing Soviet Gas in Practice

By mid-August 1968, ÖMV's engineers reported that Austria's domestic gas system was ready to receive red gas. Their colleagues in Czechoslovakia and the Soviet Union also signaled their readiness to start up the system. Then, in the night of August 20–21, 1968, Warsaw Pact forces launched a massive invasion of Czechoslovakia. The Prague Spring, the spirit of which had contributed substantially to Western Europe's enthusiasm for cooperating with the communist bloc in energy came to an end. Overnight, the illusion that the Kremlin might accept somewhat looser bonds between the Soviet Union and its Central European satellites, and that the Iron Curtain might even corrode away, was brutally crushed.

Figure 6.1 Austrian minister of transportation Ludwig Weiss and Soviet gas minister Alexei Kortunov inaugurate the Soviet-Austrian gas trade on September 1, 1968. Ten days before, Warsaw Pact forces had invaded Czechoslovakia.
Source: ÖMV. Reproduced by permission.

ÖMV feared that the turmoil and chaos that followed would make it impossible for its communist partners, which in this case were dependent on successful cooperation not only with Austria but also with each other, to maintain regular work discipline. Sabotage attempts from rebellious Czechs were also feared. The Soviet Union, however, set out to secure the arrangement by whatever measures it could, assuring the Austrians that the invasion would not be allowed to disturb the gas trade. On the contrary, the Kremlin decided to make use of natural gas for propaganda purposes by offering ÖMV deliveries ahead of schedule. In the West, voices were raised urging the Austrians to publicly show their dislike of the shocking Czechoslovak events by refusing or at least delaying the start-up of red gas deliveries. Facing the possibility of a major conflict with the Soviets, however, and by extension a need to totally rethink the country's energy supply strategy, the Austrian government found itself unable to support such a radical proposition. The Soviet offer was thus accepted, and as a result Soviet gas started flowing across the Czechoslovak-Austrian border already on September 1, 1968, that is, nine days ahead of schedule. The event was celebrated with a ceremonial festivity at Baumgarten. Moscow was also keen to point out that the invasion had by no means disturbed gas exports to Czechoslovakia itself, publicly emphasizing that "deliveries of natural gas to Czechoslovakia in the period from August 21 to September 1 of this year were 1.7–2.4 million cubic meters daily. This compares with a planned volume of 1.3 million cubic meters daily for this period." Shipments were thus not only sustained, but even exceeded the planned level by a large amount.[6]

As September drew to a close, ÖMV had already received 32 mcm of red gas. After the inauguration of the US-built compressor station at Baumgarten in

Figure 6.2 ÖMV's new compressor hall at Baumgarten, built for incoming Soviet gas.
Source: ÖMV. Reproduced by permission.

November, the import capacity rose to 3 mcm per day. By the end of December, however, cumulative imports amounted to no more than 132 mcm. Since ÖMV had asked the Soviets to deliver 180 mcm for the year as a whole, this was to a certain degree disappointing, although it was still above the lower limit of the Soviet Union's export obligations as specified in the contracts. Despite the slow start, ÖMV's management board regarded the trade as a success. Without specifying any particular cause, some "minor interruptions" were reported, but these were deemphasized in the communication with the company's government-appointed supervisory board. The management board chose to point to the initial period of importing Soviet gas as "satisfactory."[7]

ÖMV congratulated itself on having started up imports from the East in due time before the heating season. Winter 1968/1969 became unusually cold, spurring a record demand for gas. Yet during the first three months of 1969, imports from the Soviet Union remained at a low level and suffered repeated disruptions. The Soviet specialists at Baumgarten reported that only 115 mcm had been received, which was only slightly more than the 101 mcm that had been imported during the last three months of 1968. ÖMV, having been promised 745 mcm of Soviet gas for the year as a whole, had expected more, but Mingazprom proved unable to meet demand. As a consequence, ÖMV was forced to increase its domestic gas production, mainly from the large Zwerndorf field, "to a barely justifiable extent." The problems with deliveries from the Soviet Union continued during spring and summer, forcing Zwerndorf to produce at the barely justifiable level for a longer period of time.[8]

In reality, the imported volumes of Soviet gas were even lower than the ones officially reported. Unable to live up to its export obligations, the Soviets in December 1968 requested a meeting with ÖMV and the Czech foreign trade organization Metalimex. At the meeting not only Soviet but also Czechoslovak exports to Austria were discussed. For the years 1969 and 1970, Soviet exports were to reach 745 mcm and 932 mcm, respectively, whereas the corresponding Czechoslovak commitments amounted to 170 mcm and 160 mcm. However, it was now decided that the Soviet Union and Czechoslovakia would "cooperate" in supplying Austria. The cooperation consisted of a statistical manipulation: the amount of "Czechoslovak" gas arriving at Baumgarten was to be redefined as the difference between the actual amount of gas delivered to Austria and the total Soviet export volume contracted with the Austrians. This made it possible for the Soviets to hide some of its export problems, which, instead, were statistically transformed into a Czechoslovak delivery problem.[9]

The effect of the "cooperation" found its way into ÖMV's internal import statistics: for 1969, Czech exports formally amounted to a mere 94 mcm instead of the contracted 172 mcm; 76 mcm had thus been statistically transformed into "Soviet" exports. Officially reported Soviet exports to Austria amounted to 319 mcm for the year as a whole, which was far below the contracted 745 mcm. But the "real" figure, taking into account the statistical manipulations, was even lower: 243 mcm. In other words, the Soviets had failed to deliver about two-thirds of the contracted volume. Apart from that there had been eight instances of total supply disruption, each of which had lasted between 10 and 72 hours.[10]

In September 1970, two years after Austrian imports of Soviet gas had commenced, ÖMV's general director Ludwig Bauer noted that the imports continued to be subject to "repeated delivery failures." ÖMV was told that the problems were a consequence of "transport difficulties" in the Soviet pipeline network. The Soviets excused themselves for the interruptions, which were obviously not intentional. But ÖMV was highly worried, since Soviet imports, while still unreliable, were quickly becoming a key component in the company's efforts to meet national gas demand. By 1970, red gas already amounted to 35 percent of total supply, and this figure was scheduled for further growth to nearly 50 percent by 1971.[11]

Preparing for the 1971 deliveries, Mingazprom told ÖMV that it would not be able to deliver the contracted annual volume in the yearly rhythm that ÖMV had requested. Hence during the first three months of 1971, roughly coinciding with the difficult winter period, the Soviets delivered only 285 mcm, which was around 100 mcm below the volume requested by ÖMV. Again, the Austrians were forced to compensate for the delivery failures by forcing the pace of domestic production. The problems continued during the second quarter of 1971, leading to a delivery of 20 mcm less than contracted. The fact that the new storage facility at Matzen had now been taken into operation contributed to raising overall supply security, but it hardly offered ÖMV much long-term relief.[12]

According to the 1968 contract, the Soviets would have to pay a penalty fee for its delivery failures. Having discussed the issue with the Soviets, however, ÖMV chose to accept an arrangement in which delivery failures during the first half of 1970 would be compensated for by larger than contracted imports during the second half of the year. ÖMV pumped this gas into its Matzen storage facility, reserving it for future emergency deliveries. In 1971, then, the Soviets managed to deliver 1.33 bcm, which was only 5 percent below the contracted 1.4 bcm level. The next winter saw further improvements. The Soviets delivered 381 mcm in the first quarter of the year, roughly corresponding to ÖMV's request and falling within the limits of the contracted export volume.[13]

The Galician Challenge

Why did Mingazprom find it so difficult to live up to its export obligations during the first few years of deliveries to Austria? After all, the Soviet Union was in possession of the world's largest gas reserves, and no other country in Europe was even close to producing comparable volumes of gas. The exports as specified in the Soviet-Austrian contract were negligible by comparison.

The reason behind the delivery failures was that Mingazprom still suffered from a lack of long-distance pipelines that could bring abundant gas from Central Asia or Siberia to the Soviet Union's westernmost regions. The completion of such links was crucial for the realization of Mingazprom's visions of continued expansion of natural gas both for domestic use and for exports. Progress was on its way, but the task was daunting. Siberian gas would not be available in the westernmost Soviet regions before the early or mid-1970s. In the meantime, Mingazprom aimed to meet its export obligations by making use of its Galician (western Ukrainian) gas fields, conveniently located as they were near the border to Czechoslovakia. As we saw in Chapter 2, these fields had in the course of the postwar decades come to play a key role in the Soviet gas system. Not only the Ukrainian SSR, but four other union republics as well—Belarus, Lithuania, Latvia, and, to a smaller extent, Russia—were dependent on Galician gas. Small volumes were also exported to Poland along a pipeline built by Nazi engineers. In each of these user regions gas demand grew at a steady pace, and production from the Galician fields was gradually scaled up to meet the needs.

Toward the mid-1960s it started to become clear that the fields had been overexploited and that they were being emptied at an alarming rate. As of 1965, the remaining commercial reserves in western Ukraine were still sizeable by European standards, amounting to 73 bcm.[14] As of 1967, however, annual production already reached 13 bcm, a rate at which the reserves would not last long. By 1970, it was estimated that it would be impossible to produce more than 11 bcm, due to rapidly decreasing pressure in the gas wells. At the same time, aggregate regional demand plus deliveries to Belarus, Latvia, and Lithuania along with exports to Poland and Czechoslovakia were expected to *in*crease from 13 bcm in 1967 to around 15 bcm in 1970.[15] This growing imbalance between stagnating production and growing supply obligations explains why Mingazprom, in its negotiations with ÖMV in autumn 1967, was so reluctant to give in to the Austrian desire to start imports of red gas already in 1968. Although the volumes to be exported to Austria were small if seen in relation to the overall Soviet gas system, they were quite substantial from a Galician perspective.

At the time when Austria, in winter 1966–1967, initiated its negotiations with the Soviet Union, western Ukraine was already experiencing serious gas shortages. The regional communist party organization in Lvov, Galicia's old capital, in January 1967 turned to the Ukrainian branch of Gosplan with a desperate plea for improving the situation, which seemed to be looming out of control. The party representatives noted that the Ukrainian Council of Ministers had ordered Mingazprom to reserve 979 mcm of Galician gas for Lvov during the fourth quarter of 1966, but in reality only 833 mcm had been supplied. Both the population and the city's industrial enterprises experienced

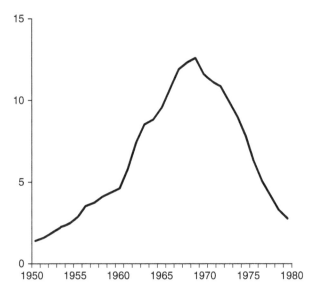

Figure 6.3 Production of natural gas in western Ukraine, 1950–1980 (bcm).
Source: Based on figures in Derzhavnyi komitet naftovoi, gazovoi ta naftopererobnoi promyslovosti Ukrainy 1997, pp. 148 and 176.

"systematic interruptions in the normal gas supply." By February 1967, the pressure in the pipelines decreased to such a low level that it became impossible to deliver gas to households. At the same time it was observed that the volume of Galician gas shipped to Belarus, Lithuania, and Latvia was higher than planned. Lvov's local party organization was outraged that gas supply to Galicia itself, the producing region, seemed to be of secondary importance.[16]

In reality the problem was much more complex. In fact, those republics to which Galician gas was "exported"—Belarus, Lithuania, and Latvia—were no less worried about their gas security than western Ukraine. Situated at considerable distance from the gas fields and lacking access to alternative supplies, they felt particularly vulnerable to the mounting production problems at Dashava and other Galician gas fields. On January 19, 1967, gas supply along the Dashava-Minsk-Vilnius-Riga pipeline broke down due to insufficient availability of gas. Mingazprom responded by bringing in Russian gas to Kiev, the Ukrainian capital, so that some of the Galician gas that normally supplied Kiev could instead be shipped to Belarus and the Baltics.[17] Even so, numerous industrial enterprises in Latvia, at the remote end of the pipeline from Dashava, were forced to a standstill as a result of gas shortages, paralyzing the republic's economic life. Latvian Premier Vitalii Ruben complained loudly about the frequent irregularities and cutoffs, which seemed to result not only from a structural shortage of Galician gas, but also from temporary shutdowns following recurring accidents at the gas fields and along the pipeline. Ruben was particularly upset about the fact that users located further upstream often used more gas than they were entitled to, leaving little or no fuel for users located downstream. Since long-term economic planning foresaw a rapid increase in Latvia's reliance on natural gas, Ruben argued that the republic's gas supply must be diversified through a new pipeline from another gas source, which would have to be independent of Galician gas.[18]

Given this already strained situation in Ukraine, Belarus, and the Baltics, Moscow's ambitions to initiate exports of Galician gas to Austria and scale up deliveries to Czechoslovakia did not give rise to any enthusiasm in the westernmost Soviet regions. The new export commitments implied that the already fierce competition for Galicia's waning gas resources was bound to further toughen. Local party organizations in the regions that depended on scarce Galician gas showed themselves very concerned with the anticipated impact of exports on regional gas security, fearing an even more "severe deterioration of gas supply to the population and to industrial enterprises."[19]

There seemed to be two principal ways of responding to the looming scarcity of Galician gas: either consumption would have to be reduced, or additional gas brought in from elsewhere. Given the wide gap between available gas resources and expected demand, the first option appeared unrealistic. Soviet economic planning for the western regions was firmly based on the assumption that ever more gas would be available. The chemical-metallurgical complex at Kalush, for example, was being expanded in the expectation that an additional 280 mcm of gas would be available for it by 1970. Kamenets-Podolsk's large cement factory, of great importance for the construction industry, counted on 250 mcm extra, and at Rovno in northwestern Ukraine a new chemical factory had started to be built following a promise from Moscow that an annual volume of 100 mcm of natural gas would be available.[20]

The choice, therefore, fell on the second option. This required construction of several new long-distance pipelines, which in turn depended on ample access to large-diameter steel pipe, compressors, and a variety of additional equipment. Gosplan, aware of domestic manufacturers' never-ending problems to deliver such items, was as always hesitant to accept Mingazprom's requests. Gas Minister Kortunov, however, managed to convince the Kremlin that the pipelines could be constructed even in the absence of domestic pipes and equipment. More precisely, Mingazprom hoped to make use of the steel pipes and compressors that would be imported to the Soviet Union in return for gas exports to Austria and Czechoslovakia, respectively. This was the reason why the Soviets, despite the highly strained gas supply situation in western Ukraine, eventually accepted ÖMV's request for gas exports to start already in 1968—under the condition that pipe deliveries would also commence early on. In the same vein, the Soviets agreed on scaled-up exports to Czechoslovakia in return for a large number of compressors.

Ukraine as a Victim

The strategy of bringing in gas from elsewhere needed to be combined with a plan for restructuring Ukrainian gas flows. New pipelines were necessary to improve transmission capacity along several routes. In particular, connections between eastern and western Ukraine would need to be strengthened

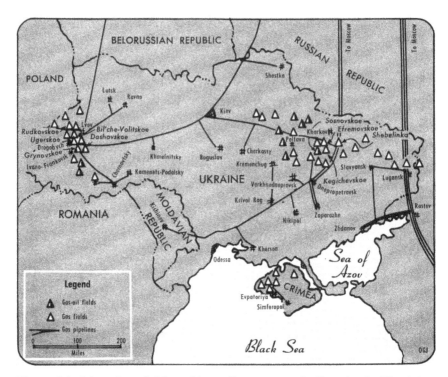

Figure 6.4 Map of gas fields and long-distance gas pipelines in the Ukrainian SSR as of the late 1960s. Before eastern and western Ukraine were more tightly interlinked, all Soviet exports of natural gas depended on deliveries from Galicia's (western Ukraine's) gas fields, here shown to the left.

Source: Oil and Gas Journal, February 17, 1969, p. 47. Reproduced by permission.

and expanded. The Soviet plan that emerged was to have the giant Shebelinka field in eastern Ukraine supply some of the regions that had so far depended on Galician gas, thus reducing the need for deliveries from Galicia to Kiev, Belarus, and the Baltics. The volumes of Galician gas that in this way were saved could be reserved, it was imagined, for local use and for exports.

Starting in 1967, at which time the Austrian contract had not yet been finalized, Mingazprom started to draw up plans for new transmission capacities on the route from Shebelinka to Kiev and onward to western Ukraine. Kiev had traditionally relied on Galician gas and still did so to a certain extent, but it had gradually embarked on a transition to eastern Ukrainian gas. For this purpose a pipeline had already been built from Shebelinka to Kiev. It would now have to be complemented by a parallel line whose purpose would not primarily be to supply Kiev itself, but rather the regions to its west. It was agreed that a new 1,020-mm pipeline be built from Efremovka in the outskirts of the Shebelinka area to Dikanka near the city of Poltava and from there in parallel with existing pipelines to Kiev. From Kiev, this trans-Ukrainian pipeline would be laid in parallel with the historical Dashava-Kiev pipeline, which since 1948 supplied Kiev with Galician gas. At Ternopol, the new pipeline would diverge from the historical line and head north, eventually merging with the important Dashava-Minsk-Riga system. Making use of Czech compressors, Mingazprom in this way aimed to pipe large volumes of Shebelinka gas not only to western Ukraine, but also to Belarus and the Baltics.[21]

The project was perceived as highly attractive for the troubled western Ukrainian regions, since it would enable them to receive gas from eastern Ukrainian sources. Belarus, Lithuania, and, in particular, Latvia were less enthusiastic. The problem, from their point of view, was that Shebelinka gas would reach the three republics by way of the already existing Dashava-Minsk-Riga system and that Belarusian and Baltic gas users would hence continue to be highly vulnerable to accidents and problems along this route. Latvian Premier Ruben, again pointing to Riga's problematic location at the remote end of the Dashava line, intensified his lobbying of Moscow for diversification of the republic's gas supply. Mingazprom was susceptible to the Latvian arguments and formed an alliance with Ruben. Together they sought to convince Gosplan that a new pipeline must be constructed through which gas from faraway Siberia could be brought in. Gosplan, however, informed Mingazprom and the Latvians that lack of material resources made the construction of such a pipeline impossible, at least for the time being.[22]

Lack of material resources also became a problem for the planned trans-Ukrainian pipeline. The reason was the delay in finalizing the contracts for pipe and compressor imports from Austria and Czechoslovakia, respectively. Although the Soviet-Austrian gas contract had essentially been completed in January 1968, tough negotiations continued during spring regarding the pipe price. Mingazprom showed considerable irritation with the Ministry of Foreign Trade regarding the slow progress in finalizing the deal. Kortunov sought to make the negotiators speed up the process and ensure that VÖEST start deliveries of pipes already in June 1968. If the pipes arrived later, it would be very difficult to complete the new pipeline in time for the winter season. This in turn would make the western Ukrainian situation extremely strained, in a way that, according to Kortunov, would most probably also affect gas exports.[23]

As for the Czechoslovak equipment, the Soviets had in September 1966 agreed in principle with its Czech partners on the import of 90 powerful compressor units. Mingazprom regarded the machines as "totally indispensible for construction of the long-distance pipelines on which increased gas exports depend." The Ministry of Foreign Trade, however, had not yet finalized the Czech contracts. Apart from the compressors, several hundred lifting cranes, to be used in the actual laying of the pipeline on Ukrainian territory, were at stake. All in all, Kortunov complained that "such a slowness in the finalization of contracts for pipes and equipment causes difficulties for Mingazprom in securing the timely launch of gas pipelines necessary for the national economy of the USSR and for meeting gas export demand."[24]

In the Soviet-Austrian contract that was eventually concluded, VÖEST and its German partners promised to have the first pipes ready by September 1968. Kortunov did his best to plan pipelaying in such a way that at least the Efremovka-Kiev section of the trans-Ukrainian pipeline could be completed in due time for winter. This would enable a reduction in deliveries from Galicia to the Ukrainian capital along the historical Dashava-Kiev pipeline and thus save substantial volumes of Galician gas, which could instead be reserved for western Ukraine itself, for Belarus, the Baltics, and foreign customers. When the first pipes from Mannesmann arrived in the port of Leningrad ahead of schedule in late August, the optimism increased. As it turned out, however, the constructors along the Efremovka-Kiev route were left to wait in vain. The reason was an acute lack of railway cars for transporting the pipes from the port to the construction site.[25]

Mingazprom was furious, stressing the necessity of "the fastest possible construction of the pipeline" and reemphasizing its key role in "securing the further growth of the USSR's gas exports." Yet by mid-October 1968, only 198 out of 316 km of pipes had arrived at the construction site, and of these only 157 km had been welded. Apart from Kortunov, the Ukrainian Council of Ministers, fearing negative consequences for the republic's near-term gas security, complained loudly. The task of guaranteeing Kiev's and western Ukraine's gas supply during the fourth quarter of 1968 was increasingly seen as a mission impossible. Out of a total western Ukrainian demand amounting to 5.5 mcm per day, it was estimated that only 3 mcm would be delivered. In other words, nearly half of the region's supply was under threat.[26]

The missing volume of 2.5 mcm per day—corresponding to 230 mcm for the fourth quarter as a whole—corresponded almost exactly to the volume Mingazprom had promised to deliver to Austria and Czechoslovakia during the same period. By refraining from exporting gas to these countries, the Soviets could thus have "rescued" western Ukraine. Yet the need to meet export obligations was prioritized over domestic needs. Deputy Gas Minister Yuli Bokserman issued a strict directive emphasizing that "export supplies to the CSSR, Poland, and Austria in the volumes specified in the approved plan must not be cut-down." Deliveries of Galician gas to Kiev were also prioritized, with the historical Dashava-Kiev pipeline transporting much larger volumes than during earlier winters. The deficit was absorbed by western Ukraine and by users along the Dashava-Minsk-Riga pipeline. By mid-December 1968, users situated on the latter route were reported to face an "acute shortage of gas." Mingazprom did its utmost to convince regional and local party organizations, municipal institutions, and industrial enterprises to lower their gas consumption so as to minimize the overall shortage, but the effect was meager.[27]

The strained situation in the regions that competed with Austria for scarce Galician gas also explains why the Soviets, though keen to live up to their export obligations, did not deliver more gas to ÖMV than what the contract specified as a minimum level in the fairly large interval defined for the fourth quarter of 1968 (120–180 mcm). During 1969, then, as we have seen, the export regime virtually collapsed as the Soviets proved able to deliver only one-third of the contracted volume. This time, the reason for the failure did not have so much to do with scarcity of gas, but rather with delays in completing a new compressor station that was under construction at Uzhgorod near the Soviet-Czechoslovak border. For the Soviet regions that competed with Czechoslovakia and Austria for scarce Galician gas, the failure to take into operation the Uzhgorod station was, ironically, good news. It prevented additional Galician gas from leaving the Soviet Union and kept the internal Ukrainian crisis during 1969 within proportions. But worse was to come.[28]

Scaling Up Exports

The Efremovka section of the trans-Ukrainian pipeline was eventually completed in time for the winter season 1969–1970. A year delayed, it boosted Kiev's gas supply by 11 mcm per day (corresponding to 4 bcm on an annual basis). The section from Kiev to western Ukraine could eventually be taken into operation in summer 1970. It was designed to handle a gas flow of 6.4 bcm per year, a target that was successfully reached in late 1970. Around half of the gas that was transported along this section was consumed in western Ukraine itself, whereas the other half was transited on to Belarus and the Baltic region.[29]

Despite this powerful injection of eastern Ukrainian gas, however, the structural supply problems in the westernmost Soviet regions continued to worsen, both because local Galician production, having peaked in 1969, had entered a phase of steep decline and because export commitments to Czechoslovakia and Austria were stepped up. The newly completed trans-Ukrainian pipeline proved too small to compensate for this development, and as a result the competition for Galicia's remaining gas intensified. The main victims in the struggle that followed were Belarus and the Baltics, whose gas supply had been in a phase of dynamic expansion ever since the inauguration of the Dashava-Minsk pipeline in 1960. Expansion now turned into contraction. For 1971, Mingazprom had originally planned to deliver 7.1 bcm to Belarus and the Baltics, but already in April that year Deputy Gas Minister Vasily Dinkov reported that it would not be possible to deliver more than 6.2 bcm. In 1972, the ministry would be able to deliver a mere 4.6 bcm.[30]

As for the exports to Austria, these were to reach the agreed pleateau level of 1.5 bcm in 1971. During winter 1971–1972, ÖMV was pleased to observe that the Soviet Union, for the first time, was more or less able to deliver what it had promised. But the price that gas users in the Soviet Union's western regions had to pay for this success was high. The winter became unusually cold, with temperatures below minus 30 degrees centigrade, and enormous problems plagued the gas supply to numerous Ukrainian cities. The pressure fell dramatically in several key pipelines. At Lutsk and Rovno in northwestern Ukraine, both of which competed with Austria and Czechoslovakia for access to scarce Galician gas, schools and other municipal institutions had to close. The crisis was deepened by the unauthorized increase in gas use from the side of many industrial enterprises. The latter explained that they were forced to use natural gas due to lack of reserve fuels (such as coal and oil). The Ukrainian Council of Ministers, referring to an "emergency" situation, accused Mingazprom of having failed to create sufficient gas storage capacities. The crisis could have been avoided had deliveries to Austria been reduced, but again Moscow chose to prioritize exports and let its own population freeze.[31]

In summer 1972, an emergency plan was worked out with measures to be taken in case of continued gas shortages during the next winter. A detailed list was compiled, consisting of 142 enterprises and power plants in Ukraine. For each of these, an exact upper limit of natural gas use was defined, along with the way in which they were to switch to reserve fuels. Some enterprises were to switch completely to reserve fuels, whereas others were to switch only partly. Predictably, the list gave rise to loud protests from a whole array of industrial gas users, many of which argued that their entire operations would be in danger if they were not allowed to use more natural gas. Judging that gas export obligations must be met at any price, however, the central authorities did not consider increased domestic deliveries an acceptable option.[32]

Unwillingness to accept the rationing plan contributed to yet another crisis during winter 1972–1973. As before, a cruel competition among user regions took place for Galician gas—and again exports were prioritized. A struggle between northwestern Ukraine and Belarus unfolded already in the early winter period, as Belarus tried to prevent Ukraine from taking out its gas needs from the Dashava-Minsk pipeline. For the Ukrainians, fall was a sensitive season because of the sugar beet harvest, which required large amounts of fuel. Deputy Gas Minister Dinkov concluded that the situation with respect to Ukrainian gas supply would remain very tense, urging the Ministry of Food Industry to ensure that its enterprises were supplied with sufficient amounts of reserve fuels. The Ministry of Chemical Industry also complained about worsening gas supply.[33]

The Unseen Crisis

It took about three years—from the first deliveries in September 1968 to the normalization of contracted flows in 1971—before the Soviet Union was able to live up to its contractual obligations vis-à-vis the Austrians. Even after that, success in meeting annual and quarterly delivery targets continued to mask problems with unplanned irregularities and unexpected delivery interruptions, which were compensated for by increased deliveries at a later point in time. ÖMV, in its communication with the Soviet side, did not hesitate to voice its dissatisfaction with the frequent delivery problems, but the company also showed considerable patience with the Soviets. Thanks to its domestic gas fields and the Matzen gas storage facility as buffers, ÖMV was able to meet national gas demand at all times.

The overall impression was that the emerging Soviet-Austrian gas trade had proven successful. The problems experienced were discarded as "childhood diseases", of a kind that could be expected in any new transnational energy system. The Austrians, and other West Europeans who eagerly followed the historical start-up of natural gas deliveries across the Iron Curtain, remained unaware of the supply crisis that the East-West gas trade gave rise to in Ukraine, Belarus, and the Baltics. West European observers did not interpret the "irregularities" in the imports from the East as signs of chaos in the East. Yet the situation in Ukraine and elsewhere in the Soviet Union was indeed both chaotic and desperate, and it appears improbable that other West European gas companies would have dared following the Austrian example, had they known of the true situation on the "other side." Lacking this knowledge, the Austrian example, interpreted as a success, opened up for radical expansion of Soviet gas exports to the capitalist world.

7
Willy Brandt: Natural Gas as Ostpolitik

The Soviet-led invasion of Czechoslovakia in August 1968 marked a watershed in Europe's postwar history. It was to have long-lasting effects on the international political climate, and it unleashed a new dynamics of East-West relations. Initially, the loud Western protests against the Czechoslovak invasion tended to generate a political reluctance to any form of cooperation with Moscow.[1] In a second phase, however, most Western governments judged that boycotts and isolation were not the way forward. Most sought, quite on the contrary, to exploit the new prospects for cooperation that seemed to result from the Kremlin's need to restore its legitimacy and reputation on the international arena. The result was a sustained and further intensified policy of détente. The issue of Soviet natural gas exports was reframed accordingly and was increasingly identified as a component of this policy by governments on both sides of the Iron Curtain. Moreover, the West European gas industry judged that the new geopolitical situation would make the Soviets more flexible in negotiating the gas price than they had been back in 1966–1967, when seemingly insurmountable differences in price conceptions had made it impossible to reach agreement with countries such as Italy and France. Gas companies now showed a new enthusiasm for (re)initiating negotiations.

First signs of renewed activity came in early 1969. In January, a possible French import of Soviet gas was taken up for discussion in the context of more general bilateral talks. The governments of the two countries decided to appoint two representatives to reconsider the gas trade prospects. This was followed in February by a Soviet agreement with Italy's ENI to reestablish a "working group", whose task would be to bring new life to Soviet-Italian cooperation in the field of natural gas.[2] In March, then, the Soviet Union and the other Warsaw Pact member states released an initiative known as the "Budapest appeal," in which a vision for long-term European peace and security was outlined. One of its key points was an emphasis on the prospects for all-European cooperation in infrastructural matters, which, according to the signatories, "can and must become the foundation of European cooperation." The idea of a trans-European natural gas grid seemed to fit this vision perfectly.[3]

Shortly afterward, the East-West natural gas talks were unexpectedly widened to include West Germany as a potential Soviet customer country. Moscow and Bonn jointly announced their intention to negotiate a major import contract. It was a remarkable step in light of the extreme political and economic tensions that had characterized Soviet-German relations up to then and which had earlier excluded Germany from the Kremlin's list of potential markets. The announcement was surprising not least in view of the Budapest appeal's explicit criticism of the Federal Republic regarding its refusal to recognize the GDR and the postwar European borders—a core foreign policy issue for Bonn. A few months later the envisaged contract was actually signed. It was the largest Soviet-German business deal ever made, and it paved the way for Germany to become Western Europe's second importer of Soviet natural gas—after Austria, but ahead of both Italy and France. This chapter explains how this could happen.

Toward a New Eastern Policy

As of June 1968, when the Soviet-Austrian gas contract was signed, the overall political relations between Germany and the Soviet Union were still extremely strained. The ambition formulated by Chancellor Kiesinger in his 1966 government declaration of reaching a renunciation-of-force treaty with Moscow and thereby launch a West German policy of détente vis-à-vis the Soviet bloc had not made much progress, and Soviet-German relations continued to be characterized by hostility and confrontation rather than trust and cooperation. As we saw in chapter 5, several attempts to improve the situation had been made during 1967 and early 1968, but the results had been meager. The Soviets felt encouraged by their talks with Foreign Minister Willy Brandt and his close advisor Egon Bahr, but they distrusted the Christian-Democratic chancellor, whose controversial Nazi past caused suspicion and who failed to establish a positive relationship with Moscow's Bonn ambassador, Semyon Tsarapkin. Kiesinger, partly as a result of this distrust, felt pessimistic about the possibilities of actually coming to terms with the Soviet side in important political matters.[4]

Whereas the chancellor saw his skepticism of Moscow's trustworthiness confirmed through the Soviet invasion of Czechoslovakia in August 1968, Brandt and Bahr drew the conclusion that efforts to establish good relations with Moscow must be intensified, not abandoned. Brandt judged that the Kremlin would now be more interested than before in actually improving its relations to Bonn, and that this opportunity must be exploited. Growing tensions between Brandt and Kiesinger, however, effectively prevented the launch of any new initiatives vis-à-vis Moscow, at least not concerning core German interests such as the contested eastern borders and the status of West Berlin and the GDR. In this internal political deadlock, which was worsened by approaching federal elections (to be held in September 1969), Brandt channelled his détente-related efforts to more concrete cooperative projects— notably in the economic field—that were not immediately linked to the core political issues. Brandt judged that the economy would have to remain "the main instrument of our East European policy for the foreseeable future." The idea of a natural gas import from the Soviet Union, preferably in combination with large-scale exports of German steel pipe, was in this connection identified as a unique opportunity.[5]

The idea had been much discussed already in 1966–1967. At that time, it had met with substantial opposition from within the federal government. Two years later, the internal support was much broader. One reason was that the popularity of natural gas had grown much greater than anyone had expected just a few years earlier, and that the need to secure additional supplies was thus greater. Another reason was that the balance of power within the responsible ministries had changed. At the Foreign Office, Egon Bahr's position had been strengthened while some of the former key officials had been removed. The Social Democratic orientation of the Ministry of Economy, which since late 1966 was headed by Karl Schiller, had also become more pronounced, particularly after the appointment of two new state secretaries in April 1967 and July 1968. The first of these, Klaus Dieter Arndt, arrived—similar to Brandt, Bahr, and Schiller—from West Berlin, where he had combined his political mission with the role as director of the important German Institute of Economic Research (Deutsches Institut für Wirtschaftsforschung, DIW). The second, Klaus von Dohnanyi, was a legal expert from Hamburg whose professional experience would prove valuable when negotiating the conditions for a German import of Soviet gas. A further change at the Ministry of Economy was the appointment of Ulf Lantzke, who had earlier been in charge of issues regarding competition in the European Coal and Steel Community, as Head of Energy Policy in 1968.

In 1967, the Ministry of Economy's main argument against a Soviet gas import had been that the gas was not really needed and that an import from the East posed unnecessary risks. As a matter of fact, the ministry took a conservative stance not only to Soviet gas, but also to natural gas in general. The main threat was seen to lie in the likely negative impact of large-scale natural gas imports on domestic coal production. Lantzke's predecessor, Gerhard Woratz, had often repeated his conviction that "the expansion of natural gas must take into account the situation of the German coal industry" and that "solving the German coal mining problems remains an urgent energy policy goal." To avoid a volatile process of creative destruction, the ministry had in this context not only opposed gas imports from the Soviet Union, but also agreed with Ruhrgas and other gas companies on a maximum level of imports from the Netherlands—13 bcm per year—above which no further shipments were allowed without Bonn's explicit approval.[6]

By late 1968, however, the ministry's stance had started to change, and the room for expansion of gas' share in the overall German energy system was deemed larger. Brandt's closest ally in the EC, Energy Commissioner Wilhelm Haferkamp, noted in November 1968 that "because of its low production costs, natural gas could replace most other energy carriers in almost all areas of use." Since large volumes of gas were available both in the Netherlands and the North Sea, this fuel must be welcomed as a way of countering Western Europe's growing dependence on imported oil. Red gas was, of course, a different matter, but the Ministry of Economy now reasoned that Soviet deliveries would stimulate the overall competitiveness of the West European gas market and reduce the danger of monopolistic abuse by the Dutch—an issue that was much debated at the time. At the same time, the rapid growth of gas demand made it possible to contemplate fairly large imports of Soviet gas in absolute terms while keeping its relative importance in terms of market share within limits. All in all, from autumn 1968, the German interest in Soviet natural gas seemed to be growing.[7]

The Soviet Union, for its part, initiated a discrete charm offensive vis-à-vis representatives of the German gas industry. At an IGU meeting held in Copenhagen shortly after the Czechoslovak invasion, the German delegates were repeatedly approached by Mingazprom's international director, Alexei Sorokin. Sorokin encouraged the Germans to raise the issue of a possible Soviet-German gas trade with the federal government in Bonn. The main German representative at the conference, Heinrich Kaun, filed a report about these informal conversations to the Ministry of Economy. The ministry did not take any action, preferring to await further Soviet moves. After several months of silence, Moscow approached the Germans again, this time by instructing Ambassador Tsarapkin to carefully raise the topic with the federal government. Meeting with Foreign Minister Brandt in February 1969, Tsarapkin stated that he had heard that "there is an interest from the side of Germany in natural gas deliveries from the Soviet Union." Brandt, who did not yet want to openly confirm that he and Bahr were actually highly interested in such a project, took Tsarapkin's comment as a sign of increased eagerness from the side of Moscow.[8]

At the next IGU meeting, held in Leningrad in March 1969, Sorokin again approached the Germans. Soviet exports of natural gas to Austria had now commenced, and Sorokin pointed to this arrangement as a model for a prospective Soviet-German gas contract. In stark contrast to the evidence presented in chapter 6, Sorokin asserted that there had been no interruptions or problems with deliveries to Austria during winter 1968–1969. He urged, again, the German delegates to approach their government and encourage it to take a formal initiative vis-à-vis the Soviet side. Heinrich Kaun reported to the Ministry of Economy that the Soviets appeared very eager to sell natural gas to West Germany.[9]

Bonn's first concrete response to the obvious Soviet interest came in mid-March. It took the form of a high-level visit from the Ministry of Economy to the Soviet trade representation in Cologne. State Secretary Arndt met informally with the representation's new director, Stanislav Volchkov, in order to probe the gas export issue and determine the seriousness of the Soviet interest. The exchange was found encouraging and was seen to open up possibilities for more formal discussions. In early April, then, Foreign Minister Brandt and Ambassador Tsarapkin came to touch upon the topic in connection with a long conversation that centered on the Budapest appeal. Brandt here chose to ignore the aspects of the appeal that were problematic from a German perspective, declaring, instead, that he fully agreed with the need, as formulated in the document, for all-European cooperation in large infrastructural projects. He suggested that the construction of a natural gas pipeline from the Soviet Union to Germany might be taken as point of departure for reviving Soviet-German trade relations, and asked Tsarapkin to forward an invitation to Minister of Foreign Trade Nikolai Patolichev to attend the German Industrial Trade Fair that was to be held in Hannover in late April, for a first tentative discussion on the topic. The Soviets were highly satisfied with this course of events, since it meant that the first formal initiative regarding exports of red gas to Germany had come from Bonn rather than from Moscow.[10]

What Role for Soviet Natural Gas?

Patolichev's visit to Hannover, with the natural gas project as the main point on the agenda, marked a turning point in overall Soviet-German relations. Similar high-level meetings at the governmental level had not taken place for several years, and the fact that the meeting could at all be organized was thus already a victory for proponents of German-Soviet détente. Although the topics discussed were strictly limited to economic issues, the significance of the meeting from a foreign policy point of view was obvious. The talks were characterized by a relaxed and positive atmosphere, which was remarkable in light of the highly strained economic relations between the two countries that had prevailed ever since the construction of the Berlin Wall in 1961. For the first time in nearly a decade a true rapprochement seemed within reach.[11]

Patolichev was accompanied to Hannover by his close advisor Andrei Manshulo, who directed the West European division at the Soviet Ministry of Foreign Trade. On the German side the meeting was attended by Minister of Economy Karl Schiller and State Secretary Klaus von Dohnanyi as well as by Ambassador Egon Emmel of the Foreign Office. No representative of Brandt's inner circle participated, probably because the Germans did not want the talks to appear linked to key foreign policy interests.

At the meeting, the two governments for the first time officially confirmed their interest in the construction of a natural gas pipeline from the Soviet Union to West Germany. Significantly, the Soviets announced that they would be interested in delivering gas not only to Bavaria and southern Germany—as envisioned by the regional Bavarian government back in 1966–1967—but to northern Germany as well. In return the Soviets wished to import large-diameter steel pipe, emphasizing that they were prepared to purchase these pipes directly from German firms, and not only indirectly through third countries such as Austria. It was agreed that another meeting be arranged for further elaborations at the governmental level. This meeting, which was to be hosted by Patolichev in Moscow, was scheduled for late May.[12]

Minister Schiller delegated the responsibility for the project to State Secretary Klaus von Dohnanyi, who in turn ordered Ulf Lantzke and his energy policy experts to come up with a proposal regarding the volumes of Soviet natural gas that could possibly become subject to negotiation and a strategy for involving the German gas industry in the project.

It was not an easy task to specify how much gas Germany could possibly import. Building on elaborations made in 1967, Lantzke thought that Bavaria would have to be the main recipient of Soviet gas. Neighboring Baden-Württemberg could possibly absorb some as well. From a pure market point of view, it was believed that the two regions would be able to absorb up to 2–3 bcm of Soviet gas by 1975, although this would necessitate a "forced sales policy." The absorptive capacity would under no circumstances be larger. Adding security considerations, it was noted that an annual inflow of 2–3 bcm of Soviet gas could be accepted only if Germany's internal transmission system was substantially strengthened. In particular, the transmission capacity between southern and northern Germany would have to be improved. Otherwise Dutch or north German gas would not be able to come to rescue in case of a major disruption in the flow of red gas to southern Germany.[13]

As for Patolichev's offer to supply not only southern, but also northern Germany, the prospects were anything but clear. The idea was in line with elaborations made in 1967 by Egon Bahr at the Foreign Office, but it had never been systematically investigated by the Ministry of Economy. The reason was that northern Germany was itself relatively rich in natural gas and that the immense Dutch gas fields were located in the region's immediate vicinity. Contractual arrangements were already in place between suppliers and distributors covering the gas needs for many years to come. Back in 1967, the Ministry of Economy had reasoned that an inflow of Soviet gas would only upset this equilibrium and disturb the government's efforts to facilitate a "harmonious" integration of natural gas into the overall German energy system. The Ministry had, therefore, discarded the idea as a naive and even dangerous suggestion by the Foreign Office.

Two years later, the ministry evaluated the situation differently. The main reason was the rapidly growing popularity of natural gas as a new, modern fuel in the still coal-oriented German energy economy. Karl Schiller's most experienced gas analysts, Gerhard Wedekind and Norbert Plesser, judged that by 1980 the share of natural gas in Germany's primary energy supply would have reached 10–11 percent, corresponding to around 40 bcm in absolute terms. This was 15 bcm above expected aggregate gas demand for 1975. Wedekind thought that around one-third of this additional demand could be covered through increased domestic gas production, whereas 10 bcm would have to be secured through additional imports. He also cited a recent report from State Secretary Arndt's economic research institute, DIW, according to which an even more rapid growth in overall energy demand could be expected for the 1970s, thus further augmenting the prospects for gas imports.

The conclusion was that although there was currently no shortage of gas in northern Germany, it was necessary to think about the needs after 1975. Everything seemed to indicate that additional import agreements would have to be negotiated. The extent to which the country's growing gas demand could be covered through imports from the East would, among other things, depend on the gas price and thus be subject to negotiations. But all in all, from a systems perspective, it was no longer ruled out that Soviet gas, starting in the late 1970s or early 1980s, might be allowed to reach regions north of Bavaria and Baden-Württemberg. The main such area that the ministry's experts had in mind was the heavily industrialized Rhein-Main district, with Frankfurt at its heart.[14]

Another difficult issue that needed to be dealt with concerned the choice of German partners with whom the Soviets were actually to negotiate. Bavarian minister of economy Otto Schedl, who had initiated the German interest in Soviet natural gas in the first place, thought it natural that he and the regional Bavarian distributor Bayerngas take the lead in negotiating and later managing the import. This followed, according to Schedl, from the fact that Bavaria, for the foreseeable future, would be the main German region to actually use Soviet gas. Since, in addition, the gas would probably enter Bavaria directly from Austria or Czechoslovakia, without transit through any other German region, the involvement of other gas companies in the import arrangement was seen unnecessary.

This reasoning made Ruhrgas, the dominant German gas company, weary. Ruhrgas had so far not taken any initiative or signaled any interest in importing gas from the East, but it was now alerted by what seemed to become a serious interest in Soviet gas from the side of the federal government. The prospects for a direct Bavarian import of Soviet gas, negotiated and controlled by the Bavarians themselves, threatened Ruhrgas' ambition to expand its activities to Bavaria, which at the time was the only German region where it had not yet any far-reaching influence. Starting from this defensive position, the company became more active and informed the Ministry of Economy that it was open for discussions with the Soviet side.

The ministry's gas experts welcomed Ruhrgas' new stance. They had come to the conclusion that Ruhrgas would have to function as the main discussion partner in the planned negotiations with the Soviet side. Ruhrgas had by far the most comprehensive experience of long-distance gas transmission in Germany, and it was in control of the most extensive grid. It was the only German gas company that disposed of "alternative possibilities to receive natural gas, which in case of a delivery interruption from the East can prevent a breakdown of supply" and of sufficient storage capacity that could be mobilized for short-term compensation in case of problems. In addition, no other German company had comparable financial muscles or management skills.[15]

Bonn's gas experts were skeptical of Bavaria's regional distributor Bayerngas, which they regarded as a weak player without experience of big business. They did not trust the management board of this company to lead negotiations with high-level Soviet negotiators such as Alexei Sorokin or his counterpart at the Ministry of Foreign Trade, Nikolai Osipov. On the other hand, the ministry's advisors had great respect for Otto Schedl, and thought it wise to try and exploit the good relations that Schedl had built up with the Soviets since 1967. In addition, von Dohnanyi thought it "good if Mr. Schedl in this instance achieves a certain independence for the regional gas distribution company in Bavaria, since this would contribute to a loosening of the dominant position of Ruhrgas" on the German market. The conclusion was that it would be advantageous if Ruhrgas and the Bavarians could be made to cooperate in a constructive way, preferably through the formation of a consortium of prospective importers. Given Schedl's ambition to make Bavaria independent of north German energy interests, however, cooperation with Ruhrgas would not come easily.[16]

In mid-May, Ulf Lantzke traveled to Munich in an attempt to persuade the Bavarians to take an active role in the project but let Ruhrgas play the main part in the negotiations. Schedl agreed with Lantzke that a coordinative effort would be advantageous, but was not willing to accept a leading role for Ruhrgas. Sensing that his position in a joint consortium with Ruhrgas would be weak, Schedl was disappointed that von Dohnanyi did not envisage any noteworthy role for Bonn in the planned Soviet-German talks. Such a federal participation, Schedl reasoned, might have provided a more effective counterweight to Ruhrgas' dominance.[17]

Ruhrgas, for its part, sought to exclude the Bavarians from the negotiations altogether. The company did not consider it necessary to make use of Bavaria's already existing relations with Moscow, but initiated its own contact with the Soviet side. Convening with Stanislav Volchkov at the Soviet trade representation in Cologne, Ruhrgas President Herbert Schelberger announced that he supported a German import of Soviet natural gas somewhere in the interval 1–5 bcm per year. Like the federal government, Schelberger reasoned that the final volume that could possibly be accepted would depend on the gas price that the Soviets were able to offer. He suggested that negotiations be initiated between a Ruhrgas delegation, led by himself, and a Soviet delegation, preferably under the lead of Sorokin. Schelberger argued that the Bavarians should not be directly involved in the Soviet-German talks, but that Bayerngas could later on negotiate access to Soviet gas indirectly by way of Ruhrgas. Schelberger stressed that Ruhrgas intended to talk with the Soviet Union about a volume that would be large enough for Ruhrgas to easily cover any Bavarian needs.[18]

The Ministry of Economy was not happy with this course of events. Ulf Lantzke concluded that the conflict between Ruhrgas and Bayerngas would have to be quickly overcome so that "a potent discussion partner" was ready to meet the Soviets. Gerhard Wedekind at the gas division agreed that the "danger of fragmentation is, in light of the partly very different interest positions in the German gas industry, relatively big."[19]

From Politics to Business: Negotiating Price and Volumes

On May 23, 1969, State Secretary Klaus von Dohnanyi flew to Moscow to continue the intergovernmental elaborations on a possible Soviet-German gas trade. Soviet minister of foreign trade Nikolai Patolichev and his deputy Nikolai Osipov were eager to hear whether von Dohnanyi and his colleagues in Bonn had already worked out any "concrete proposal". Von Dohnanyi explained that the issue was being thoroughly evaluated by the federal government. He made clear, however, that it was not the government itself that would be the buyer of the gas, but private companies, that the capacity to absorb Soviet gas was limited, and that there were already delivery contracts with the Netherlands in place. Against this background, the most principal issue from a German point of view was whether the Soviet Union would be able to offer the Germans a competitive price; only then would it be possible to seriously discuss possible pipelines to southern and northern Germany. Patolichev responded that the price could not be discussed in isolation, but would have to be dealt with in close relation to other key issues such as gas volumes and duration of the contract.[20]

The parties agreed that an annual delivery of 1–5 bcm, that is, the same volume that Ruhrgas had already signaled its interest in, would be an appropriate point of departure for the negotiations. Von Dohnanyi informed the Soviets that as long as the talks concerned volumes in this interval, the federal government would not raise any objections from a foreign or energy policy point of view. It was decided that von Dohnanyi and his ministry "provide

Figure 7.1 Herbert Schelberger, Ruhrgas' chairman and main negotiator in the Soviet-German gas and pipe talks.

Source: gwf. Reproduced by permission of Oldenbourg Industrieverlag GmbH.

assistance" with the formation of a German consortium of potential customers of Soviet gas, which would then be responsible for the actual negotiations. This consortium would in turn form part of a larger German delegation that would include not only the gas industry, but also the steel companies, with which the Soviets would negotiate the intended import of large-diameter gas pipes, and a group of German banks, whose task would be to design an appropriate credit scheme. On the Soviet side, a similar delegation would be formed, to be led by Deputy Minister of Foreign Trade Nikolai Osipov. Von Dohnanyi and Patolichev agreed that the gas, pipe, and credit talks would all start in Vienna on June 20, 1969.[21]

Two days before the talks were to commence, however, it was still unclear who would actually negotiate with the Soviets. Ruhrgas and the Bavarians were still unable to agree on a joint negotiation committee. Volchkov at the trade representation, through which the Germans communicated with Moscow, received contradictory messages. A telegram arrived from the Bavarian Ministry of Economy, containing a "final list of German participants" from which representatives of Ruhrgas were totally missing. Ruhrgas, for its part, contacted Volchkov independently, confirming the company's intention to participate. In the end all German actors that the Ministry of Economy had wanted to involve did meet the Soviet delegation as planned in Vienna on June 20, but to the Soviets it was still not clear whether there were actually one or two delegations on the other side of the table. As a temporary solution to the intra-German disagreement, the negotiations were from the German side led partly by Schedl and partly by Schelberger.[22]

The first part of the talks focused on basic technical and infrastructural aspects. Here it proved easy for the two sides to come to terms. Issues such as the chemical quality of the gas and technical standards for measuring it could quickly be settled. Similarly, agreement was easily reached about the preferred transit route, as all stakeholders advocated a transit through Czechoslovakia. Back in 1966–1967, the possible involvement of Austria as an additional transit country had been up for discussion, but this was no longer seen to offer any advantages. Not only would an additional transit country make the arrangement more complicated from an institutional and political point of view, but since northern Germany was now envisioned as an additional market for Soviet gas, it was seen inconvenient to have the gas enter Germany from the south. Transit through Czechoslovakia opened up for a more northerly point of entry. The choice of border-crossing point eventually fell on a location near the village of Waidhaus high up in the Bavarian Forest.[23]

Having settled the main technical and transit issues, the negotiators turned to a much more difficult nut: the gas price. This was in turn closely linked to the volumes to be traded. Schelberger stressed that in northern Germany the Soviets would face competition from Dutch gas and that they, if they wished to reach the north German market, would have to accept the Dutch gas price as a point of departure. Otherwise Soviet natural gas would have a chance on southern German markets only, which meant that the total volume that could possibly be traded would be much smaller.

Schelberger noted that the Dutch gas price was currently 0.56 Pf/Mcal at the German border. The Soviet price, however, would have to lie significantly below this, for two reasons. First, the Soviets were not prepared to offer the same degree of flexibility in delivery as the Dutch, implying that the German gas industry would be receiving the gas at a more or less uniform rate throughout the year. Ruhrgas would have to take measures to match these uniform deliveries with a nonuniform pattern of consumption, notably through the construction of expensive underground storage facilities. Second, Schelberger argued that the gas "must be brought to the heat value of the German network". Otherwise it would not be interchangeable with domestic and Dutch gas. Since the heat values differed substantially—Dutch and domestic German gas was chemically classified as "L-gas" whereas Soviet gas belonged to the "H-gas" category—this was bound to necessitate further investments on the

German side. Taking into account the total investment needs, Schelberger argued that the Soviet border price must be below 0.50 Pf/Mcal.

Osipov and his colleagues, for their part, reasoned very differently, arguing that the point of departure must be the price for Soviet gas as agreed upon a year earlier with the Austrians. The Soviets referred to the fact that ÖMV paid 0.614 Pf/Mcal for its gas at the Czechoslovak-Austrian border, and suggested that the Germans would have to pay this price plus an additional fee of around 0.05 Pf/Mcal to compensate for the longer transit through Czechoslovakia. The Soviet side, therefore, regarded a price level of 0.66 Pf as the absolute minimum.[24]

Schelberger thought this price "unacceptable for the German gas industry." Given the prices of competing fuels in the Federal Republic (in particular oil), it would prevent Ruhrgas from selling any significant amounts of Soviet gas in Germany. But the Soviets refused to give in, and for the moment no agreement could be reached. On June 30, after ten days of intense talks, the negotiations were adjourned.[25]

State Secretary Klaus von Dohnanyi, who did not participate in the negotiations but was informed on a regular basis of their progress by both Schelberger and Schedl, concluded that the difference in price conceptions made it difficult to tell whether the attempt to negotiate a contract would eventually succeed or not. Given the perceived political importance of the talks, von Dohnanyi took up the issue for discussion with ambassador Tsarapkin, though in vain. Von Dohnanyi was shocked by Tsarapkin's—and the general Soviet—unwillingness or perhaps inability to appreciate sound market economy arguments. He concluded that it had "not yet been made sufficiently clear to the Soviets that they arrive with their natural gas in an intensely competitive market and that the price conceptions presented by the German side are justified." The Bavarians, in contrast, showed themselves more flexible. Fearing that the price gamble might ruin the prospects for an eventual agreement, Schedl's close advisor Hans Heitzer confidentially informed von Dohnanyi that the Bavarians would, as a matter of fact, be happy to consider a somewhat higher price than the one demanded by Schelberger. Munich regarded a border price of 0.55–0.57 Pf/Mcal as sufficiently low.[26]

The Soviets, for their part, made their best to persuade the German government of the necessity to take the Soviet price arguments seriously. On July 9, 1969, Osipov and Sorokin were back in Germany. Before resuming the negotiations with Ruhrgas and the Bavarians at the former's Essen headquarters, they paid a visit to the Federal Ministry of Economy in Bonn, complaining loudly about the tough stance from the side of Ruhrgas. The Soviets obviously hoped that the federal government might intervene in one way or the other, perhaps through a state subsidy. Von Dohnanyi, however, skilfully defended Schelberger's price demands. Emphasizing that a number of new gas sources were just about to be taken into operation in Germany, he argued that the current trend was toward lower gas prices, and that the Soviets would have to adapt their negotiation strategy to this situation. In addition, an increasing number of suppliers—such as Algeria, Libya, and possibly Britain—were allegedly seeking access to the German market, which further increased the competitive pressure. At the same time, however, von Dohnanyi stressed that he was very much in favor of "connecting Eastern and Western Europe." Osipov thought the Germans overestimated the prospects for Libyan and Algerian gas, given the enormous difficulties in the area of LNG technology and submarine pipelining. He was, therefore, not as certain as the Germans that prices were bound to fall in the future. He did acknowledge, though, that this was currently the trend.[27]

A few days of intense talks with the gas companies followed, though without much progress in the price negotiations. A positive turn was that Schedl and Schelberger, realizing that they shared an interest in negotiating a favorable contract for German access to Soviet gas, had started to accept each other's roles in the talks. Schedl, impressed by Schelberger's negotiation skills, agreed that the discussions be led by Ruhrgas rather than by himself. Gradually he also came to accept Ruhrgas' idea that Bayerngas would buy Soviet gas indirectly from Ruhrgas rather than directly from Soyuznefteexport, at least during the first contractual phase.[28]

The next negotiation round was held in Moscow. For the first time, the Federal Ministry of Economy, sensing a need for its presence, now sent a representative to take part in the talks: Norbert Plesser from the ministry's gas division. Plesser had the official status of an "observer," but this did not prevent

Figure 7.2 Bavarian minister of economy Otto Schedl (far right, with interpreter) and Soviet minister of foreign trade Nikolai Patolichev in Moscow, August 1969. Schedl was the original initiator of the German-Soviet gas negotiations and had initially hoped to conclude an independent deal with Moscow. Bonn, however, anticipated that the Soviets would prefer to deal with the federal government and with Ruhrgas, rather than with Schedl and the Bavarian gas industry. Schedl was thus largely outmanoeuvred.

Source: RIA Novosti.

him from repeatedly intervening in the discussions. Taking up the difficult price issue again, Schelberger repeated Ruhrgas' demand, 0.50 Pf/Mcal. Osipov offered 0.61 Pf/Mcal, which was lower than the initial Soviet bid, but still far above what the Germans expected. Schelberger and his deputy Jürgen Weise spent long hours trying to persuade the Soviets to accept the Dutch gas price, 0.56 Pf/Mcal, as a point of departure. They also elaborated in great detail on the flexibility argument: whereas in the Dutch case Ruhrgas could choose to use the export pipeline anywhere in the interval 3,500–8,760 hours per year and purchase only 70 percent of the contracted volume if it wished to, the Soviet side insisted on 7,000 hours of use and a minimum take-or-pay volume of 91 percent. Ruhrgas thought this lesser degree of flexibility must be compensated for by 0.02 Pf/Mcal. In addition, a further price reduction of 0.03 Pf/Mcal was needed as compensation for the need to adapt red gas to Dutch and north German gas quality.

The Germans also noted that the average distance from the Dutch border to the users in Germany was only 75 km, whereas for Soviet gas—taking into account potential users not only in Bavaria, but also further north—the corresponding average distance was 150 km. As Ruhrgas saw it, this gave rise to additional transmission costs of 0.05 Pf/Mcal. Finally, Ruhrgas argued that the long transit through Soviet and Czechoslovak territory increased the probability of supply disruptions due to technical failures along the pipeline route. Ruhrgas would have to take into account this risk when planning for emergency situations, estimating that it would give rise to additional costs of 0.02 Pf/Mcal. All in all, the extra costs amounted to 0.12 Pf/Mcal. Subtracting this amount from the Dutch gas price, Schelberger and Weise argued that a reasonable price for red gas at the German border would be 0.44 Pf/Mcal. From this perspective, the Germans argued that their 0.50 Pf/Mcal bid was actually a very generous one.

Osipov replied that he could not take Ruhrgas' arguments seriously. His experts had already investigated the German arguments carefully and found them unsustainable. Showing themselves particularly sensitive to the supply risk argument, Osipov and Sorokin explained that although the plan was to ship Soviet gas all the way from Siberia to Germany, other (notably Ukrainian) gas could easily come to rescue in case of unexpected delivery problems. Having said this, however, Osipov added that he was prepared to accept the Dutch price as a point of departure, and that he could possibly imagine to go a little bit below that level.

At this point the Ministry of Economy's "observer" Norbert Plesser entered the discussion. Plesser spurred the parties by noting that Osipov's recognition of the Dutch price level as an appropriate point of departure meant that there was now a much better foundation for further discussion. He also urged the Soviets to take into account current trends regarding the Dutch gas price, explaining that the Ministry of Economy had recently paid several visits to the Hague to discuss the problem of what the Germans regarded as unjustifiably high prices for Dutch gas. Plesser emphasized that the Dutch gas price was being "discussed within the framework of the EC Commission" and that it must be expected that the attempts to force the Dutch to lower their prices would "sooner or later yield success." He explained that the Dutch gas price had been subject to severe criticism both in the parliaments at the federal and regional levels in Germany and from the side of municipal customers. A similar criticism could be expected in the case of imports from the Soviet Union if the price was not significantly below the Dutch price. Otto Schedl, for his part, stated that 0.6 Pf/Mcal at the Bavarian border would be much too high, but that 0.6 Pf/Mcal at Munich would be fine.

The price continued to be discussed between the two of Schelberger and Osipov. This generated a breakthrough. Osipov came up with a new Soviet offer of 0.53 Pf/Mcal, while Schelberger declared that he would accept 0.525 Pf/Mcal—though only for a volume of 3 bcm per year. Osipov had hoped to sell 5 bcm per year, but eventually agreed on the smaller volume in order to avoid having to accept a lower unit price. Ruhrgas was given a one-year option for an additional 2 bcm per year.[29]

At the next negotiation round, held in late August, the parties were full of optimism regarding the prospects for a soon finalization of the contract. The price had not yet been finalized in a definite sense, but the parties were now so close to each other that it seemed highly unlikely that they would not come to agreement. The remaining items to be negotiated were considered less troublesome: this concerned issues such as a price adaptation formula and a clause defining the German consortium of gas companies involved in the ongoing talks as "most favored customer." Plesser reported from Moscow that a positive outcome of the negotiations could be expected before the end of the year.[30]

Finalizing the Contract

To what extent was the emerging deal politically important for Bonn and Moscow? Foreign Minister Brandt and strategist Bahr followed the project with great interest. In mid-July, Brandt emphasized in his communication with the Ministry of Economy that he "would not have any objections against a comparatively large volume of Soviet natural gas," hoping that "the clarification of the commercial aspects of the natural gas deliveries...will come about in a timely manner and in such way that the results can be used as a possible basis for German-Soviet negotiations on a long-term governmental treaty." This perceived potential of natural gas as an instrument and catalyst for improving overall Soviet-German relations appears to have grown as the negotiations progressed. Chancellor Kiesinger, in contrast, showed himself skeptical of the natural gas project. At a summit with US president Nixon in early August, Kiesinger made clear that he did not see any prospects for red gas to have any positive effect on Soviet-West German relations more generally.[31]

Minister Schiller and State Secretary von Dohnanyi conveyed Brandt's message that natural gas was not only about energy, but also about foreign policy, to Ulf Lantzke, Norbert Plesser, and the other energy policy experts. In his communication with the German delegation of negotiators, von Dohnanyi increasingly stressed the political importance of actually coming to agreement with the Soviets. The gas contract was expected to become "an important component" of the government's overall Eastern policy. The ministry, whose energy policy advisors had earlier been highly suspicious of Soviet natural gas, thus gradually moved from a position of light support to Soviet natural gas to forceful insistence that the negotiations must become a success. The "uncontested" fact that, in von Dohnanyi's words, "political and foreign trade policy considerations speak for an agreement with the Soviet Union" was also communicated to Ruhrgas. Von Dohnanyi put pressure on Schelberger by stressing that "for political reasons an interruption in the negotiations must, at all events, be avoided."[32]

Brandt's argument that the deal had better become as large as possible was also conveyed to Ruhrgas. The ministry sought to convince the company to take on 5 or even 6 bcm of Soviet gas per year, while explicitly warning Schelberger and his colleagues from concluding a deal for less than 3 bcm. In the early phase of the Soviet-German talks, the ministry's gas experts feared that the Soviets might step out of the negotiations if the gas companies were unwilling to import less than 5 bcm, since the construction of a pipeline all the way from the Soviet Union might in that case become uneconomic.[33]

Von Dohnanyi and his advisors elaborated on complex arrangements through which the project in such a case might be "rescued." Under discussion was, for example, a possible subsidy to Ruhrgas for importing Soviet gas in case the price would not be competitive on north German markets. The idea met with opposition, however, from within other divisions of the ministry, whose advisors argued against it "both for general economic policy and energy policy considerations." Von Dohnanyi also developed the idea that the price adjustment clause, which was one of the last things to be negotiated, could be coupled to the volume issue in such a way that the Germans agreed to import more gas in case overall German gas demand exceeded a certain level. Gas expert Norbert Plesser, for his part, thought that the ministry might put pressure on Ruhrgas to accept a larger volume by pointing to the fact that if the volume was too low, then very little Soviet gas would flow to northern Germany. In such a situation, it would be more difficult for Schelberger to gain support for his argument that Ruhrgas and not Bayerngas should be in charge of the project.[34]

Von Dohnanyi actively sought to prevent the Soviet side from letting the talks collapse due to disagreement on price and volumes. Meeting with Volchkov from the trade representation, he uttered his conviction that Soviet gas had very good long-term prospects in Germany even though the first contract might be small. The Soviets must simply see the prospects for exports to Germany in a more long-term perspective. Von Dohnanyi wanted Volchkov to understand that large gas markets were about to take form in both Bavaria and, from the mid-1970s, Baden-Württemberg, and that the Soviet Union would have excellent opportunities to win a large share of these markets.[35]

There were clear signs that the Soviet Union, too, viewed the gas deal as being of great political importance. Egon Bahr noted that Premier Kosygin and Foreign Minister Gromyko had personally received the entire Soviet delegation of negotiators upon their return to Moscow after the mid-July negotiation round. Bahr concluded that the issue had thus been lifted from a departmental to a political level and that the conclusion of a long-term gas contract had "become a political test" for the Soviet Union.[36]

In the end the German government did not manage to convince Ruhrgas to take on more than 3 bcm per year. Even so, the impression is that the political pressure from the side of the Foreign Office and the Ministry of Economy had a certain effect on Ruhrgas' negotiators. Brandt, Schiller, and von Dohnanyi were relieved by the preparedness of Ruhrgas to accept 3 bcm rather than 1–2 bcm and by the Soviet willingness to accept an annual volume below 5 bcm.

The political dimension of the project took on a new turn following the federal elections that were held in Germany on September 28, 1969. The Social Democrats won a historical victory, paving the way for Willy Brandt to become new chancellor. For the first time ever in the history of the Federal Republic, the Christian Democrats were not represented in the federal government that was formed. Instead, a majority coalition government was formed by the Social Democrats and the Free Democrats, whose chairman Walter Scheel took over Brandt's earlier position at the Foreign Office. Karl Schiller stayed on at his position as minister of economy, whereas Klaus von Dohnanyi, who as state secretary at the same ministry had played a key role in the Soviet-German gas negotiations, moved on to a new position at the Ministry of Education and Science.

Brandt's victory meant that his and Bahr's "Ostpolitik" could start to be implemented in a much more forceful way than had been possible under Kiesinger's and Brandt's "grand coalition." As a result, the efforts to improve overall relations between West Germany and the Soviet Union entered a new phase, the emphasis shifting from economic and technical cooperation to key foreign policy issues linked to core national interests. Brandt and Bahr immediately set out to probe the prospects for a governmental treaty with the Soviet Union that would regulate, in particular, the sensitive border issues and, by extension, the Federal Republic's relations to the GDR.

In this situation it became even more crucial to actually bring the natural gas negotiations to a successful end. A failure here might very well disturb the attempts to resolve key foreign policy issues. Conversely, a successful outcome of the gas talks would demonstrate to the Soviets that it was indeed possible to cooperate with West Germany and its new government. A German preparedness to accept a dependence on Soviet natural gas deliveries would testify to the new government's commitment to détente.

In the meantime, however, several new problems had arisen that seemed to endanger the expected success of the negotiations. The first had to do with the security of supply. Ruhrgas, which had so far been happily unaware of the Soviet problems to build the export infrastructure and guarantee a stable gas supply to domestic and foreign customers, had received new information from ÖMV, the Austrian oil and gas company, according to which there had been a whole array of "technical difficulties" with gas imports from the Soviet Union, including several serious supply disruptions that had lasted for up to three days. This contradicted the earlier German expectation that the Soviets would be a fairly reliable supplier. Ruhrgas had so far considered outages lasting longer than three hours improbable. The new information forced Ruhrgas to adapt its risk management strategy in such a way that it would be able to deal with longer cutoffs. This was seen to increase the overall costs of the project. For this reason, Ruhrgas considered annulling the tentative agreement on the gas price that had been reached in August.[37]

A further problem stemmed from an expected revaluation of the D-mark, a much-discussed topic both in Germany and internationally at the time. Kiesinger's and Brandt's grand coalition had not been able to agree on a revaluation, but the new political landscape increased the prospects for carrying it out. If the D-mark were revalued, it would make Dutch gas, which was paid for in Dutch guilders, cheaper for the Germans and thus more competitive vis-à-vis Soviet gas, the price for which was negotiated in D-marks. Schelberger expected that this would make Dutch gas cheaper by around 3.5 percent. In addition, he stated that the Dutch exporter intended to lower their export prices very soon, which in combination with the revaluation effect would likely amount to a total price reduction of around 10 percent.[38]

When the gas negotiations were resumed in mid-October the parties thus faced a partly new situation. Nikolai Osipov, who had hoped to finalize the price issue, was disappointed by the new stance from the side of Ruhrgas. He acknowledged that there had been a few technical problems with the supply of gas to Austria, but argued that fears of similar difficulties with the supply of gas to Germany were not motivated. The problems in the Austrian case allegedly stemmed from the fact that the gas partly had to pass through an old, vulnerable pipeline, whereas in the case of exports to Germany a brand new line would be built. Regarding the expected reduction of the Dutch gas price, Osipov was surprised that Ruhrgas had not taken into account this possibility already at an earlier stage. After the negotiations were adjourned on October 22, Osipov paid a visit to Germany's Moscow ambassador Helmut Allardt, uttering his concern about the slow progress. Allardt reported to Bonn that the Soviets had started to suspect that Ruhrgas was not seriously interested in coming to agreement with the Soviet side, but merely kept the negotiations alive in order to put the Dutch under pressure to reduce their export price. The ministry informed Ruhrgas about these Soviet concerns and urged the company to use every possibility to make clear to the Soviets that its interest in red gas was real and serious.[39]

A month later, the parties met again for a marathon negotiation round that would turn out to be the final one. The talks, which started on November 27 and continued without interruption into the next morning, generated the final breakthrough. Although the Dutch exporters had not yet announced any new, lower gas price, the parties managed to agree on the Soviet export price. It was set at 0.5198 Pf/Mcal for a volume of 3 bcm per year. This level would be reached after a six-year build-up phase. As expected, Ruhrgas was awarded an option to increase the annual volume to 5 bcm. Deliveries were to start on October 1, 1973. Agreement was also reached on a price adaptation formula. The price could be renegotiated after April 1, 1975.[40]

A draft contract for the pipe business was also finalized. The main producer was to be a new joint subsidiary of Mannesmann and Thyssen—Mannesmann-Röhrenwerke—with possible aligned contracts for two other pipe producers, Hoesch and Salzgitter. The pipes, intended for the construction of pipelines from Siberia to the Soviet Union's main industrial regions, were to be 1,420 mm in diameter and 17–20 mm thick. Production was to take place at Mündelheim, where Europe's only plant for manufacturing thick steel plates up to 5 meters wide was located. The total volume of pipes to be delivered amounted to 1.2 million tonnes, which was equivalent to about 2,000 km of pipelines. The value was 895 million DM. After the revaluation of the D-mark, which was eventually carried out in late October, this corresponded to $245 million. In addition, various equipment for the gas industry was to be exported, boosting the total value of German counterexports to 1.32 billion DM. The pipes were to be delivered in the period from July 1970 to December 1972.[41]

The formal signing of the contract was scheduled for February 1, 1970, pending timely completion of a credit arrangement. Most components of such an agreement were settled following an additional negotiation round held in mid-December 1969. The Soviet Union was awarded a credit amounting to 85 percent of the total value of the German pipe and equipment exports. The credit was to run for 10 years counted from the last pipe deliveries. Importantly, the federal government agreed to provide a guarantee for half the credit. The only critical issue that remained to be discussed concerned the interest rate. It was later set at 6.25 percent, which was seen highly favorable in view of market interest rates, which at the time amounted to around 9 percent. The federal guarantee and the generous credit, which gave rise to severe criticism from other EC member states, was yet another confirmation of the perceived political importance for Bonn of actually coming to agreement with the Soviets. It showed, again, that the gas deal was thought of as a vital component in improving overall Soviet-German relations and in launching a new Eastern Policy.[42]

Shell and Esso: Lobbying against Unwelcome Competition

A remarkable aspect of the Soviet-German gas deal was that the contractual partner on the German side, Ruhrgas, was largely controlled by interests that seemed to have everything to lose from the planned imports. This concerned, in particular, Shell and Esso, the oil and gas companies that together controlled much of West Germany's domestic gas production and were in charge of the competing Dutch exports. Each of them held 12.5 percent of Ruhrgas' shares, but since they were also indirect shareholders their total influence in Ruhrgas amounted to 40 percent. In addition, the German coal industry held major interests in Ruhrgas.

Bonn had the impression that Ruhrgas' board of directors were genuinely interested in a positive outcome of the Soviet-German talks. Schelberger and his colleagues had initially been skeptical of the project, but as the negotiations progressed the board became increasingly eager to make use of the perceived opportunities. The potential advantage for the company of gaining access to natural gas not only from domestic and Dutch sources grew clear: a diversified import structure would strengthen Ruhrgas as a transmission company vis-à-vis the gas producers, and boost the overall competitiveness of natural gas vis-à-vis coal on the overall German energy market.

The problem was that the company, as a result of the large stakes held in it by Shell, Esso, and the coal industry, was "in a double conflict of interests," as von Dohnanyi put it. Fearing that the conflict might prevent a successful outcome of the negotiations, von Dohnanyi at an early stage urged Schelberger to deal with potential resistance from the minority shareholders. Schelberger followed this advice, seeking to personally convince the CEOs of Shell's and Esso's German subsidiaries, Dirk de Bruijne and Emil Kratzmüller, that the Soviet project would eventually be of great benefit for all. The companies, however, showed themselves hostile.[43]

Two days before the gas negotiations were to start in June, the Ministry of Economy was approached by Hans Carsten Runge from Deutsche Shell's division for natural gas. Paying a personal visit to Bonn, Runge informed the ministry's gas experts Plesser and Wedekind that there was a "rare unanimity" among Germany's domestic gas producers regarding Soviet natural gas: the industry considered an import from the East inadvisable and would prefer if the negotiations did not take place. This statement irritated Plesser, who made clear that the talks were supported by the ministry "for political reasons," explicitly warning Runge from trying to prevent and disturb the process. At the same time, however, it was important for the ministry to avoid unnecessary trouble in its relations with Shell and Esso, since a clause in the statutes of Ruhrgas gave its large minority owners the right of veto in decisions that were of "vital importance" to the company.[44]

In July, von Dohnanyi and his advisors were again approached by Kratzmüller and de Bruijne, who together with Schelberger and the chairman of Ruhrgas' supervisory board, Friedrich Funcke, demanded a meeting with the Ministry of Economy. Kratzmüller, de Bruijne, and Funcke all argued that Germany had sufficient gas resources of its own and that it would be easier and safer to compensate for German pipe exports to the East through imports of Soviet oil, rather than gas. Von Dohnanyi replied that the Soviets seemed to have problems increasing their oil exports to Germany, and that only gas could hence be used to balance the pipe deal. In addition, it would be advantageous for Germany to diversify its gas supply. This position irritated Kratzmüller, who stated that if this was the case, then Shell and Esso might no longer be willing to accept any self-imposed limit on their supplies of oil to the German market, in the way they had done so far in mutual agreement with the federal government. He thus hinted that Esso and Shell might deliberately flood the German market with oil, thereby jeopardizing the competitiveness of both Soviet gas and German coal—both of which were considered politically important for the federal government. Von Dohnanyi got furious when hearing this, and the meeting ended in open conflict.[45]

Herbert Schelberger, for his part, had to strike a balance. Meeting on his own with government representatives, he explicitly distanced himself from the stance taken by Funcke, Kratzmüller, and de Bruijne. He stressed that Ruhrgas had a genuine interest in coming to agreement with the Soviet side and that his tough stance in the price issue, which at times was seen to threaten the envisaged gas deal, must under no circumstances be taken as a sign that Ruhrgas merely negotiated with the Soviets for tactical reasons. He stressed that most members of Ruhrgas' supervisory board were unambiguously in favor of the conclusion of a gas import contract with the Soviet Union. Von Dohnanyi was full of respect for Schelberger's efforts to deal with his company's internal dilemma, judging that Ruhrgas had "with great fairness tried to resolve its conflict of interests."[46]

As the Soviet-German negotiations progressed, Shell and Esso, along with other gas producers in Germany, sharpened their arguments against an import of red gas by carrying out a number of in-depth quantitative analyses. The companies stressed that they were making strong efforts—spending 138 million D-marks in 1968—on expanding domestic gas reserves but that it would be very difficult to sell this gas, should Soviet gas suddenly appear on the market. This in turn would have a negative impact on further domestic exploration activities, so that rich domestic gas reservoirs might remain undiscovered. The gas producers ensured the Ministry of Economy that an "import of Russian natural gas at the present time is not necessary, since the domestic firms, with certainty, are able to cover the estimated demand in the Federal Republic." Arguing that Germany's gas demand "up to the 1980s can easily be covered from German deposits," they recommended the government to wait a few years and see if there would really be any need for Soviet imports. A thick report submitted to the Ministry of Economy in August developed these arguments further.[47]

At the same time, US-based Esso lobbied Washington to put pressure on the German government—though without much success. Norbert Plesser at the gas division was contacted by the US Embassy, which merely wished to inform the ministry that Esso had become active presenting its interests to the State Department. The company had made clear to the Nixon administration that it strongly opposed German gas imports from the Soviet Union. However, the Embassy had received no instructions to intervene, and it did not expect any such instructions to appear in the future.[48]

Apart from lobbying the German and American governments, Shell and Esso launched a media campaign to spread their view that Soviet gas deliveries to Germany would make life difficult for the companies and that this would have severely negative consequences for their willingness to continue investing in the German gas industry. In a series of newspaper articles it was stated that the Federal Republic's own gas deposits would be completely sufficient to cover gas demand in Germany not only through the 1970s, but all the way "up to the end of the century." Hence, there was no need for Soviet natural gas, an import of which would have only negative consequences.[49]

Responding to press statements, the ministry consulted information from the Federal Geological Survey (Bundesanstalt für Bodenforschung), on the basis of which it was concluded that domestic gas production might reach 19 bcm by 1980. The German natural gas market was growing at an unprecedented rate, with increments of 30–60 percent per year, and this growth was expected to continue. The ministry expected gas consumption in Germany by 1975 to have reached 25–28 bcm and perhaps 37–40 bcm by 1980; 11 bcm of the latter had already been contracted with the Netherlands (i.e., Shell and Esso), but 7–10 bcm still remained to be sourced. Hence there was ample room for Soviet gas on the German market.[50]

The oil companies continued to oppose the attempts to conclude an import agreement with the Soviet Union, whereby Schelberger's insistence on a relatively small volume—3 bcm as opposed to the Soviet wish to export at least 5 bcm—may be interpreted as an attempt to appease Shell and Esso. Had it not been for the latter's resistance, it appears probable that a larger contract had come about.

Seeking Coordination with Italy and France

The Soviet-German gas negotiations took place in parallel with analogous talks held with Italy and France. This was the result of a deliberate strategy from the Soviet side, as Mingazprom did not intend to build separate export pipelines to each importing country, but rather exploit economies of scale through the construction of a single, integrated export infrastructure. The importing countries were also highly interested in such a coordination. Most prospective system-builders agreed that it would be uneconomic to build a Soviet-West European export pipeline for a gas flow below 10–12 bcm per year, which was much more than any one of the individual importers could possibly absorb. Italy was the country that had expressed the largest interest—6 bcm per year—but even this was too low a volume to motivate the construction of a pipeline all the way from the Soviet Union. The success of the Soviet-Italian negotiations was thus seen to depend on the progress of the Soviet Union's negotiations with other countries. In the case of Germany and France, this need for coordination with others was even more pronounced.

Coordination among prospective West European importers was also considered desirable for security reasons. Ulf Lantzke reasoned that with three or four West European countries connected to the same pipeline, the Soviets would come under strong pressure to maintain a reliable supply regime. Especially if Germany would function as a transit corridor for Soviet gas on its way to France, Lantzke thought that "the danger of supply interruptions for political reasons" would be substantially reduced. Back in 1967, France had been elaborating on an import of Soviet gas by way of Italy, but the initiation of the Soviet-German talks in spring 1969 opened up for a German transit of the French volumes as an alternative. To judge from information received by Bonn following a French-Soviet meeting held in April 1969, both the French and Soviet governments took great interest in this opportunity. Lantzke even thought that the strong Soviet interest in exports of natural gas to France was an underlying reason for Moscow's eagerness to come to terms with Germany.[51]

The Soviet-Italian talks made good progress in spring and summer 1969, and as of early fall finalization of a contract seemed within reach before the end of the year. The Soviet-French talks, however, turned out to be anything but straightforward. Paris initially showed only a moderate interest in actually importing Soviet natural gas, at least as far as the near future was concerned. The country's gas supply was seen covered at least up to 1976 through contracts signed with the Netherlands and Algeria. Domestic gas still dominated the French gas market at the time, contributing 8 bcm in 1969. LNG from Algeria contributed 0.5 bcm, a volume that would increase by 3.5 bcm from the early 1970s as a result of a new arrangement agreed upon in 1967 (see chapter 4). Meanwhile in northern France, Dutch gas was about to become the main source of supply, with annual imports expected to grow to 7 bcm by 1976. All in all, although the Soviet option was clearly interesting as a diversification opportunity, France was not at all in the same hurry to secure additional supplies as Bavaria or Italy. Yet the perceived need for coordination with other prospective West European importers made it difficult for the company to postpone the Soviet talks.[52]

The German Ministry of Economy aimed for cooperation with both Italy and France. West European cooperation would be particularly critical for the Germans in case Ruhrgas dropped out of the talks, which the ministry thought it might well do as a consequence of opposition from Shell and Esso or its insistence on a gas price that the Soviets could not accept. If Ruhrgas withdrew from the negotiations, the Bavarian interests would be left alone, and the potency of the German delegation would be lost. If this happened, State Secretary Klaus von Dohnanyi intended to "encourage the Bavarian side to negotiate a separate deal with the Soviets and, as far as possible, reach a cooperation between Bavarian natural gas supply and deliveries to Italy and France."[53]

As a matter of fact, von Dohnanyi had initiated preparations in this direction already upon his return from Moscow in late May. His first step had been to contact Otto Wolff von Amerongen, an experienced industrialist with far-reaching political connections who had been a key figure in organizing and expanding Germany's trade relations with the Soviet Union. A respected figure both internationally and in Bonn, Wolff was at the time on his way to Italy. On von Dohnanyi's request, he agreed to pay a visit to ENI's headquarters and probe how the Italians currently viewed their prospects for Soviet gas imports. Having discussed the issue informally with ENI's management board on behalf of the German government, Wolff reported back to Bonn that a connection between the Italian and the Bavarian gas systems "could have certain chances."[54]

Meanwhile von Dohnanyi met with the general secretary for energy in the French Ministry of Industry, Jean Couture, seeking to persuade the French about the benefits for all of a coordination between the French and German interests. In addition, Axel Herbst of the Foreign Office's division for trade policy met with his counterpart at the French Foreign Ministry, General Director Brunet. It was agreed that the responsible gas advisors in Bonn and Paris—Plesser and Herbin, respectively—were to "hold coordination talks about the natural gas situation and the opportunities for a German-French partnership in the import of Russian natural gas."[55]

In the case of Italy, which in its 1966–1967 negotiations with the Soviets had contemplated an import of red gas by way of Austria, Bonn's energy experts developed the bold alternative vision of a Trans-European Pipeline routed through Czechoslovakia and Bavaria and from there to France and, by way of transit through Switzerland, to Italy. If realized, this arrangement would turn southern Germany—rather than Austria—into the main hub for Soviet gas distribution in Western Europe.[56]

As the Soviet-German negotiations progressed, signs came that the French were getting more interested in Soviet gas. The Germans actively sought to promote this interest, and it was suggested that the topic be taken up for

Figure 7.3 Alternative transit vision for Soviet natural gas. Its realization would have turned Germany into the central hub for Western Europe's import of Soviet gas. The vision may be contrasted with the one visualized in figure 4.3.

Source: Oil and Gas Journal, October 13, 1969, p. 56. Reproduced by permission.

discussion at a planned summit to be held in September between Chancellor Kiesinger and President Pompidou. Minister Karl Schiller also took up the issue at a meeting with his French counterpart, François-Xavier Ortoli. Head of Energy Policy Ulf Lantzke further supported the coordination efforts by encouraging the French government to establish a contact with Ruhrgas in order to investigate the economic feasibility and technical possibilities of a transit pipeline. The Germans informed the French on a regular basis about the progress in their Soviet negotiations—though without disclosing the gas price that was up for discussion. By late August, the Ministry of Economy had become so eager to coordinate the German and French efforts that it asked its French counterpart to let the German delegation take up the French interest on its behalf in the next Soviet-German negotiation round. The French side, however, thought this premature.[57]

Regarding cooperation with Italy, it soon became clear that ENI did not favor a pipeline route through Germany and Switzerland to Italy, and this reduced Bonn's interest in a direct cooperation with the Italians. The French government, however, suggested that the three countries might still join forces in order to strengthen the general West European negotiation position vis-à-vis the Soviets. The director of foreign economic relations at the French Ministry of Economy and Finance, Jean Chapelle, thought the best way of organizing Western Europe's import of Soviet natural gas would be to involve Brussels, linking the issue to the "common energy policy" that the EC at the time was preparing.[58] Ulf Lantzke, however, judged that it would be very difficult to develop a common EC strategy for imports of red gas. The Dutch, in particular, were in a position that differed markedly from that of the other member states:

> In the natural gas area, the unambiguous seller's interest of the Netherlands could in case of a Community-based treatment of the [Soviet gas] issue make itself disadvantageously noticeable. Therefore one should not burden the Community's energy policy, which stands at its beginning, with this issue. It must be feared that the already hesitant stance of the Netherlands to a common energy policy would be further strengthened.[59]

Yet Lantzke thought it important not to give the impression that Bonn wished to avoid discussing the issue with Brussels. Although he thought it advisable to avoid community bindings, he suggested that the federal government remain open for consultations both with the EC Commission and individual member states. Lantzke thought this was well in line with the principles codified in an EC document from 1964 known as the "energy protocol." Klarenaar at the Foreign Office endorsed Lantzke's judgment.[60]

In the end the negotiations continued on a bilateral basis between the Soviet Union and Germany, Italy, and France. Of the three countries, Italy made the fastest progress, with Germany not far behind. The French-Soviet talks commenced in earnest only in September 1969. An agreement-in-principle was soon reached according to which deliveries to France would start in 1977 and reach a plateau level of 2.5 bcm per year by 1980. The gas would be shipped through Czechoslovakia and Germany. In late October, the French declared that they, although they saw no "real" need for Soviet gas before 1977, were prepared to adjust their planning to Soviet preferences by agreeing to start up the imports already in 1975.[61]

This flexibility from the French side was welcomed by the Germans, since the addition of Gaz de France as a Soviet customer made it less problematic that Ruhrgas only wished to import 3 bcm per year. For the same reason the Germans were happy to receive the news that ENI on October 15, 1969, had initialed a contract with the Soviets for an annual import of 6 bcm per year. Adding Austria, which already had contracted 1.5 bcm per year, Western Europe's total import of Soviet gas was now expected to reach a level of at least 13 bcm per year. Such a volume was clearly large enough for a Soviet-West European pipeline system to make economic sense. This made it much easier for the German delegation to finalize its own contract with the Soviets in late November without a binding German commitment to larger imports.[62]

The Italian contract, according to which ENI would start importing Soviet gas in January 1973, was formally signed on December 10, 1969, at a ceremony in Rome. Imports would grow from 1.2 bcm in 1973 to 2.5 bcm in 1974, 4 bcm in 1975, and 6 in 1976, a level that would then be maintained until 1993. In return the Italians were to export pipes and equipment worth $270 million, mainly in the form of 1 million tons of 1,420-mm mm steel pipe, 3 large compressor stations, and communication equipment. Several Italian newspaper claimed that Italy's contract was more favorable than the

German one, but ENI argued that the price difference, which was not officially disclosed, vanished if Italy's costs for transit through Austria were taken into account. Ulf Lantzke similarly defended the German deal by emphasizing that ENI, in contrast to Ruhrgas, would have to contribute financially to the transit arrangement, since the Soviets did not take any responsibility for the part of the system that would traverse Austria. Moreover, the most important Italian gas users were situated several hundred kilometers from the border-crossing point, whereas in Germany the main prospective end users were in Bavaria near the Czech border. Domestic distribution costs would thus be lower in the German case.[63]

The Significance of the Soviet-German Natural Gas Deal

On Sunday February 1, 1970, the Soviet-German natural gas contract was formally signed. Ruhrgas president Herbert Schelberger gave the welcome address at the main ceremony, which took place at Hotel Kaiserhof in Essen. German minister of economy Karl Schiller gave a speech in which he applauded the successful completion of the negotiations in only nine months counted from the Hannover spring fair. He concluded that the stage had now been set "for a tight cooperation in an important and future-oriented area between the Federal Republic and the Soviet Union."[64] Soviet minister of foreign trade Nikolai Patolichev also praised the achievement. Nikolai Osipov, who together with Alexei Sorokin had led the negotiations on the Soviet side, placed the deal in a broader European context, reminding the guests that Soviet natural gas was expected to reach not only Germany, but other West European countries as well:

> The concluded agreement for the construction of a trans-European gas pipeline is a good example of economic cooperation between Europe's countries. This example also indicates that other economic problems in Europe can be successfully solved, under the condition that all European countries manifest their good will and their efforts regarding the development of mutually advantageous cooperation. I can add, gentlemen, that what I have said also fully concerns the solution of all *political* issues in Europe. It is precisely from this perspective that we view the signed contract: we are completely confident that the spirit of realism and mutual trust is stronger than all obstacles and that this development can lead to all-European cooperation.[65]

In the afternoon, Ruhrgas, together with the steel companies and the banks, gave a press conference, while Patolichev and Schiller met for a one-hour talk. Patolichev told Schiller that the Soviet Union now thought the time ripe to think about further steps in improving Soviet-German trade relations. The two ministers agreed that the natural gas deal would have a positive influence on the solution of many political issues. Otto Wolff von Amerongen acknowledged that it had a "demonstrative character." Indeed, some analysts did not think it a coincidence that the contractual ceremony in Essen took place in parallel with the crucial political talks that at precisely the same time were held in Moscow between Egon Bahr and Soviet foreign minister Andrei Gromyko. These talks centered on core foreign policy issues such as the status of Germany's postwar borders. The natural gas deal was seen to give "added impetus" to the successful conclusion of these high-level elaborations.[66]

Viewing the deal from an energy policy perspective, Ulf Lantzke prophetically argued that the importance of the gas deal must be seen

> not so much in the now concretely negotiated contract of 3 billion m^3, but in the fact that this contract for the first time has led to a successful connection between Eastern Europe's huge natural gas deposits and the West European network. On this basis, developments of decisive importance for the energy structure of Europe could, on the long term, come about.[67]

The Federal Ministry of Economy was keen to emphasize that the gas deal would not make the German energy system more vulnerable. Before the formal signing of the contract, the ministry had been approached by NATO's Economy Directorate, which wanted Bonn to report about the result of the negotiations, particularly regarding the extent to which the deal would lead to a dependence on deliveries from the East. Plesser at the gas division stated that Germany's natural gas consumption was expected to grow to 40 bcm by 1980, of which 3 bcm would come from the Soviet Union. Hence Germany's dependence on Soviet gas would be less than 10 percent. Moreover, he ensured the critics that "Soviet natural gas can in case of emergency be replaced by natural gas deliveries from north German or Dutch sources." More precisely, this would be possible through the creation of a unified pipeline network for the Federal Republic as a whole. Such a network did not yet exist, but Plesser promised that the necessary pipeline links would be completed in due time for first deliveries of red gas.[68]

In response to a question posed by a critical member of the Bundestag, the ministry further stressed that

> the Soviet Union has an extraordinarily strong economic interest in a friction-free fulfilment of the contract, since it is not only anxious to pay for the pipe deliveries that it needs from the revenues of the gas sales, but also because its economic intentions are directed at coming to terms with other European countries about natural gas deliveries. In this connection it is of particular importance that Soviet natural gas deliveries to the Federal Republic will attain an additional security momentum if, as intended, agreements are concluded between the Soviet Union and France concerning natural gas deliveries, which will be realized through the territory of the Federal Republic.[69]

Security issues aside, the economic effect of the deal was a major issue. As it turned out, a positive impact of the first Soviet gas contract on the competitive dynamics of the German gas market was soon felt, as Ruhrgas and Thyssengas were able to renegotiate their import price for Dutch gas from 0.56 to 0.49 Pf/Mcal. Italy's ENI, having negotiated its first Soviet contract, was similarly able to attain an extremely attractive contract with the Dutch, signed in spring 1970. Although no single cubic meter of Soviet gas had yet entered Germany or Italy, the expected deliveries from the East thus already played an important role in shaping the European market.[70]

From European to American Imports of Soviet Natural Gas?

In what followed, a whole array of Western countries became increasingly interested in linking up with the envisaged Soviet gas export regime. Gas companies that had already concluded first contracts with Moscow set out to negotiate additional imports, while prospective importers from countries such as Finland, Sweden, Switzerland, Belgium, and Spain stepped up their efforts to access red gas. Ruhrgas had earlier been pressed by its minority owners not to accept an annual Soviet import larger than 3 bcm, but once the contract had been signed it did not take long before the company indicated its interest in additional supplies. The main reason were the new forecasts indicating that Germany needed much more gas than earlier anticipated. Shell and Esso, both of which had vehemently opposed the first Soviet-German contract, no longer objected to an increased import from the East. The two companies even initiated their own contact with the Soviet Ministry of Foreign Trade. Through their joint German subsidiary, Brigitta und Elwerath-Betriebsführungsgesellschaft mbH (BEB), Shell and Esso met with the Soviets in Helsinki in January 1971 to probe the issue. The talks ultimately failed due to difficulties to come to agreement on the gas price, but the very attempt to negotiate with the Soviets was remarkable, to say the least, in view of the aggressiveness of the companies' earlier anti-Soviet lobbying campaigns.[71]

Ruhrgas was more successful in negotiating additional imports. The first deal had awarded the Germans an option of 2 bcm in addition to the contracted 3 bcm, but Ruhrgas soon came to the conclusion that this would not suffice. In mid-December 1970 Klaus Liesen, who shortly after the conclusion of the first contract had taken over the company's Soviet business from Herbert Schelberger, traveled to Moscow to negotiate additional supplies. The Soviets first hesitated, signaling that "technical transport problems" would make further increases in exports difficult. Liesen, finding this argument implausible, thought the Soviets deliberately downplayed its export ability in order to strengthen its negotiation position.[72]

In any case a second import agreement was soon worked out. The talks were finalized in April 1971, paving the way for an additional German import of up to 4 bcm of Soviet gas per year. Deliveries were to be successively stepped up to this plateau level during 1973–1980. The price agreed upon was 12 percent higher than in the first contract (somewhere just below 0.6 Pf/Mcal), but since world fuel prices had increased markedly after the conclusion of the first Soviet contract, it was still considered favorable. In particular, it was seen to be "clearly below the prices that are demanded today by the Dutch in new contracts." The contract would run for 20 years and the gas was to reach Germany by the same pipeline route. Due to problems to come to agreement on a credit arrangement for the countertrade in pipe and equipment, the formal signing of the contracts had to wait until July 6, 1972.[73]

Eventually the Soviet negotiations with France, which had been initiated already in 1966, also yielded success. The agreement-in-principle reached in fall 1969 did not immediately translate into a commercial contract, mainly because of Gaz de France's judgment that it was not in a hurry to conclude a contract. Yet rapidly growing domestic demand made the French increasingly interested, and in August 1972 a 20-year contract could eventually be signed, covering 2.5 bcm per year starting in 1976 and reaching 4 bcm in 1980. During the first four years of deliveries, however, GdF was not to receive Soviet gas in a physical sense. Instead, the French agreed on a complex "switch" involving four countries. In material terms, Italy's ENI was to import the Soviet gas that the French had contracted, and in return France would import an equivalent amount of Dutch gas on which ENI had an option. Only from 1980 would GdF start importing Soviet gas by pipeline.[74]

Meanwhile Austria's state-owned oil and gas company ÖMV, which had pioneered imports of Soviet natural gas to Western Europe, sought access to additional volumes. ÖMV took up the topic for discussion in June 1970 at the Soviet-Austrian Commission for Economic Cooperation, in which ÖMV's chairman Ludwig Bauer was an active member. An agreement-in-principle was soon reached that seemed to pave the way for a near doubling of imports.[75]

Shortly afterward, in December 1971, Moscow was able to add Finland to its list of export markets. As foreseen in an intergovernmental Soviet-Finnish agreement signed in April 1971, Soyuznefteexport agreed with the Finnish oil company Neste on the delivery of natural gas starting in January 1974 at 0.5 bcm per year and rising to a plateau level of "at least" 1.4 bcm by 1979. The long-term target was to reach an annual volume of 3 bcm. The gas, which Neste aimed to transmit mainly to industrial users and electric power plants, was to be imported through an extension of the domestic Soviet pipeline that supplied Leningrad. Measuring 720 mm in diameter on the Finnish side of the border, the line would be able to carry substantially more gas than the actually contracted volume. A major reason for this overdimensioning of the link was that the Soviet Union also hoped to export natural gas to Sweden, which might be supplied by way of transit through Finland. For the time being, however, the private Swedish actors involved in these negotiations failed to come to agreement with the Soviet side.[76]

Other West European countries that for the time being were left without Soviet contracts included Switzerland, which took its first contacts with the Soviet side in July 1970, Belgium, which, as of spring 1971, was reported to be negotiating with the Soviets for imports by way of Germany, and Spain, with which loose talks were initiated in the aftermath of the successful completion of Germany's and Italy's first Soviet deals.[77]

More surprising—and controversial—were the attempts from the side of several American gas companies to access Soviet natural gas. In sharp contrast to the seemingly limitless gas riches of the Soviet Union, US gas production was expected to reach a peak in the early 1970s. Several distributors were already importing large volumes of Canadian gas, but these supplies were not expected to be sufficient to meet future demand. Imports in the form of LNG from further away were suggested as a solution. The main interest was in imports from Algeria, but in November 1971 the Nixon administration confirmed that several US gas companies had approached the White House on the issue of possible imports from the Soviet Union as well. Washington appeared to take a certain interest in the industry's wish to move "Siberian LNG from Baltic Sea ports to the US East Coast," prompting US secretary of commerce Maurice Stance to take up the idea for discussion with the Kremlin. The real driving actor in the proposed project appears to have been the French firm Gazocean, which specialized in liquefied gas transports and which was one of two owners of Distrigas, a large US-based supplier.[78]

Two other US companies, El Paso Natural Gas and Occidental Petroleum, proposed to import Soviet LNG from the Soviet Far East to the American West Coast. These talks, which took place in cooperation with Japan, were also supported by Washington, and in 1972 El Paso was already reported to be near a deal with the Soviet side. As for imports to the US East Coast, Distrigas' interest soon waned, but its vision was taken over by a consortium of three other US gas companies, Tenneco, Texas Eastern Transmission, and Brown & Root. Whereas Distrigas had proposed an annual import of 6 bcm, the three firms expanded the proposal to a staggering 20 bcm per year. Nixon took up the consortium's proposal for Soviet LNG imports during his visit to Moscow in May 1972. The State Department was reported to see opportunities for improvement of Soviet-US political relations arising from the envisaged gas trade. Still, the proposal was highly controversial, and fears were raised that the Soviets might "close the valves and leave the East Coast to shiver." The government insisted that deliveries would most probably be "safe."[79]

In June 1973 the three-company US consortium signed a preliminary agreement with the Soviet Ministry of Foreign Trade. The "North Star" project, as it was now referred to, was expected to become operational in 1978–1980. A feasibility study suggested that the gas might be sourced from the supergiant Urengoi field in northwestern Siberia. From there, a 2,400 km pipeline would have to be built to Murmansk, where an LNG terminal and tanker harbor would have to be constructed; 20 LNG tankers would be needed for moving the gas to the United States.[80]

By 1973, then, at which time Germany and Italy eagerly awaited their first red gas supplies, five capitalist countries—Austria, Germany, Italy, Finland, and France—had already signed contracts with the Soviet Union, while at least six more—Sweden, Switzerland, Belgium, Spain, Japan, and the United States—were in the process of negotiating with Moscow. This extraordinary dynamism had come about in only a few years' time and was essentially the combined result of two key trends: an unprecedented growth in the popularity of natural gas and a new wave of attempts to improve East-West relations. The perceived economic and political opportunities linked to a prospective import of Soviet natural gas made a range of actors highly enthusiastic about the possibility of interconnecting the Soviet pipeline infrastructure with its emerging counterpart in Western Europe. The risks, for their part, were downplayed. Austria's imports of red gas, which started in late 1968, played an important role as a positive reference case, supposedly proving—in stark contrast to the evidence presented in the preceding chapter—that Soviet gas deliveries were secure and reliable. The negotiations themselves, as explored in depth for the case of Germany, functioned as an important arena for establishing additional trust and generating resonance. Crucially, the talks also became a key vehicle for the formation of a transnational coalition of system-builders.

Imports from the East were strongly opposed by a number of powerful actors, though not so much for security reasons. West Europeans were generally optimistic about the possibilities to deal with the security issue by technical means, particularly through improved intra-Western grid integration. The danger, from the viewpoint of the opponents, was rather seen to lie in a restructuring of power relations on domestic and intra-Western energy markets, and in unwelcome and unfair competition from beyond the Iron Curtain. The latter aspect was in turn seen to threaten Western Europe's internal gas exploration activities. Some actors also opposed imports from the East because the time did not seem ripe. From a foreign policy point of view, however, it was seen absolutely crucial to make effective use of the one-time opportunity that the 1968 invasion of Czechoslovakia had opened up. In other words, the negotiations could not wait. Not surprisingly, then, gas companies were strongly pushed by governments, and it is unlikely that any negotiations would at all have come about, let alone been successfully completed, without the vigorous facilitating role played by ministers, state secretaries, and their energy policy advisors. In this sense, the story told in this chapter highlights the making of European energy dependence as a complex process, involving a variety of interests and stakeholders, that needed not only political support, but also active political coordination.

8
Constructing the Export Infrastructure

Through the addition of West Germany, Italy, Finland, and France to the Soviet Union's list of capitalist markets, Moscow's gas export commitments jumped to completely new levels. Whereas Austria, the only capitalist country so far to actually import red gas, received 1.5 bcm annually, Italy and Germany would together import 9 bcm per year according to the contracts signed in 1969–1970. The second German contract and the first French contract signed in 1972 added another 8 bcm. Finland was to receive 1.4 bcm or more. At the same time, exports to Poland (1.5 bcm), Czechoslovakia (2.5 bcm), East Germany (at least 3 bcm), and Bulgaria (3 bcm) were to be initiated or expanded. Total Soviet deliveries were scheduled to grow steeply from less than 5 bcm per year in the early 1970s to 7.4 bcm in 1973, 17.2 bcm in 1974, and 24.7 bcm in 1975. The first big test would come in 1973, when exports to Italy and the two Germanies would commence.

The stakes were high, especially for exports to Western Europe. Failure to live up to contractual obligations in the sensitive formative period of the East-West gas trade would ruin Mingazprom's international reputation and jeopardize its prospects for a major long-term role on the West European gas scene. Moreover, political leaders such as German chancellor Willy Brandt had linked the gas trade to a much wider foreign policy agenda. A collapse of the arrangement might thus turn out politically hazardous and might conceivably infect East-West relations more generally. This chapter takes a closer look at how the Soviets, against this background, took on the challenge of living up to what they had promised.

Siberian Megalomania

In order to pipe additional volumes of natural gas to Western Europe, the Soviet Union had to expand its export infrastructure. The Bratstvo pipeline, which supplied Czechoslovakia and Austria, was no longer sufficient, but had to be complemented by additional lines. Similarly, it was no longer possible to rely on Galicia's gas fields, in western Ukraine, for the exports. Though conveniently located near the Soviet Union's western border, the Galician reserves were much too insignificant if seen in relation to the country's new contractual obligations. Gas destined for export markets would increasingly have to come from farther away, and for this purpose substantial new capacities would have to be added to the domestic Soviet pipeline grid.

Mingazprom's export strategy was based on an arrangement in which deliveries from Siberia's giant gas fields would play the main part. Export deliveries, however, constituted only a small share of a much more far-reaching Siberian system-building vision. The gas, it was imagined, would be brought westward by way of two main pipeline routes: a "northern" route that would carry gas from northwestern Siberia's Tyumen region to European Russia, the three Baltic republics, the Belarusian SSR, western Ukraine, and onward to foreign customers, and a "southern" system that would supply the Urals and other regions. The first plans, presented in late 1965, envisaged an annual transmission of 50 bcm along the northern route and 40 bcm along the southern route.

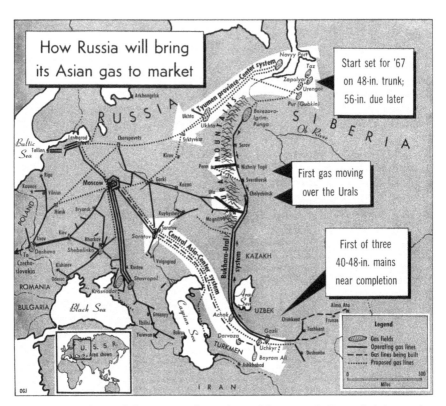

Figure 8.1 Planned Soviet gas flows from Siberia and Central Asia to the European part of the USSR. Mingazprom's large-scale system-building activities were followed with great interest in the West. This illustration, published in the American Oil and Gas Journal, shows the overall Soviet scheme for bringing "Asian" gas to the Soviet Union's main industrial and population centers.

Source: Oil and Gas Journal, February 27, 1967, p. 84. Reproduced by permission.

These figures were subsequently scaled up in response to new spectacular gas finds in Siberia and upward revisions of the available reserves.

As of spring 1966, the annual volumes to be transported had been raised to 85 and 45 bcm, respectively, and a goal of reaching an annual transmission of 15 bcm along the northern route in 1970 had been set. In 1967, then, the discovery of several new giant Siberian gas fields stimulated a further scale-up. The volume of confirmed Siberian reserves grew from 714 to 3,020 bcm and the total Soviet ones from 4,381 to 8,013 bcm in a single year, prompting Mingazprom to elaborate on a supply of more than 200 bcm of Siberian gas annually along the northern and southern routes. Deputy Gas Minister Yuli Bokserman noted that the volumes of gas production in Tyumen, and particularly the Yamal-Nenets national region, did no longer "depend on the gas resources, but on the transport possibilities."[1]

By January 1969, the Siberian reserves had grown to 4,400 bcm, which was equivalent to no less than 50 percent of the total probable reserves of all capitalist and developing countries. This immense growth was celebrated as an extraordinary achievement by the regional party organization in Tyumen, whose first secretary Boris Shcherbina argued that "such natural gas reserves are available in Tyumen region that the area can provide the country with 600–700 bcm of gas annually." A year later *Izvestiya* claimed that "discoveries made to date will enable western Siberia to produce 1,000 bcm of gas annually" and develop "hydrocarbon reserves in this new region on a scale unknown in any other country in the world." No West European country could dream of matching the Soviet Union's extreme ambitions. For example,

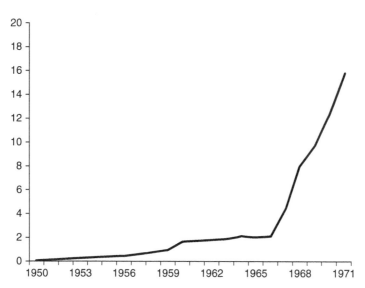

Figure 8.2 Soviet gas reserves, 1950–1971 (tcm) (A+B categories in Soviet terminology).

Source: Based on figures in Gazovaya promyshlennost, various issues.

West Germany, as we saw in the preceding chapter, planned for an increase of its gas supply to 40 bcm by 1980—an enormous growth from Germany's horizon but almost negligible in the Soviet context.[2]

In quantitative terms, exports of red gas would thus play a very minor part in Mingazprom's overall system-building efforts. Siberian gas would contribute to exports both directly and indirectly. In an initial phase, its role in the export scheme would be to supply fuel-thirsty regions in Latvia, Lithuania, and Belarus that had so far received their gas from Ukraine. This would make it possible to save fairly large volumes of Galician (western Ukrainian) gas, which, instead, could be piped to customers abroad. In a second phase, Siberian gas would itself be exported. For this to be possible, however, the pipelines from Siberia would have to be extended all the way to western Ukraine and onward through Czechoslovakia—and possibly Poland—to Central and Western Europe. Finland was a special case, as it would be supplied via Leningrad, whose gas users would similarly switch to Siberian gas in a near future.

The early plans developed in 1966 in preparation for the negotiations with Austria, Italy, and France foresaw that Siberian gas reach Belarus in 1971. Belarusian transition to Siberian gas would free 6 bcm of Galician gas, a large share of which could instead be exported to the West. From 1973, export deliveries from Galicia, whose reserves were very limited, would then be phased out and replaced by direct deliveries from Tyumen.[3] The failure to reach agreement with Italy and France in 1966–1967 meant that the timetable had to be revised, but the overall scheme was retained as the physical fundament of Mingazprom's export strategy. It was reactivated following the finalization of export contracts with Italy, Germany, and France in 1969–1972.

Arctic System-Building

The "northern" pipeline route, which was expected to become the most important route for transmission of Siberian gas, was popularly referred to as the "Northern Lights" (Siyanie severa). Originating in northwestern Tyumen region, it stretched via Nadym on the Ob and Ukhta in Komi ASSR—itself a gas-rich region—to the town of Torzhok, located halfway between Moscow and Leningrad. From there, the gas would be piped through partly existing and partly to-be-built lines to Leningrad, the Baltics, Belarus, and onward across the country's western border to eagerly waiting customers abroad.

The Council of Ministers gave Mingazprom green light to start construction of the first section of the Northern Lights system already in 1967, that is, before any West European contracts had been concluded. Stretching from Ukhta to Torzhok—a distance of 1,050 km—this first link would initially not be used for transmission of Siberian gas proper, but for tapping the large Vuktylskoe gas field in Komi. With probable reserves of some 500–600 bcm, Vuktylskoe was one of the largest gas fields in the Soviet Union at the time, and the possibility of integrating it into the new Siberian system was one of the main reasons behind Mingazprom's enthusiasm for the Northern Lights route.

Uktha-Torzhok could be taken into operation in late 1969. As for its upstream extension to Siberia, however, progress was slow in coming. Gas

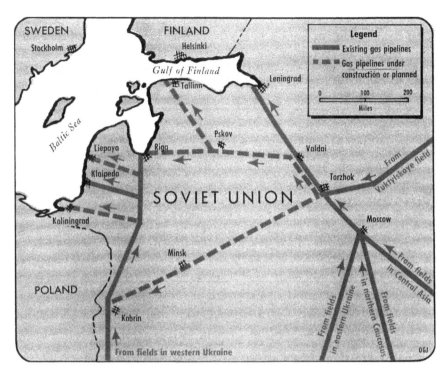

Figure 8.3 Planned pipeline routes for the transmission of Komi and Siberian gas to Leningrad, the Baltics, and Belarus. The plan was that these lines would later on be extended into western Ukraine and across the border to Czechoslovakia, whereby the south-north flow along the existing route from western Ukraine to Belarus and the Baltics would change direction. Moscow, for its part, was expected to rely more on Central Asian than on Siberian gas.

Source: Oil and Gas Journal, May 29, 1972, p. 12. Reproduced by permission.

Minister Alexei Kortunov pointed to uncertainty regarding the availability of large-diameter pipes and equipment. The ministry thought that the use of 1,220 mm wide pipes, on which the Uktha-Torzhok section had relied, would be uneconomic for the intended transmission of vast volumes of Siberian gas westward, and that construction of the Siberian lines would, therefore, have to wait until 1,420 mm pipes became available. Domestic pipe makers, however, were not yet able to produce such large pipes. The pipe industry faced enormous difficulties not least as a result of the harsh climate in Russia's far north, which meant that the steel had to be of a particularly high quality.

From this perspective, the successful completion of the gas and pipe negotiations with Italy and West Germany in late 1969 were of crucial importance. The two countertrade agreements specified that the Soviet Union would receive 2.2 million tons of high-quality 1,420 mm pipes. Finalization of the contracts allowed Mingazprom to work out a "programme for accelerated development of the fields in northern Tyumen region," which was submitted to the Central

Committee of the Communist Party and the Council of Ministers immediately after the initialing of the German gas and pipe contracts on November 29, 1969. On the basis of this program, the Central Committee and the Council of Ministers issued a special directive that paved the way for faster Siberian system-building.[4]

The new program suggested that efforts concentrate on exploiting the giant Medvezhye field, which was the southernmost of the recently discovered Siberian giants, and enabling its gas to be shipped westward. In a second phase, supergiant Urengoi, located a hundred kilometers further north, would be phased in. A goal was set to produce 60 bcm per year of Siberian gas by 1975. In order to transport such large volumes westward, a double 1,420 mm pipeline would be laid on the 2,500-km-long Nadym-Ukhta-Torhzok route. From Torzhok to Minsk and western Ukraine, the line would continue as a 1,220 mm wide and 1,135 km long line. The system would subsequently be expanded through the construction of numerous additional lines that would be laid in parallel with the first ones. By 1980 the transmission capacity would reach 230 bcm along the "northern" route and 50 bcm along the "southern" route. At such a system-building pace, the Soviet dream of producing and delivering 1,000 bcm of Siberian gas per year seemed within reach before the end of the 1980s.[5]

On Christmas Eve 1969, Mingazprom's board of directors convened to distribute responsibilities for the new undertakings. In January, then, the ministry informed Gosplan that it intended to start constructing the first Nadym-Ukhta-Torzhok pipeline immediately upon reception of the first batch of 1,420 mm pipes from Germany, and that the first Siberian pipeline would be ready by 1973. In February 1970, the ministry organized a large internal meeting, where all managers of the ministry's enterprises and organizations came together to discuss the challenge of accelerating the Siberian undertakings. In the course of the following months similar meetings were held at the regional level in the provinces involved.[6]

When the first German pipes arrived in July 1970, construction of the first pipeline could start. The practical problems, however, turned out to be enormous. A major challenge was the virtual absence of roads and railways for transporting Mannesmann's huge pipes to the pipeline route. Mingazprom planned to lay the pipeline in parallel with the old railway track that had been built for shipping coal from Vorkuta in the Far North—where a number of Gulag camps had been established for mining purposes—to central Russia. But the railway did not always follow the most optimal gas transmission route, and the ministry thus had to decide whether to favor a longer but more comfortable route—at the expense of additional pipe demand—or a shorter and less comfortable one—which would worsen project logistics. Given the scarcity of pipe, Mingazprom initially tended to favor the shortest possible route. The first Siberian pipelines thus followed the railway line only along the 450 km section from Ukhta to Kotlas, whereas the remaining 2,000 km were built straight through the Russian wilderness.[7]

In this wilderness, the huge German pipes had to be transported on water or on snow. Water transport was possible on navigable rivers, but only during the short summer season. Land transports, in contrast, were possible only in winter, when the labyrinth of swamps, lakes, and rivers that made up northern

Figure 8.4 Bear cub found along the Northern Lights pipeline route. The new lines from Siberia and Komi went through some of the wildest and most inaccessible parts of Russia.
Source: RIA Novosti.

Russia lay frozen, enabling specially designed trucks to operate. Since pipelaying was carried out under strong time pressure, the Soviets used a combination of summer and winter transports.

From Kotlas to Rybinsk, a distance of 500 km, Mingazprom sought to lay the pipeline close to two navigable rivers, the Severnaya Dvina and the Sukhon. The route here went through an extremely swampy area that during summer could be accessed only by barge. Pipes destined for the region near the Siberian gas fields, for their part, were transported by rail to the northernmost railway station, Labytangi on the Ob, just on the Polar Circle. From there, 73,000 tons of pipes destined for the Medvezhye-Nadym pipeline segment were brought in on road-less snow, the average distance to the construction sites being 500 km.[8]

The overall challenge turned out to be much greater than anticipated even by Gosplan's stern pessimists. Although both pipes and people were available in sufficient quantities, Mingazprom's lack of experience of pipelaying in permafrost regions caused delays. The German pipes were designed to withstand temperatures down to minus 60 degrees centigrade, but the quality of the pipes was only one of many aspects that needed to be dealt with. Because of the permafrost, the pipes had to be constructed above ground, and this made it crucial to prevent the frozen land on which they rested from melting under the influence of the warmer gas. Various methods for chilling the gas were developed, but this took time and delays were inevitable.

Another problem area was the construction of compressor stations. Powerful compressors were seen crucial for transmitting large volumes of Siberian gas over a distance of several thousand kilometers. Soviet compressor makers had always lagged far behind their Western counterparts, and Western export regulations made it difficult to compensate for the shortage through imports from the West. The Soviet-Italian countertrade deal did include delivery of three very large (25 MW) compressor units, but Mingazprom needed hundreds of units. Domestic manufacturers did not yet produce 25 MW units, but were still struggling to master 10 MW and 16 MW ones, and the quality and reliability of these was much lower than the corresponding Western machines. Mingazprom noted that foreign compressors were able to work without renovation for up to three years, whereas the units produced in the Soviet Union had to be renovated after less than one year. "In this way," Gas Minister Kortunov complained, the country's metallurgy and machine-building enterprises "restrain the development of the gas industry."

At the 24th Congress of the Communist Party, held in March 1971, Brezhnev was proud to inform the delegates that "a gas pipeline of unique dimensions is being laid to carry natural gas from Siberia to the country's European part." The rise of Siberian natural gas was pointed at as a key instrument in making "the USSR's relations with the countries of the capitalist world...fairly active and diverse." Premier Kosygin emphasized that during the new five-year plan, 1971–1975, "an enormous increase in the output of gas will be achieved." In reality, however, the problems in Siberia and elsewhere had already put a troublesome brake to the industry's spectacular development. The Soviets had earlier elaborated on a doubling of gas production from around 200 bcm in 1970 to around 400 bcm in 1975. It was this predicted increase in the overall availability of gas, to be achieved mainly through rapid exploitation of Siberia's newly discovered gas riches, that had inspired the Soviets to aim for contracts as large as possible in their negotiations with Germany and other countries in 1969. The Siberian setbacks forced the Soviets to lower their ambitions, and as a result the formal directives adopted by the 1971 Party Congress specified a much lower growth rate, from 200 bcm in 1970 to 300–320 bcm in 1975. This was still pointed at as impressive, but it was a far cry from the original goal. It was now also deemed unrealistic that natural gas from faraway Tyumen would reach the country's westernmost regions in 1973. Only in 1976, it was now believed, would Siberian gas be available.[9]

Kortunov vehemently opposed the new planning targets. At a speech delivered at the Supreme Soviet in May 1971, he acknowledged that the 1973 target for first Siberian deliveries would probably not be met, but assured the audience that Mingazprom would be able to finish construction of the first Siberian lines before the end of 1974. This would make it possible to start up gas deliveries from Medvezhye in January 1975. Kortunov argued that the problems in Siberia could be mastered, and that all he needed was ample support from Gosplan and the political leadership.[10]

A month later, Mingazprom submitted an updated Siberian plan to the Council of Ministers. To avoid further delays, the radical decision was taken to change the main pipeline route from northern Tyumen to the western regions. As noted, scarcity of large-diameter steel pipe had so far led Mingazprom to favor the shortest possible routes. In April 1971, however, the second Soviet-German gas and pipe deal was finalized, which meant that Mingazprom could suddenly count on the availability of another 800,000 tons of 1,420 mm pipes, to be delivered by Mannesmann in 1972–1974. This allowed the ministry to rethink its Siberian transmission scheme and opt for a route that to the greatest possible extent avoided the difficult permafrost areas. The old, straight route was thus abandoned in favor of an alternative one. Stretching from Nadym down to Punga and from there to Vuktylskoe in Komi and further on to Ukhta and Torzhok, it would require more pipes, but in return the section of it that went through permafrost areas could be radically shortened from 495 km to only 28 km. The main drawback was that large quantities of pipes had already been delivered to the original route. These pipes now had to be picked up for transfer to the new construction sites further south.[11]

Apart from problems with pipelines and compressors, a main source of delay stemmed from unexpected problems at the gas fields. New drilling and production technologies needed to be developed for the extreme Siberian conditions, and this took longer than expected. Having constructed its first operational wells at Medvezhye, Mingazprom drew the conclusion that it was "necessary to revise the current technological approach." Similarly, the harsh climate in combination with the enormous distances from population centers seemed to negatively affect the work morale. The ministry called for "new forms of work organization, strict technological and productive discipline, and the dissemination of best practice in the fast-drilling of wells."[12]

The difficulties in developing the large Vuktylskoe field were also seriously underestimated. As of early 1970, only 10 out of 20 planned wells had been constructed, and out of these 10, 3 were reported to have broken down. Moreover, the field contained huge quantities of condensates that needed to be separated out and processed.[13] The situation did not improve much during the following years, and earlier targets according to which 15 bcm of Komi gas in 1970 and 35–40 bcm in 1972 would have been be produced and shipped westward could not be met. Kortunov concluded that it was meaningless to try and quickly expand the Vuktylskoe-Uktha-Torzhok section of the Northern Lights system, since the transmission capacity already exceeded Vuktylskoe's production.[14]

The Ukrainian Crisis and Kortunov's Death

How would the Soviet Union be able to live up to its new export commitments without Siberian natural gas? Mingazprom's 1969 program had counted on the arrival of first Tyumen gas in 1973, just in time for it to contribute to first export deliveries to Italy and the two Germanies. Siberia and Komi would help boost the overall availability of gas in the Soviet Union's westernmost regions and thereby prevent domestic customers from being left without fuel as western Ukrainian gas was increasingly earmarked for exports. Later on, Siberian gas would also start to be exported in its own right.

The delays in Siberia meant that the plan had to be revised. A new strategy started to be worked out in 1971, whereby the focus shifted from solving the Siberian problems to speeding up Ukrainian system-building. This was a daunting task, as became clear from a report submitted by Mingazprom to Gosplan in January 1971. Regional gas demand in western Ukraine, through which most exports were to pass, was expected to grow only slightly during the next few years, and this was also the case regarding Ukrainian deliveries to Belarus and the Baltics. Yet since export commitments were expected to grow from 4.8 bcm in 1971 to more than 20 bcm in 1975, Galicia's gas fields and the western Ukrainian pipeline grid would have to handle an enormously much larger flow than before. Gas destined for export was about to become the most dominant flow post. Total gas demand in western Ukraine and the regions dependent on supplies from it was expected to increase from 17.5 bcm in 1971 to 35.5 bcm in 1975, while at the same time local production, following rapid depletion of Galicia's gas fields, was expected to *de*crease from 10.8 bcm to 6.7 bcm. A huge deficit in western Ukraine's gas budget was thus in the making.[15]

The original Soviet strategy for covering the deficit had been to first bring in gas from the giant Shebelinka field in eastern Ukraine, in fairly modest quantities, and then increasingly Siberian gas. Following the Siberian delays, Mingazprom judged that Shebelinka gas would have to play a much larger role than earlier anticipated. Moreover, since production at Shebelinka already showed signs of decline, it was deemed necessary to arrange for Central Asian gas to make a contribution. Such an arrangement had been proposed by Gosplan as a permanent export solution already in 1966. At that time it had not found support from the side of Mingazprom, whose managers had firmly believed that they would be able to supply Ukraine with Siberian gas within only a few years' time. The problems in Siberia now forced Kortunov and his colleagues to change their mind.[16]

To enable large-scale flows of Shebelinka gas to Central and Western Europe, the transmission capacity between eastern and western Ukraine would have to be strengthened. Trans-Ukrainian flows had been negligible up to 1970, when the completion of a new 1,020 mm pipeline for the first time enabled significant volumes of eastern Ukrainian gas to be moved to western Ukraine (see chapter 6). Mingazprom's new plan was based on a scaling-up of this arrangement. The ministry planned, first, to build additional compressor stations along the existing pipeline and thereby boost its transmission capacity to 10 bcm per year. Much of this gas would be piped to Belarus and the Baltics, which had so far received western Ukrainian gas only. In this way more of Galicia's rapidly waning gas resources could be saved for exports. Second, the ministry planned to construct a new, larger pipeline from Shebelinka that for most of its length would follow the same route. At Ternopol in western Ukraine, this second trans-Ukrainian line would diverge from the first one, turning southward to Dolina, and from there continue in parallel with the 820 mm "Bratstvo" pipeline to Uzhgorod

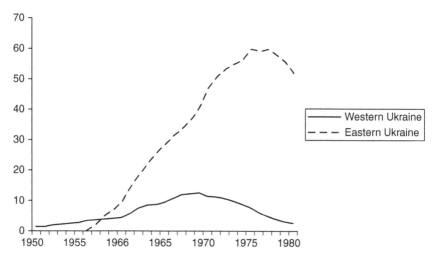

Figure 8.5 Production of natural gas in western and eastern Ukraine, 1950–1980 (bcm).

Source: Derzhavnyi komitet naftovoi, gazovoi ta naftopererobnoi promyslovosti Ukrainy 1997, pp. 148 and 176.

on the Czechoslovak border. This would make it possible for Shebelinka gas to make a direct contribution to Soviet gas exports, all of which had so far depended on Galician deliveries.[17]

As for Central Asian gas, the first pipelines from Uzbekistan to the European part of the Soviet Union had come on-stream in late 1967, just in time for the 50th anniversary of the October Revolution. As of 1971, two pipelines already delivered Central Asian gas to the central regions of European Russia, and a third line was about to be completed. The latter was to terminate at Ostrogozhsk near the Russian-Ukrainian border, where it was to link up with the gas flows from the Caucasus to Moscow and Leningrad. Mingazprom had for some time already elaborated on an extension of the Ostrozgozhsk line westward into Ukraine, where Central Asian gas would play an important role in countering the expected decline of eastern Ukrainian gas production. The perceived need to compensate for the Siberian troubles speeded up the decision-making process, clearing the way for construction of the Russian-Ukrainian connection.[18]

The revised export plan looked fine on paper, but its actual implementation turned out to be anything but straightforward. Again, problems with compressors became a main headache for Mingazprom. Kortunov was furious about the low quality of the machines that his ministry received from the country's main compressor manufacturers. The first station along the first trans-Ukrainian pipeline was to have been taken into operation already in December 1970, but the ministry found that the units started vibrating dangerously when switched on, and as a result the station could not be taken into regular operation. The problems were investigated and the mistakes corrected, enabling the units to be connected to the grid in July 1971. But since a total of 11 analogous units were to be installed on other sections of the same line, further delays seemed inevitable. For 1971 as a whole, the Ministry of Heavy Machine Building in the end delivered only 40 out of 60 promised units to Mingazprom, and out of these 40 compressors, 15 turned out to lack certain key components.[19]

The problems continued in 1972. The ministry now set out to construct the second trans-Ukrainian pipeline, which would be of crucial importance for moving larger volumes of Shebelinka gas across the Soviet-Czechoslovak border. Again, progress was painfully slow. Pipelaying suffered from lack of equipment, logistical failures, and organizational meltdowns, and only a small fraction of the 1972 targets, as measured in kilometers, was eventually met. The construction of compressor stations also progressed at an alarmingly slow pace. According to the revised export system-building timetable, 5 out of 13 compressor stations along the export route were to have been completed before the end of 1972, but in reality none of them had yet been taken into operation as the year drew to a close.[20]

Mingazprom's managers and engineers were deeply worried in view of the approaching start-up of exports to Italy and the two Germanies. A gift from above was that ENI, which according to the original agreement was to have received its first shipments of red gas in January 1973, was facing its own system-building problems and wished to postpone the start-up of its Soviet imports until spring 1974. Even so, the export commitments grew rapidly. The addition of East and West Germany to the list of Soviet gas recipients from January and October 1973, respectively, meant that exports would have to grow by 50 percent from just below 5 bcm in 1971 and 1972 to 7.4 bcm in 1973.[21]

Mingazprom was not the only Soviet stakeholder that worried about the slow progress. Gas users in Belarus, Lithuania, and Latvia were also upset by the delays. In these republics, dependent as they were on deliveries from Ukraine, industries and households had traumatic experiences of what increased competition from foreign customers for scarce Ukrainian gas might mean, particularly in winter. They had been strong supporters of Mingazprom's ambitions to quickly bring in Siberian gas to the westernmost regions of the country, as such deliveries promised to diversify their overall gas supply. Accordingly, they were deeply worried by the looming Siberian problems and even more so by the delays along the Shebelinka-Uzhgorod route, which was to compensate for the nonavailability of Siberian gas.[22]

Of the Soviet republics that depended on western Ukrainian gas, Latvia was the most vulnerable. Located at the remote end of the Dashava-Minsk-Vilnius-Riga pipeline, the republic had pushed hard for access to gas from elsewhere in due time before the start-up of exports to Germany and Italy. The Latvians had early on secured Mingazprom's support for the construction of a new, 558-km pipeline from Valdai (near Torzhok) to Riga, which would give the Latvians convenient and direct access to both Komi and Siberian gas. Following the delays in Siberia, Mingazprom wished to postpone the construction of the new line, arguing that it would be more efficient, on the mid-term, to cover Latvia's gas demand through increased deliveries of Ukrainian gas. Latvian premier Vitalii Ruben, however, insisted on the project's early realization and eventually succeeded in convincing Moscow's central planners to make available the necessary resources and have Mingazprom construct the pipeline as originally planned. Hence the new connection was built as anticipated and taken into operation in May 1972.[23]

In Latvia the start-up of the new line was celebrated as an important victory for the republic. Yet as a consequence of the problems to ship sufficient volumes of gas from Komi, the diversification was of limited practical value. During its first year of operation, the Valdai line brought only half as much gas to Latvia as planned. Neighboring Lithuania, which hoped to profit from the new line through existing connections to Latvia, was also affected. From January 1974, moreover, both Latvia and Lithuania would have to compete with Finland for scarce Komi gas. Lithuania and Belarus were, in addition, affected by the Ukrainian system-building problems.[24]

The increasingly tense situation in the Soviet natural gas system was discussed in depth at the September 1972 session of the Supreme Soviet. Several speakers at this high-level political meeting criticized the slow pace of construction, and a number of critical articles addressing the problems appeared in leading Soviet media. The Soviet Council of Ministers, backed up by the Central Committee of the Communist Party, eventually opted to deal with the crisis through a major administrative reform. Mingazprom's system-building activities were transferred to a brand new Ministry for Construction of Oil and Gas Facilities (Minneftegazstroi), to which parts of the Oil Ministry were also added. The smaller Mingazprom that remained would from now on first and foremost focus on the operation of the already existing pipeline network, including the management of gas exports. The reform also included major personnel changes. Gas Minister Kortunov left his position to become head of Minneftegazstroi. As new minister for the gas industry, Premier Kosygin appointed Sabit Orudzhev, a 59-year-old Azeri who had earlier served as first deputy oil minister.

Kortunov literally was short-lived on his new post. The extremely stressful and difficult situation with which he had to cope in terms of Siberian and Ukrainian system-building had a severely negative impact on his health. His physical strength failed, making it impossible for him to continue directing the new ministry. His illness came at the worst possible moment, precisely when his previous energy, persuasiveness, and enthusiasm were more direly needed than ever before. He passed away on November 17, 1973.[25]

Desperation and Chaos

The organizational reforms of 1972 notwithstanding, it appeared improbable that the Soviet Union would be able to complete its export infrastructure on time. The problem was that Mingazprom's export commitments were now growing so large that transmission of Shebelinka gas through Ukraine and onward across the border was absolutely necessary. New compressor stations would have to be added to the existing East-West Ukrainian pipeline, and the new Shebelinka-Uzhgorod line would have to be started up. Gas Minister Orudzhev, pointing to his predecessor's failure to meet the planning targets, in February 1973 warned Veniamin Dymshits, chairman of the State Committee for Material-Technical Supply (Gossnab), that "if serious measures are not taken to secure the completion of these pipelines in the first quarter of the year, then during the second quarter the non-delivery of gas for export will amount to 400 million cubic meters." This corresponded to about one-third of total contracted exports.[26]

By early March, it was clear that the measures taken were not sufficient, and that the Soviets would thus not be able to deliver more than two-thirds of the contracted exports. In this situation, deliveries to Austria, which was still the only capitalist country to import red gas, were prioritized, and ÖMV was thus able to report that it received the volumes of Soviet gas it had requested. The shortage was absorbed by the Soviet Union's communist customers, notably the GDR, to which exports had started in January. For 1973 as a whole, the East Germans received only 0.7 bcm out of its contracted 1.7 bcm. The Czechs were more lucky, receiving 2.4 bcm out of 2.5 bcm.[27]

In early May, Orudzhev reported that export system-building was still in a "desolate" state throughout the pipeline route from Shebelinka to Uzhgorod. The construction of compressor stations was "carried out with deviations from design." Housing modules, machinery, and people were missing. In order to meet the approved deadlines, Orudzhev argued, Minneftegazstroi would have to "immediately increase the number of workers by at least 4,000 and organize the work in 2–3 shifts."[28] The construction workers who were already in Ukraine, for their part, were reported to be in a bad mood, being squeezed into small portable dwellings. The work morale was low. The Ukrainian Council of Ministers, fearing consequences for the republic's own gas supply of any further delays, begged Moscow to increase the pecuniary incentives by awarding all workers exceptional bonuses. The Ukrainians emphasized that it was a matter of compressor stations "on which the fulfilment of gas exports depend," but their real fear was clearly that of another severe winter crisis in Ukraine itself. Kiev's request was turned down by the responsible authorities in Moscow.[29]

By May 20, 1973, when, according to the plan, 408 km of new pipelines along the route from Shebelinka to Uzhgorod were to have been taken into operation, only 68 km had actually been laid. The situation with respect to compressor stations, necessary for boosting exports up to contracted levels, was equally discouraging. The station at Dikanka, for example, was to have been taken into operation in June 1973. However, a few key components that should have arrived several months earlier from an Estonian machine-building factory were not yet available, and, as a consequence, construction of the station could not proceed. This was hardly an unusual problem in the Soviet Union, but the weakness and unreliability of its centrally planned economy now threatened to spill over into the realm of energy exports and jeopardize Mingazprom's efforts to meet its contractual obligations vis-à-vis Western Europe.[30]

The Kremlin urged Mingazprom to take whatever measures it could to complete the export infrastructure on time. Deputy Gas Minister Sidorenko ensured the Council of Ministers that "Mingazprom understands the full seriousness of the question of securing export deliveries of gas and is extremely worried about the state of pipeline and compressor station construction in the Ukrainian gas transport system." Sidorenko argued the the delays were beyond Mingazprom's control, pointing, instead, to the inefficiencies of Minneftegazstroi, the Ministry of Energy, the Ministry of Petrochemical Machine Building, and the Ministry of Heavy Machine Building.[31]

The Ukrainian Council of Ministers tended to agree with Orudzhev and Sidorenko that Mingazprom was not the main source of the problems, noting that Minneftegazstroi and Minenergo carried out their work "in an extremely unsatisfactory way." Minneftegazstroi, however, was furious about the slow pace at which Mingazprom carried out necessary tests on completed pipeline sections. The Commission for Foreign Trade, a body that sorted under the Council of Ministers, blamed all the involved branch ministries, remarking that "despite repeated instructions from the Council of Ministers, Mingazprom, Minneftegazstroi, and Minenergo permit a significant stagnation in the construction of pipelines and compressor stations on the territory of the Ukrainian SSR, necessary for securing export deliveries of Soviet natural gas to West European nations."[32]

Minneftegazstroi, which now had to operate without Kortunov, struggled desperately to complete the Shebelinka-Uzhgorod export pipeline, but the work was hampered by the discovery of several pipe ruptures on sections that had already been completed, forcing the ministry to channel resources to their reconstruction. As a result, system-building was further delayed.[33]

The most difficult task in terms of pipelaying was to traverse the scenic but logistically challenging Carpathian mountains, near the Czechoslovak border; 42 km of pipes were to be laid there. The Soviets had already gained some experience from constructing the "Bratstvo" system along the same route, but the much larger diameter of the new line—1,420 mm as compared to 820 mm for the Bratstvo—made the challenge in terms of transportation and pipelaying much greater. The section would preferably have to be completed before mid-August. Construction after that would be hazardous due to heavy rainfalls. Orudzhev, therefore, asked the Council of Ministers to double the amount of welding brigades on the Carpathian section. At the sites of compressor stations, none of which had yet been completed, an additional 2,000 workers were similarly needed.[34]

On July 27, 1973, the main responsible managers convened for a crisis meeting at Dymshits's Moscow office. It was acknowledged that Minneftegazstroi and Mingazprom had worked out suitable measures for accelerating the pace of construction in Ukraine. Unfortunately, however, they had not been satisfactorily implemented. Minneftegazstroi was instructed to take "systematic control" of implementation. Gas Minister Orudzhev was ordered to complete pipelaying and launch all remaining compressor stations by the end of September 1973, in time for first export deliveries to West Germany. After lengthy discussion, Orudzhev's plea for a massive addition of working cadres found approval. The Ministry of Interior was instructed to make additional numbers of construction workers available; 1,150 probationers and conditionally released prisoners were to be sent to the construction sites in Ukraine; 200 people from the "working reserves" were also to be drawn upon. The Ministry of Trade was instructed to arrange for supplying these new workers with food while on the construction sites.[35]

The additional supply of workers did generate some progress in pipeline and compressor station construction. Yet by the end of September the difficult Carpathian section of the export pipeline had not yet been completed. Given the bottleneck in physical transmission capacity, this meant that it was still unclear whether the Soviet Union would be able to launch exports to West Germany as planned on October 1, 1973.

To sum up, the efforts to make real the proud export project had turned into a nightmare for Soviet system-builders. The efforts had started out in the late 1960s with breathtaking pipeline plans that would bring Siberian gas to the European part of the country and onward to Central and West European markets. The challenge turned out much greater than anticipated, and by 1970 it had become clear that a Plan B was needed to prevent a collapse of the emerging export regime. This alternative plan centered on Ukrainian system-building and the transmission of eastern Ukrainian gas to western Ukraine, whose once very large gas reserves were no longer sufficient to meet growing foreign demand, and onward to export markets. Western Ukraine was now gradually transformed from a producing region into a transit corridor for gas from elsewhere on its way to Central and Western Europe. This was a painful process that gave rise to concern throughout the westernmost Soviet republics, whose gas users rightly feared that the chaotic transformation of the whole system would make them more vulnerable.

The restructuring of gas flows in western Ukraine that took place in preparation for first export deliveries to Italy and the two Germanies in 1973 was essentially a continuation of a process that had started a few years earlier in connection with first exports to Czechoslovakia and Austria, but the scale of the projects was now much larger and the stakes, accordingly, higher. A number of high-level political initiatives were launched to come to grips with the export system. Yet the Soviets failed to complete the new export infrastructure on time, and the parts of it that did meet planning targets were often of alarmingly low quality. The countertrade deals signed with several West European countries meant that Mingazprom did no longer have to worry about access to high-quality steel pipe, but this could not protect system-builders against logistical breakdowns, low work morale, and other internal Soviet problems. It remained to be seen whether the situation could still be mastered.

9
Trusting the Enemy: Importing Soviet Gas in Practice

Enabling Transit through Czechoslovakia and Austria

While the Soviets were struggling to complete the new export infrastructure, West European gas companies were preparing to receive Soviet gas. This involved intense activities in Bavaria, northern Italy, and Finland's Karelian province, where Western Europe's main soon-to-be users of red gas were located, as well as in the countries and regions through which the gas was to be transited.

Regarding the transit, Czechoslovakia played the key role. Since 1968, small volumes of Soviet gas were already being transited through Czechoslovakia to Austria, but the challenge of transiting red gas to Italy and West Germany was an undertaking of completely different proportions. In the contracts signed it had been agreed that Italy and West Germany would not be involved in arranging the Czechoslovak transit, but that the Soviet Union and Czechoslovakia would handle it on a bilateral basis. Negotiations in this matter started immediately after the conclusion of the Soviet export contracts with Italy and West Germany in late 1969 and early 1970. Agreement was delayed, however, by the initial failure to finalize a gas export contract with France. The addition of France to the list of importing countries would necessitate a larger transit capacity in Czechoslovakia, but as of 1970 it was still uncertain as to whether a Soviet-French gas contract would actually materialize. Eventually the two communist countries decided to go ahead anyway, designing a transit system in which a possible flow of 2.5 bcm per year to France was taken into account.

On this basis, a formal agreement for the transit of Soviet gas through Czechoslovakia to no less than four capitalist and one communist country—Austria, Italy, France, and the two Germanies—could eventually be signed in late 1970. The system was to consist of a double pipeline, to be laid in parallel to the already existing "Bratstvo" line. The new pipelines were to measure 1,220 mm and 820 mm in diameter, of which the larger one was to be reserved for exports across the Iron Curtain and the smaller one for shipments to East

Figure 9.1 Construction of the first transit pipeline through Czechoslovakia, July 1971.
Source: RIA Novosti.

Germany. With compressor stations the system would be able to handle an annual gas flow of no less than 30 bcm. All in all, some 600,000 tons of pipes would be laid on Czechoslovak territory for the purpose of transiting red gas, and along them compressor stations with a total capacity of 400 MW would be installed. The project was seen to pave the way for "the largest construction project of the new Czechoslovak five-year plan" (1971–1975). The investment volume was around three times as large as for the much-publicized "Druzhba" oil pipeline.[1]

The way in which countries other than Czechoslovakia would contribute as transiteers was subject to dispute. In Austria, it was initially taken for granted that ÖMV would be awarded responsibility for transiting Soviet gas to both Italy and Germany. The Austrians were shocked when they got to hear, in summer 1969, that Ruhrgas and the Soviets were negotiating an arrangement that would circumvent Austria. ÖMV had hoped to use the Soviet-German transit pipeline to strengthen regional gas supply, particularly in Upper Austria with its sizeable chemical and steel industries. Austrian minister of trade Leo Mitterer was severely criticized by leading Austrian industrialists for not ensuring that Austria's economic interests were defended in the Soviet-German gas talks, while Germany was accused of aggressively investing in Austria without really opening up its own markets. Mitterer reacted to the internal criticism by approaching the German government, urging the Germans to change their transit plans, but it was too late. The transit route, through Czechoslovakia only, had already been firmly settled with the Soviets. The Austrians were thus sidestepped.[2]

The transit to Italy was also contested. The Soviet-Austrian gas contract from 1968 had obliged ÖMV to ensure friction-free transit of red gas to Italy and France. In the course of the Soviet-German negotiations in 1969, however, the energy experts at the German ministry of economy had challenged this plan by elaborating on a route through Czechoslovakia to Bavaria and onward to Italy by way of transit through Switzerland—without any Austrian involvement. Moreover, immediately after the finalization of the Soviet-Italian gas contract in December 1969, the Hungarian government approached the Kremlin with a proposal for transiting Soviet gas to Italy through Hungary and Yugoslavia, rather than through Czechoslovakia and Austria. This route had been discussed as an option already in 1966, at which time the prospective transiteers had not shown much interest. Since the Hungarians had now signed their own import contract with the Soviet Union and needed to build a pipeline for this purpose, they now evaluated the possibility differently. The Soviets showed themselves interested in the updated proposal, although the Council of Ministers appears to have used it mainly as a way to put pressure on the Austrians to speed up transit negotiations with Italy. Soviet gas minister Alexei Kortunov thought the Hungarian proposal worth taking into account "as one of the variants" for transiting Soviet gas to Italy. A committee was formed on the Soviet side for the purpose of further investigating the Hungarian proposal.[3]

The Austrian-Italian transit negotiations were delayed by an internal Austrian debate about the optimal routing of the transit line and by Austrian-Italian disagreement regarding ownership of the system. Nearly all Austrian provinces wanted the Soviet-Italian pipeline to pass through their territory, hoping to use the transit for strengthening their own gas supply. ÖMV sought to maximize the national gain from the Italian transit by involving as many Austrian regions as possible, but this irritated the Italians. Seeking to speed up the talks and ensure timely completion of the transit system, ENI pointed to the Hungarian transit option as a viable alternative, thus putting ÖMV under pressure.[4]

Eventually the Austrian route won out, although it took until August 1971 before an agreement-in-principle for the transit could be signed. According to this agreement, ÖMV would build, own, and operate the transit pipeline. A joint Austrian-Italian company, to be set up in Vienna, would have exclusive rights to conclude transit agreements. ÖMV was to own 51 percent of this company and ENI 49 percent. The transport costs were to be carried to 85 percent by the Italians and to 15 percent by the Austrians, whereby the Austrian share corresponded to the volume of gas that was to be reserved for domestic use, that is, in the regions through which the pipeline passed. ÖMV and ENI were both satisfied with the arrangement. However, the delays in finalizing it had consequences for Italy's imports of Soviet natural gas. The two state-owned companies no longer thought it realistic to aim for completion of the transit pipeline by January 1973, as originally intended. May 1974 was set as a new deadline.[5]

The Italian transit agreement marked a watershed for ÖMV's gas business, paving the way for the company's transformation from a production into a transmission enterprise. Although the Germans had opted to receive its Soviet gas directly from Czechoslovakia, the Italian contract promised to turn Austria into an important hub for the transit of red gas to southern Europe. In addition, the transit pipeline was of great importance for supplying Austria itself. Especially the southern Austrian provinces of Styria and Carinthia would benefit from the arrangement.[6]

The transit pipelines through Czechoslovakia started to be laid in early 1971. The project was widely publicized, being regarded by the government in Prague as "a priority interest for our society as a whole." Broadcasts and newspaper articles stressed the enormous economic significance of the system, as well as its relevance to Czechoslovakia's "foreign political links." But actual construction ran into difficulties at an early stage. Aversion among Czech workers, in the aftermath of the traumatic 1968 events, to large-scale cooperation with the Soviet Union contributed to repeated delays and inefficiencies. To speed up construction, the communist government sought to attract additional working cadres, offering 20 percent bonuses and exceptional "separation allowances" for workers while away from home. The project was also complicated by the second Soviet-German gas trade agreement reached in mid-1971. Soviet deputy minister of foreign trade Nikolai Osipov complained that it would have been much better if Ruhrgas immediately had signaled its interest in importing 7 bcm per year, rather than first concluding a 3 bcm and then a 4 bcm contract, since the Soviet Union and Czechoslovakia could then have dimensioned the pipeline accordingly. Now it had already started to be built in a smaller dimension. This was cited as one of the reasons for the higher price Ruhrgas had to pay in its second Soviet deal.[7]

Laying the Austrian transit pipeline, which started at Baumgarten on the Austrian-Czechoslovak border and ended at Tarvisio in northeastern Italy, was particularly challenging because it traversed the Alps, climbing to a record height of 1,470 meters above sea level. Pipelaying started in late 1972. By autumn 1973, nearly all of the pipes had been welded and the pipeline trenches filled in. In winter, then, two Italian-made compressors were installed at Baumgarten. Further compressors were to be added later on, both at Baumgarten and at other to-be-built stations further south. The first two units gave the pipeline a transmission capacity of 3 bcm per year. Having installed them, ÖMV was ready for transiting its first cubic meters of natural gas from the Soviet Union to Italy.[8]

Doubts in Bavaria

Meanwhile in Germany, Ruhrgas and Bayerngas were busy preparing for the arrival of red gas. According to an arrangement agreed upon between the two companies in connection with the Soviet-German negotiations, Ruhrgas was to take full responsibility for importing Soviet gas in practice. Bayerngas, and later on other regional distributors, would buy the gas from Ruhrgas and then transmit it to municipal distributors and major industrial customers.

At the time, Bavaria's gas supply was in a phase of structural change. The popularity of natural gas was growing at a pace that would have been hard to imagine a few years earlier. As of 1970, it was estimated that Bavaria's gas demand would grow to a level of 3–4 bcm in 1975 and 9–10 bcm in 1990. Since the province's own gas reserves—around 6–7 bcm—were about to be depleted, deliveries of such large volumes were almost totally dependent on access to gas from external sources. Soviet gas would have to play the main role in this context. Most of it would be used in southern Bavaria, which accounted for around 75 percent of the total Bavarian gas market. In the smaller north Bavarian gas market, coal and coke gas of north German origin had traditionally been most important. However, plans were being devised for a transition to natural gas, whereby it appeared probable that the western part of northern Bavaria would be supplied from the Netherlands and the eastern part from the Soviet Union.[9]

It was thus of crucial importance for Bavaria that Soviet gas arrived as planned. Bonn's gas experts assured the Bavarians that "there is no reason to see any doubts in the preparedness of the Soviet Union to fully and completely fulfil its trade-contractual agreements." In April 1970, however, Germany's leading business daily *Handelsblatt* published an article in which it was argued that internal Soviet failures to meet planning targets, along with delays in pipeline and compressor station construction, posed a threat to the technical security of deliveries from the East. Not only were the Soviets reported to be far behind schedule, but it was also stated that low welding quality made the Soviet pipelines "untight." A second article published a few months later conveyed the same message.[10]

In Munich the press reports gave rise to a certain nervousness. The Bavarian Ministry of Economy approached Ruhrgas with a query concerning the reports and whether they should be taken seriously. Ruhrgas immediately downplayed the Bavarian fears, pointing to the lack of evidence. The company emphasized that it had far-reaching knowledge of the Soviet natural gas system and great respect for Mingazprom's system-building skills. Hans Geilenkeuser of the management board confidently stated that "the Russians put a very particular value on the security of gas transport," adding that he had personally visited several Soviet gas facilities and that he had the clear impression that the quality of these "corresponded to the general [international] level."[11]

In autumn 1971 the reliability of Soviet gas supply was again questioned by West German media, as a much-cited broadcast by the Bavarian Radio reported that the Soviets faced technical problems in their export system-building efforts. The report was followed up by a series of newspaper articles in which reference was made to American gas experts who had recently visited the Soviet Union. The Americans had painted a gloomy picture of Soviet gas system-building, particularly in Russia's far north, indicating that hardly any planning target or deadline for project completion could be counted on and that the Soviets had seriously underestimated the territorial and climatic conditions of Siberia.[12]

Since it was precisely Siberian gas that the Bavarians, according to the initial plans, were to receive, this was alarming news. Hans Heitzer at the Bavarian ministry of economy spent considerable effort collecting whatever information he could about Soviet pipeline construction. Although much of the information obtained on Siberian and Ukrainian system-building relied on secondary and tertiary sources, it seemed to confirm that the Soviets were in trouble. Heitzer's conclusion was that the export targets could theoretically be met, "but only if we do not take into account the domestic needs of the USSR." This was, as a matter of fact, quite a precise summary of the real Soviet situation. However, since the Bavarians found it hard to imagine that the Soviets would actually sacrifice domestic needs for the sake of exports, they were deeply worried.[13]

The Bavarians once again tried to raise the issue with Ruhrgas, whose top managers, however, repeated that they had "no concerns" about the timely start of red gas deliveries, and that the company did not want to discuss the topic further. Heitzer argued that Ruhrgas must "provide a detailed comment on this issue," adding that Munich considered it necessary to investigate "how preparations can be made which, in case of longer technically conditioned interruptions of deliveries from the Soviet Union, guarantee a minimal supply of the south Bavarian area with Western gas."[14] Klaus Liesen of Ruhrgas insisted that the Bavarians must not take the media reports seriously, since they, according to the company, were based on unreliable information. In addition, Ruhrgas had been informed by the Soviet side that exports to Germany could be realized even if the much-publicized Siberian project was delayed:

> Possible delays with this pipeline project are of no relevance for the start-up of deliveries to Bavaria. We have alerted the Bavarian Radio about this and other falsities in its report, and asked it to carry out more careful researches in future reports on this theme, in order to avoid unnecessary worries among the population.[15]

In Case of Emergency

The media reports, which were widely circulated and discussed, put the Bavarian government and the regional gas industry under pressure to assure the general public that the province's gas supply would remain secure. Although the regional government considered this to be a task mainly for Ruhrgas to handle, the perceived need to counter what seemed to be major uncertainties in red gas deliveries forced Bayerngas and the government in Munich to take appropriate measures. The most important task was to make sure that reserve capacities could be called upon in case of unexpected supply disruptions. This concerned in particular the construction of new pipelines through which Bavaria, in case of problems, would be able to draw on emergency supplies from the Netherlands or northern Germany.

The center-piece of the emergency plan was a proposed link between the regional network of Bayerngas and that of its counterpart in neighboring Baden-Württemberg, Gasversorgung Süddeutschland (GVS). GVS was an interesting partner for Bavaria because it had recently become interlinked with the Dutch export system. Starting in 1968, this had allowed GVS to import small volumes of Dutch gas, and if Bayerngas linked up with Baden-Württemberg it would thus be able to access Dutch gas as well. Bavaria intended to use such an interconnection not primarily for regular imports of Dutch (or north German) gas, but as a back-up to be called upon in extraordinary situations. GVS, for its part, interpreted the proposed connection as an excellent way to prepare for possible future imports of Soviet natural gas to Baden-Württemberg via Bavaria.

The "only" problem was finance. Connecting Ulm with Augsburg, a distance of 72 km, the link was expected to cost 22.5 million D-marks. In addition, a gas treatment facility would have to be built at a cost of 13.5 million D-marks. The latter was necessary because the heat value of Soviet and Bavarian gas (high-calorific gas or "H-gas") differed markedly from the heat value of Dutch and north German gas (low-calorific gas or "L-gas"), a circumstance that made the two gas types incompatible unless one of them was transformed to the heat value of the other. This was an expensive procedure requiring costly investments.[16]

In its 1969 negotiations with the Soviets, Ruhrgas had stated that all Soviet gas would have to be transformed to the heat value of domestic and Dutch gas, which at the time was the predominant gas type in Germany. For this purpose a conditioning facility was planned at the Czech-German border-crossing. By 1971, however, it had become clear that H-gas rather than L-gas would grow dominant in the Federal Republic. This was because a second large contract had been signed with the Soviets, while at the same time the prospects for large-scale supplies of North Sea gas had increased, the calorific value of which was roughly similar to that of Soviet gas. Against this background, Ruhrgas grew hesitant. By late 1970, the gas division at the Federal Ministry of Economy in Bonn reported that the issue of whether or not Soviet gas would undergo transformation was "still not clarified."[17] Eventually, Ruhrgas opted to relinquish its plans for a conditioning facility at the border. Since the need for calorific transformation had been an important argument in the tough price negotiations with the Soviet side, it is also possible that Ruhrgas in reality never intended to build the plant, but merely used the alleged need for it for tactical purposes in the negotiations.

For Bayerngas it had always appeared favorable not to transform Soviet gas to the heat value of Dutch and north German gas, for the simple reason that Bavaria's own gas deposits were of roughly the same calorific value as Soviet gas. However, when the security connection to Baden-Württemberg—which had become part of the L-gas system—started to be discussed, the problem of incompatible gas qualities became a major issue. Bayerngas was clearly unable to finance, on its own, a conditioning facility of the kind that was needed for transforming L-gas into H-gas. As Bayerngas managers Presuhn and Kolb pointed out, both the pipeline and the conditioning facility would be uneconomic, since it was uncertain as to whether they would ever be used for more than emergency deliveries. Yet both were needed to guarantee Bavarian security of supply in case of long-term disruptions in deliveries from the East.[18]

Another project that the Bavarians hoped would increase security of supply was a proposed new pipeline in northern Bavaria. It would stretch from Würzburg to Nürnberg, a distance of 110 km, and like the Baden-Württemberg link it would function as an interconnecting line between the Bavarian gas system and that of northwestern Germany. This line was originally proposed by Ruhrgas, which, however, did not think about it primarily as a guarantor of Bavarian supply security. Rather, Ruhrgas planned to use the link for large-scale flows of gas in the opposite direction—from Bavaria to northwestern

Germany. More precisely, this gas would be of Soviet origin, the aim being to enable its transmission to markets beyond Bavaria. Since such deliveries would not commence in earnest before the late 1970s, however, Ruhrgas was not in a hurry to build the line. Bayerngas, for which the pipeline filled a different function, pushed Ruhrgas to speed up construction. Ruhrgas acknowledged that the technical possibilities for emergency supplies to Bavaria would be "considerably improved" once the pipeline was completed and signaled to Bayerngas that the project was progressing. Yet the Bavarians felt that Ruhrgas did not prioritize the project. In this situation the regional government in Munich turned to Bonn with a plea for "financial participation by the federal government," without which the Bavarians feared that the 50 million D-mark project would "not be built in the near future."[19]

The gas experts at the Federal Ministry of Economy were initially reluctant to any direct involvement in either of the projects. It referred to established liberal traditions in the German gas industry, whose laudable development was "mainly the result of entrepreneurial initiative and a readiness to assume risk," rather than of state contributions. It was noted that the German gas industry had survived very well so far without federal subsidies. The advisors acknowledged that the necessary investments were substantial in the Bavarian case, but emphasized that both Bayerngas and the other regional gas companies had "rich mothers and gas suppliers." In other words, the experts expected Ruhrgas and Bavaria's regional government to finance the interconnecting projects.[20]

Not everyone in Bonn, however, subscribed to this reasoning, and the gas experts were eventually forced to give in to higher-level political considerations. In August 1971, following negotiations between Bonn and Munich, the decision was taken to support the two emergency projects through a public subsidy amounting to no less than 50 percent of the total project costs. Two-thirds of this amount were to come from Bonn and one-third from Munich. The decision was motivated partly by energy policy considerations, particularly with regard to fuel diversification targets, and partly by regional policy concerns. Support to the Bavarian gas industry was seen as a way of strengthening the competitiveness of the Bavarian economy, which was still lagging behind northern Germany. Access to priceworthy natural gas, it was argued, would help attract industrial investment and reduce unemployment.[21]

Meanwhile Bayerngas sought to strengthen Bavarian supply security by diversifying regional supply in terms of geographical origin. In March 1972, the company joined forces with a consortium of European gas companies—consisting of Gaz de France, Belgium's Distrigaz, Saarland's distributor Saar-Ferngas, und GVS—in an attempt to bring about imports of liquefied natural gas (LNG) from Algeria. Toward the end of the year, a precontract was concluded between Algeria and the consortium, which, if realized, promised to give Bavaria access to 2 bcm per year of Saharan gas, starting in 1977–1978. Austria Ferngas and Swissgas joined the consortium in early 1973. Bavaria would receive Algerian gas from a new LNG harbor, to be built at Monfalcone near the Italian-Yugoslavian border, and from there by pipeline through Italy and Austria.[22]

When the Bavarian government in December 1972 was called on to report to the regional parliament on its preparations for receiving Soviet gas, it was largely satisfied with what it had achieved. Minister of Economy Anton

Jaumann, who in 1970 had succeeded Otto Schedl, pointed to the successful measures taken to integrate Bavaria's gas supply into the overall German and European system. The creation of links with Soviet, Dutch, and Algerian supplies, as foreseen under current agreements, was seen to decisively strengthen Bavaria's supply security. It was seen "improbable that foreign policy crises or technical disturbances would coincide so unfortunately that several supply sources, at the same time, would no longer be available for a longer period of time."[23]

The remaining preparations for receiving Soviet gas became a matter for the managers and experts of Bayerngas and Ruhrgas to handle. This concerned in particular the completion of a receiving station at Waidhaus on the Czech-German border, where the gas, having passed through Czechoslovakia, was to be measured and dried. A compressor station was also erected, equipped with three large 11 MW units. From Waidhaus, a large-diameter pipeline was built to the town of Weiden, some 40 km to the west. There, the import pipeline was split into two, one branch heading for Nuremberg and the other for Munich.[24]

On the Verge of Breakdown

On October 1, 1973, the historical day had come. Soviet gas minister Sabit Orudzhev, who had spent his last few weeks in constant crisis meetings, arrived at Waidhaus to take part in the formal inauguration of the Soviet-German natural gas trade. The opening ceremony, of which Ruhrgas was in charge, was also attended by Federal German minister of economy Hans Friderichs, who together with Orudzhev pushed the "red button." Although the Soviets had failed to complete the Shebelinka-Uzhgorod export pipeline on time, the gas could start flowing thanks to an arrangement by which the Ukrainian section of the already existing "Bratstvo" pipeline, built for the purpose of exports to Czechoslovakia, was temporarily used for exports to Germany as well. The first cubic meters of Soviet gas thus flowed into Bavaria without disturbances. Since there was hardly any spare capacity on the Bratstvo, however, the arrangement meant that deliveries to the Soviet Union's communist customer countries had to be reduced. Since the bottleneck was in Ukraine, it did not help that Czechoslovakia's powerful transit system was already in place.

In the evening, the Soviet representatives were invited to a reception at Nuremburg's famous Kaiserburg. Bavarian minister Jaumann in a speech proudly noted that "two peoples dared the step to integrate, on the long-term, their national economies with each other in a sustainable way." Friderichs spoke of "a further important cornerstone in the relations between our countries." The Soviet-German gas trade allegedly showed that "the politics of détente and normalization" was the way forward. Orudzhev spoke of a memorable day. Soviet gas had crossed the border "without passport and visa" and from now on it "burnt like a torch" at Waidhaus. This torch was of "great symbolic significance," providing light and heat as a sign of the good relations between the German and Soviet peoples. Ruhrgas president Herbert Schelberger, who had been the key negotiator on the German side back in 1969, pointed at the new German-Soviet cooperation in natural gas as a "trend-setting model for new forms of economic cooperation." Jaumann further reminded the guests of Czechoslovakia's important role in the project. Josef Odvarka, who had been in charge of constructing the Czech transit pipeline, did not deliver any speech. Instead, the participants at the ceremony were cordially invited to the screening of a film that documented the hard and heroic work carried out on Czechoslovak soil to enable the transit of red gas to the West.[25]

But not everyone had reason to celebrate. In Ukraine, the population was already preparing for a cold and chaotic winter. By September 1973, it had become clear that the looming export commitments, which had priority over domestic gas supply, would make it impossible for Mingazprom to cover the gas needs of many municipal institutions and electric power stations. Given scarce gas resources, Gosplan issued a new regulation according to which the municipal and electricity sectors would receive only one-third of normal supplies, and in many cases even less. Supplies to thermal plants that heated dwellings in Kiev, Lvov, and other cities had to be cancelled altogether.[26]

The Ukrainian Council of Ministers contemplated covering the municipal sector's gas needs by sacrificing supplies to some of the republic's industrial enterprises. Steel producers, cement factories, chemical combines, and sugar makers were identified as candidates in this context. However, disrupting supplies to these industrial gas users would be a desperate measure, inevitably causing "disorganization, destruction of technological processes, and large losses in production that cannot be made up for." The Ukrainians, therefore, begged Moscow to reconsider the new, harsh regulation.[27]

Meanwhile Mingazprom and Minneftegazstroi continued to struggle with the new export infrastructure. Temporary usage of the Bratstvo was a short-term solution only. On October 11, 1973, Deputy Minister of Foreign Trade Mikhail Kuzmin turned to Gosplan and the Council of Ministers pointing to the utmost importance of completing the new export pipeline and stressing that the export targets must under all circumstances be met. The situation was not completely hopeless. Thanks to additional workers brought in—mainly probationers and conditionally released prisoners—some progress could be reported. Since the country's export commitments grew steeply from month to month, however, the overall challenge continued to grow. In early November, Gas Minister Orudzhev noted that uncertainty prevailed over the extent to which the ministry would be able to meet near- and mid-term export targets.[28]

Internal Soviet complaints about gas shortages started arriving in Moscow from late October 1973. From Belarus, whose gas users competed directly with Mingazprom's foreign customers for Ukraine's increasingly scarce gas resources, Premier Kosygin received a letter in which Belarusian party secretary Piotr Masherov and premier Kisilev noted that "if in earlier years great difficulties were experienced during the coldest months as a result of non-availability of gas, then currently an extremely difficult situation has come about already now, in October." Consequently, the Belarusians were forced to make use of reserve fuels that were normally used only under the most extreme winter conditions. Frustrated complaints also came from Minister of Power and Electrification Piotr Neporozhnii, who noted that Mingazprom had failed to deliver 99.3 mcm of natural gas to a number of electric power stations during October 1973. During the second half of October, total gas supply to the ministry's gas-fueled power stations had been reduced from 169 to 135–145 mcm per day. To compensate for the shortage, the power stations had been forced to burn large amounts of valuable reserve fuels in the form of oil and coal.[29]

The Ministry of Chemical Industry also complained, particularly concerning the difficult situation at Rovno's large chemical combine in northwestern Ukraine and the important fertilizer plant at Jonava in Lithuania—both of which were direct competitors with Bavaria, Austria, Czechoslovakia, and East Germany for scarce Ukrainian gas. In early November, the pressure in the pipeline to Jonava decreased from 12 to 6 atmospheres, jeopardizing ammonia production. Similar emergencies troubled the Ministry of Ferrous Metallurgy, whose Ukrainian enterprises received their gas at rates far below what they had been promised.[30]

On December 6, 1973, Ukrainian party secretary Alexei Titarenko informed Gossnab chairman Veniamin Dymshits that the situation in Ukraine, despite the measures taken by Gosplan and Gossnab in October, had become extremely strained.[31] Leonid Brezhnev, for his part, began to receive desperate letters from inhabitants in the regions most severely hit by the looming crisis. One of the most troubled towns was Galicia's historical oil capital, Drogobych, which was now located on the export pipeline route to Western Europe and thus had to compete for its gas with Czechoslovakia, Austria, and the two Germanies:

> We, inhabitants of Drogobych, Lvov region, turn to you with a big request for cooperation concerning the increase of the upper limits for gas. Since four years already we endure a disastrous situation during the autumn-winter period. At this time of the year the amount of gas delivered is insufficient for supplying dwelling houses, children institutions, medical and administrative facilities. Houses are very cold, and since apartments are not designed to be heated with firewood or coal it is impossible to cook. As a result of these conditions grown-ups, not to mention children, often fall ill. Under such difficult conditions we have to live and work. We have repeatedly turned to the municipal government and the municipal party organization, though without result. The municipal powers answer us that they are not in the state to supply the town with gas, due to the low limit. It is impossible to continue living like this.[32]

At about the same time first reports came that Mingazprom had been unable to meet its export obligations to West Germany. The issue was discussed in detail at the December plenum of the party's Central Committee. Mingazprom and Minneftegazstroi were both severely criticized for the long delays with "building gas transport capacities in the Ukraine and the non-delivery of Soviet gas to the Federal Republic of Germany." The ministries accepted the criticism but tried to calm the party leadership down by assuring them that progress was on its way. On December 12, Minneftegazstroi could report that the most critical sections of the Shebelinka-Uzhgorod line had been completed. The first compressor station was also ready to be taken into operation, whereas five other stations were "almost" ready. Resources were now channelled to completing the last few linear sections of the export system, in particular the difficult Carpathian part. This was the last bottleneck that needed to be removed for increased gas exports to Germany and, from spring 1974, to Italy to be physically possible: 38 km of the Carpathian segment still remained to be laid. In the meantime, exports to Western Europe continued to hinge on the "Bratstvo."[33]

The lack of physical transmission capacity along the export route should have been good news for the Ukrainians, since it implied that some of the gas destined for export could not leave the republic. Indeed, gas shortages in Soviet Ukraine during winter 1973/1974 would most probably have been more severe, had the new export pipeline been completed on schedule. Even without the new lines, however, exports were large enough to cause far-reaching damage in the westernmost Soviet regions. In January 1974, facing extremely cold weather, Ukrainian premier Lyashko turned to Kosygin in Moscow with a desperate plea for more gas, stating that the situation in the republic was extraordinarily problematic. Most severely hit were the cities of Kharkov, Kiev, and Lvov with surroundings, all of which competed with export customers for scarce Ukrainian gas. Lyashko reported that

> technological processes in metallurgical and chemical plants have been destroyed, factories have stopped working, and heating plants in residential areas are on the verge of breakdown. The situation is aggravated by the fact that many enterprises lack reserve fuel, making it impossible, on particularly cold days, to transfer gas to municipal needs.[34]

Dymshits at Gossnab was frustrated by the chaotic situation in Ukraine, over which the government and the party seemed to have lost all control. Eastern Ukrainian gas destined for export failed to reach the western border regions of the country, because gas users along the trans-Ukrainian route consumed more gas than they were entitled to according to Gosplan's and Gossnab's strict regulations. Dymshits ordered Lyashko to take "immediate measures for stricter observation of the established limits on gas use, and for the provision of gas exports."[35]

By mid-February 1974, the situation had still not improved. Deliveries to industries and residential areas in Ukraine continued to be severely undercut, and the temperature in houses, schools, and kindergartens repeatedly fell "below the permissible limit." The attempts to establish "strict control" over gas consumption had failed. Gas Minister Orudzhev acknowledged "serious mistakes," but promised to do his utmost to improve the situation in due time for the next winter season.[36]

Perceived Success

In Bavaria, gas users were happily unaware of the deep Soviet winter crisis. Ruhrgas and Bayerngas noted a few disturbances in their operations, but the companies were well prepared for a period with start-up problems. The Austrian experience had made clear to the Germans that the Soviets might face problems in the early export phase, but similar to ÖMV, Bayerngas was able to compensate for these problems by temporarily expanding local gas production. The local gas fields in southern Bavaria (with remaining reserves estimated at 6.4 bcm as of January 1973) effectively functioned as a buffer. Moreover, the emergency link between Baden-Württemberg and Bavaria had been completed and taken into operation as planned. Ruhrgas and Bayerngas were thus able to show a certain patience with the initial Soviet failures. What they did not know was that the Bavarian imports contributed to a severe gas supply crisis on the other side of the Iron Curtain.[37]

The start-up of exports to Finland was somewhat less troublesome than scaling up the flows through Ukraine. The Soviet-Finnish gas trade was ceremoniously inaugurated as planned on January 10, 1974. Finland's president Urho Kekkonen and prime minister Kalevi Sorsa attended the ceremony, as did several high-level Soviet officials. Finland had not been using natural gas at all prior to the first Soviet shipments and had no access to alternative gas supplies. The Finns were, therefore, somewhat nervous as the pipeline was taken into operation. During the first year of deliveries, however, only 0.5 bcm of Soviet gas was to be imported, and all of it was to be consumed by energy-intensive industries that had prepared for the worst by bunkering large volumes of reserve fuels, mainly fuel oil. The Finns wanted to make sure that the import arrangement was reliable before proceeding to feed Soviet gas into municipal heating systems. From a Soviet point of view, the only problem with deliveries to Finland was that the Siberian gas that had been earmarked for the Finns was not yet available. As a result, Finland's gas-consuming industries competed with gas users in Leningrad and the Baltics for scarce Komi and Caucasian gas.[38]

As for the main, Ukrainian export route, a positive turning point came in early 1974 when the Soviets finally were able to take the last few sections of the troubled Shebelinka-Uzhgorod export pipeline into operation. Since the Soviet Union's exports along the Ukrainian route were to nearly double from the last quarter of 1973 to the first quarter of 1974, further delays with taking the line into operation would have caused far-reaching supply disruptions to Central and Western Europe. This was now avoided. By April, winter loosening its grip on Ukraine, shipments to Germany reached the planned levels and the earlier delivery failures could even be compensated for by increased summer deliveries. In this way the Soviets, statistically speaking, were able to fully meet their export obligations during the first contractual year. Both the Soviets and the Germans were relieved. Yet since exports were scheduled for a further steep increase during 1974 and 1975, the responsible Soviet agencies were given no time to rest.[39]

The next task was to ensure a timely start-up of deliveries to Italy. Luckily, these were to start in spring rather than in fall, when the stress on the Soviet transmission system was much more intense. On May 20, 1974, the Austrian-Italian transit pipeline was ceremoniously inaugurated at Baumgarten. ÖMV's chairman Ludwig Bauer, Soviet deputy minister of foreign trade Nikolai Osipov, ENI's president Raffaele Girotti, and Austrian ministers Erwin Lanc and Josef Staribacher all attended the ceremony. Osipov and the other Soviet representatives gave a nervous impression, but to everyone's relief the system started working without disturbances. Back in Moscow, however, the Soviets received a new crisis report from Ukrainian Premier Lyashko, who noted that the Ukrainian gas system was in a mess and that it continued to be plagued by serious gas shortages and construction delays. As of June 1974, the prospects for timely completion of additional capacities looked anything but rosy. Of the 527 km of pipelines scheduled for completion during the year as a whole, only 122 km had yet been built. Similarly, only 3 out of 13 compressor stations had been completed. Lyashko pushed Kosygin to urge all involved ministries to take their responsibilities seriously. While the Ukrainians were certainly most worried about the supply situation in their own republic, Lyashko emphasized that improvement was clearly in the Kremlin's interest, since the complex intertwinement between internal and external supply schemes meant that any delay was bound to have a negative impact on gas exports.[40]

Once again, extra work forces had to be brought in to help the responsible organizations complete the key objects. In a huge effort, thousands of workers were called on to complete pipelines and compressor stations in due time for winter. Out of the logistical chaos of the campaign, some progress could eventually be discerned. Kilometer by kilometer, a functional system did take form. Eventually, the year 1974 became a success, as all pipelines and compressor stations serving the export system could eventually be taken into operation as planned. Relieved, Minister Orudzhev in early 1975 reported to the Central Committee that all contractual obligations vis-à-vis the Soviet Union's foreign customers had been met.[41]

Given the enormous problems faced just a few months earlier, it was all but a miraculous achievement. Western observers, unaware of the extreme measures that had been necessary to make it happen, took the Soviet feat as a confirmation of Moscow's ability—and willingness—to live up to its export commitments. The overall impression was that the most critical phase in realizing the vision of an East-West gas trade had stood up to the test. This conclusion would prove to be of immense importance for the future. From now on, what was widely perceived as a positive experience could be rhetorically used as a powerful argument against opponents to Western Europe's integration with the East.

In reality, the emerging East-West gas system was a shaky construct, built in a haste and based on a capricious blend of inferior Soviet methods and technologies and Western pipes and equipment. Keeping the new export system operational became as challenging a task as constructing it in the first place. Emergency events were reported on a more or less continuous basis throughout the route both in the Soviet Union and in Czechoslovakia, and it did not take long before the impact was felt in Western Europe. In October 1974, for example, seasonal floods in eastern Czechoslovakia caused a collapse of the supports on which the new 1,220 mm export pipeline rested. As a result, the gas flow to Austria, Italy, and the two Germanies was completely interrupted for six days. The Soviets rushed in to help the Czechs repair the line, but the work was carried out in a haphazard way and the line continued to jeopardize exports. In February 1975, the pipeline exploded at exactly the same place and once again the gas stopped flowing. The Bratstvo pipeline, which went in parallel with the new export line but had been built in a lesser hurry, was not affected by the accidents.[42]

Meanwhile an increasing number of German, Finnish, and Italian gas users were connected to the Soviet export system. In Bavaria, the first municipalities to receive Soviet gas included Neustadt, Ingolstadt, and Munich. Another early user was Gebersdorf's electric power plant in Nuremberg's western outskirts. In summer 1974, when gas demand was low, Soviet natural gas also started to be fed into underground storage facilities near Munich and Nuremberg. During winter 1974/1975, the transition process continued. Regensburg, which until then had relied on refinery gas, and Landshut, which had earlier been supplied from Bavaria's own gas fields, belonged to the localities that from now on would be dependent on deliveries from the East.[43]

Soviet gas also started to be delivered to northern Bavaria. This was a troublesome process, since the area had undergone transition from coke gas to Dutch natural gas only two years earlier. Since Dutch and Soviet gas were not interchangeable, customers who had just purchased and installed new gas burners or updated their old ones to accommodate for Dutch natural gas were forced to change equipment once again. Consumers were upset. The problem was taken up in the Bavarian parliament by the Social Democrats, who were outraged about this seeming lack of clear long-term strategy from the side of Ruhrgas and the Bavarian government. Minister of Economy Anton Jaumann defended himself by arguing that up to the conclusion of the second Soviet-German gas deal in 1971 it had still appeared probable that L-gas would remain the dominant natural gas standard in Germany. In the meantime, however, H-gas from the Soviet Union as well as from the North Sea had become available in such large quantities that a transition to H-gas in northern Bavaria had become "unavoidable."[44]

The Würzburg-Nürnberg pipeline, which Bayerngas regarded as a key security link, was completed in December 1974. In addition to serving emergency purposes, it made it possible for customers in the Würzburg area to receive Soviet natural gas. By 1976, all of northern Bavaria had switched to Soviet gas. The Nürnberg-Würzburg line was further extended in the direction of Fulda in Hessen, enabling Soviet gas to penetrate further north. A new compressor station was also built near Würzburg to boost supplies. The area in Germany supplied with red gas was thus rapidly expanded.

For Bavaria, deliveries from the Soviet Union came about precisely at the right moment, as local Bavarian gas reserves were about to be depleted. Thanks to red gas, the Bavarians were able to scale back local production by half already during the first delivery year, from 1.65 bcm in 1973 to 0.87 bcm in 1974. By 1976, at which time Bavaria had already imported 10 bcm of Soviet gas, its reliance on local sources had been reduced to 18 percent. Nearly all the remainder was supplied by the Soviet Union.[45]

Otto Schedl's original vision of natural gas imports as a powerful tool in strengthening Bavaria's economic competitiveness also seemed to materialize. Natural gas became cheaper in Bavaria than in any other part of Germany—including the regions located next to the huge Dutch gas fields. As for the reliability of Soviet deliveries, Minister Jaumann could report to the Bavarian parliament that the imports had been realized "without noteworthy disturbances." The contracts had been "precisely fulfilled." Although there had been a number of "technical problems and short-term delivery interruptions," these "could be eliminated conjointly without adverse effects."[46]

The cited short-term disruptions and irregularities, which were usually compensated for at a later point in time and thus remained invisible in the annual supply statistics, were certainly unwelcome both for the Soviets and the West Europeans. Precisely because the West Europeans had been so skeptical about the Soviet Union's trustworthiness, however, importers had taken a variety of measures to protect themselves against supply disruptions and other potential problems in the gas trade. As a result, end users were never affected more than marginally by the problems that did occur.

In the case of Bavaria, Soviet deliveries were regarded as secure because of the region's deepening integration with the rest of the German gas system and the growing local availability of underground storage capacity. In 1976, the perceived security was further boosted as H-gas from the North Sea came on-stream. This gas was interchangeable with Soviet gas, and it could thus come to rescue in case of unexpected problems with imports from the East. North Sea and Soviet gas were able to mix following the completion of the Nürnberg-Würzburg pipeline and the extension of this system northward. Bavarian minister of economy Anton Jaumann concluded that the "last gap between the natural gas supply systems in Eastern and Western Europe" had thus been closed.[47]

To sum up, the debate before and after the arrival of first supplies of red gas to West Germany, Italy, and Finland in 1973–1974 showed that there was considerable disagreement within the West about the Soviet Union's reliability as a gas exporter. At focus was now the Soviet Union's technical and organizational abilities, rather than its political intentions. The disagreement here stemmed from the fact that the actual characteristics and problems of the Soviet gas industry were largely unknown to most West European actors—including those who were part of the East-West system-building coalition. The perceived uncertainty prompted West European system-builders to speed up the implementation of various technical measures designed to strengthen the intra-Western gas grid and lower Europe's vulnerability to unexpected supply disruptions. The Soviets, for their part, did everything to show that supplies from the East were secure. After a problematic start-up phase, Mingazprom actually managed to deliver the annual volumes that the West Europeans had been promised in the contracts signed a few years earlier. But the price paid for this apparent success was high. Apart from alarming gas shortages in the Soviet Union itself following Moscow's decision to prioritize exports over domestic supply, the export system was built in a hurried, chaotic fashion that was bound to make itself reminded in the future through numerous pipeline breaks, explosions, and accidents. It was a highly unstable system from which anything could be expected.

10
Scale Up or Phase Out?

A Turbulent Energy Era

By the early 1970s, Europe was seen to have "fallen in love with natural gas."[1] Gas use grew at an unprecedented pace, unmatched by any other energy source. Despite the complexity and high capital costs of the pipeline infrastructure, and the perceived uncertainties regarding imports from far away, exporters and importers rushed to expand the emerging transnational system and connect new users to it. "Blue gold," as natural gas was nicknamed in Russian, was identified as an alternative to oil, the supply of which was no longer as trouble-free as it had once been, and as a way of diversifying overall energy supply. Moreover, the excellent environmental properties of natural gas were increasingly pointed at. In 1972 the issue of "acid rain" was brought up on the agenda at the United Nations' Stockholm Environmental Conference, and in some countries restrictions were put in place to restrict sulfur contents of fuels. Since the combustion of natural gas hardly produced any sulfur dioxide, but only water and harmless—as it was believed at the time—carbon dioxide, it profited markedly from this trend.

Yet it was not easy to predict what long-term role the Soviet Union would attain on Europe's increasingly dynamic gas market. As of 1970, at which time only three countries—Austria, Italy, and Germany—had signed Soviet gas contracts, many analysts still believed that Siberia's blue gold would merely be used as a "supplementary supply," and that "Soviet gas definitely will not become a major factor in the overall supply picture."[2] The follow-up contracts signed in 1971–1972, along with declarations of interest from several other potential importers, challenged this analysis. In fact, there was hardly any European country that did *not* take an interest in the possibility of gas imports from the East.

Then, in 1973–1974, came the first oil price shock, dramatically pulling Europe and the whole world into a new, turbulent energy era. In this situation, depending on perspective, it was possible to view Soviet natural gas both as part of the problem and as part of its solution. On the one hand, it contributed to Europe's troublesome dependence on imported fossil fuels; on the other, it offered a pathway to diversification away from oil. The oil crisis reminded importers of the deeply political nature of the international fuel trade, and of the potentialiaty of energy exports being abused for political purposes. From this angle, the dismay expressed in connection with the Arab oil embargo resonated well with the voices that for political reasons had opposed the first Soviet gas deals. If, on the other hand, diversification away from oil was the overarching goal, then the emerging East-West gas trade seemed to be a step in precisely the right direction.

Most actors and analysts tended to regard red gas as part of the solution. The thesis that red gas constituted a political weapon was largely rejected. Instead, actors emphasized the political *opportunities* linked to gas from beyond the Iron Curtain, interpreting it as an integral component of East-West détente. As for the risks, there seemed to be good reason to interpret these as low. One reason was that natural gas was still a new fuel that did not yet play any major role in Europe's energy system. At the aggregate EEC level, the share of natural gas in primary energy supply amounted to a mere 8 percent in 1971. Another reason was that, precisely because the West Europeans did not fully trust the Soviets with regard to their willingness and ability to live up to their contractual obligations, gas companies had taken far-reaching precautionary measures that would enable them to respond effectively to any short- or long-term supply disturbances. The experiences gained from first imports showed that such measures were certainly needed, though not so much to counter political moves by the Kremlin as to deal with unintended irregularities and breakdowns of a technical nature. The Soviets excused themselves for these mishaps, and always compensated for them through deliveries at a later point in time. As long as annual export obligations were met, and the importers' crisis tools did their job, Western Europe's gas companies preferered to interpret the emerging transnational system as a success.[3]

Soviet gas was thus welcomed in Europe. A long-term strategic question, however, was how large a role it could possibly be allowed to play. Upon conclusion of Ruhrgas' second contract with the Soviet Union in July 1972, the federal German government estimated that by 1980 "the import dependence of the Federal Republic on the USSR would, with 14 percent, not yet be worrisome." However, the question was seen to arise "as to when German gas supply, through further Soviet gas deliveries, gets into an import dependence on the Soviet Union that can no longer be considered sustainable."[4]

Europeans widely agreed that a diversified supply structure was to be encouraged. Imports from the Soviet Union would thus need to be complemented through imports from the Netherlands, Algeria, Norway, and possibly a few other sources. Bonn's advisors thought that imports of Soviet gas might be allowed to increase more or less indefinitely as long as the country retained a diversified import structure in which different exporters balanced each other. Basically there was no need to define any *absolute* upper limit for imports from the East, only a *relative* one. In other words, "to the extent that the German gas industry can get hold of additional natural gas volumes from other areas, imports from the Soviet Union could also be increased without further ado." Some analysts, moreover, argued that it was meaningless to define any maximum share of red gas in a certain regional or national context. What mattered, they argued, was the availability of tools for countering potential disruptions.[5]

For the Germans, the main challenge was to respond to rapidly growing gas demand at a time when the country's own gas fields were being depleted. In 1971, domestic production was still able to cover 70 percent of overall West German gas demand, whereas only 30 percent needed to be imported. Already in 1974/1975, however, domestic supplies and imports were roughly equal, and by 1980 imports were expected to reach 70 percent of total supply. This rapid and fundamental shift contrasted sharply with the forecasts made by Shell, Esso, and several domestic gas companies in connection with the 1969 Soviet-German negotiations. At that time, an important argument against Soviet gas had been that domestic gas would be able to cover Germany's needs "up to the end of the century."[6]

Italy and France were in a similar position. Not long ago, they had been Western Europe's leading gas producers, and France had even been identified as a potential gas exporter. The dynamic development of the domestic gas industry in both countries had started out as an attempt to strengthen energy autarky. The very popularity of the new national fuel, however, generated a dynamic expansion with which domestic producers proved unable to keep pace. System-builders then turned to supplies from abroad. By the time of the first oil crisis, French and Italian gas imports already exceeded domestic production. Like Germany, however, the two countries were able to diversify their imports by signing long-term contracts with several foreign suppliers. Apart from the Soviet Union and the Netherlands, these included Algeria and Libya.

Austria and Finland found it more difficult to attain a diversified supply; both imported Soviet gas only. The dependence on red gas was particularly noteworthy in the case of Finland, which lacked domestic gas deposits that might be used as a national buffer in case of problems. For this reason, the Finns opted not to introduce natural gas for space heating, but only for industrial purposes and electricity generation, where reserve fuels could more easily be mobilized.

Austria, for its part, had made some attempts to import Algerian LNG, which, according to the most advanced plans, was to be shipped to Yugoslavian or Italian harbors, and as of 1973/1974 it appeared probable that at least western Austria would succeed in accessing Algerian gas. In the meantime, Austria relied on its remaining domestic deposits and a well-developed gas storage capacity for countering vulnerability. In addition to the already existing storage facility at Matzen, which had commenced operation in 1969, ÖMV in 1973 started up a new, larger facility at nearby Tallesbrunn. Together, the two facilities ensured the Austrians an emergency stream of 2.2 mcm per day in case of supply disruptions, corresponding to a sizeable share of total demand. ÖMV could also compensate for import disruptions by temporarily increasing domestic production; as we have seen, this strategy was used successfully during the start-up phase of imports from the Soviet Union.[7]

Clearly, the security of Soviet gas imports had to be put in relation to the security of imports from elsewhere. In fact, the oil crisis generated uncertainties regarding both Dutch and Algerian gas. The Netherlands, along with the United States, had been singled out as a main target by the Arabs in their 1973 oil embargo. Controversially, Dutch prime minister Joop den Uyl sought to force his country's West European neighbors to resell Arab oil to the Netherlands by threatening to disrupt gas exports to Germany, France, and Belgium. At the same time, an internal debate was initiated in which a growing number of Dutch actors suggested that the country's vast natural gas riches be reserved for domestic needs, and that no further export contracts be signed.[8]

Algeria, for its part, early on earned a reputation as a partner from whom anything could be expected. Immediately after the outbreak of the Arab-Israeli war in October 1973, Algerian LNG exports to the United States were disrupted. The state-owned Algerian oil and gas company Sonatrach initially referred to technical problems at the LNG plant, and exports were soon resumed. In December, however, Algerian energy minister Abdessalam told American journalists that "future shipments of LNG to the US on a continuous basis may depend on the satisfactory settlement of the Arab-Israeli conflict." This was hardly an encouraging message for those European gas companies that held high hopes in the possibility of large-scale imports of Algerian gas for diversification and security purposes. In addition, the turbulence on the world oil market made the Algerians rethink their pricing strategy. An immediate result was that a large preliminary export contract that had been concluded in 1972 with a European consortium of gas companies was declared invalid. At the time, far-reaching preparations to receive Saharan gas had already been initiated in the prospective importing nations. Several large investments linked to the deal, including a first German LNG terminal (at Wilhelmshaven), had been designed specifically for handling incoming Algerian gas. Their fate now became highly uncertain.[9]

A long-term gas supply strategy based on imports from the Soviet Union was thus not necessarily the most risky one. Yet it would have been naive to expect the emerging East-West gas trade to remain unaffected by the turbulence on the oil market. The most immediate effect of the oil price shock was to make Soviet gas more attractive from an economic point of view. In the next phase, however, Moscow signaled that it would not accept this state of affairs for long. Approaching their Western partners, the Soviets offered additional gas supplies, but only under the condition that previous deals were renegotiated. In this way both Germany and Austria in 1974 contracted additional Soviet supplies at prices several times higher than in the previous deals. This was seen to reflect the new harsh realities.[10]

France was also approached with a request for renegotiation of its Soviet contract. But the French case was different since imports, scheduled for start-up in 1976, had not yet commenced. Given the turbulence on the global energy market, upward adjustment of the gas price agreed upon in 1972 was reported to have been "a constant bone of contention" between the two sides ever since the contract was originally signed, and at one point it even seemed that the deal would be annulled. The issue was finally resolved in December 1974, when Soviet leader Leonid Brezhnev and French president Giscard d'Estaing ceremoniously signed a new, updated contract.[11]

In the case of Germany, Austria, and France, as well as Italy, which had signed the largest single contract with the Soviets so far, the post-1973 trend thus served to reconfirm and further expand a commitment to imports from the East. By contrast, those countries that at the time of the oil crisis had been interested in Soviet natural gas, but had not yet signed any contracts, tended to move further away from an actual import. This was so with regard to Sweden, Switzerland, Belgium, and Spain, as well as the United States. America's "North Star" stakeholders, who since 1970 had elaborated on a major LNG project with the Soviet Union, in June 1974 reported that they had reached agreement on everything with Moscow except the gas price. Accord on this final point, however, was complicated both by the oil price shock and by the appointment of Gerald Ford, who was more skeptical to the project than his predecessor Richard Nixon, to the US presidency.[12]

Looming shortages of natural gas in large parts of the United States, leading to harsh winter curtailments and other emergency measures, served to keep the interest in Soviet gas alive in the period that followed. But little progress was made in the actual negotiations, and in the meantime overall Soviet-US relations started to worsen. In early 1976, following Moscow's much-criticized intervention in Angola, Secretary of State Henry Kissinger declared that although he had initially favored the red gas proposals, he now believed that "political conditions have reached a point where right now would not be the most opportune moment to produce or come forward with projects of large-scale economic cooperation." In the period that followed, the grand visions of the early 1970s gradually faded away. It thus appeared that America would be left without access to Siberia's blue gold.[13]

The countries of Central and Eastern Europe, for their part, headed toward a radically scaled-up import of Soviet natural gas. As of 1973, Poland, Czechoslovakia, and East Germany were already major Soviet gas customers. In August 1974, then, Bulgaria received its first deliveries. When Hungary in 1975 linked up with the Soviet-Czechoslovak "Bratstvo" system through the construction of a 130-km branch line from Uzhgorod across the Soviet-Hungarian border, all Soviet satellites except Romania had become importers.[14]

In June 1974, moreover, the Soviet Union and its Central European neighbors decided to embark on a highly prestigious, multilateral project aimed at exploiting southern Russia's giant Orenburg gas field for export purposes. A 2,750-km pipeline was to be built from Orenburg to Uzhgorod through the joint efforts of workers and engineers from Czechoslovakia, Poland, Hungary, Bulgaria, and East Germany. Each participating country was assigned the task of constructing a section of the pipeline. In return, each country would receive an annual 2.8 bcm of Orenburg gas. Romania, which initially opposed the project, eventually became part of the arrangement as well. The gas was to start flowing in 1978 and reach a plateau level in 1980. The "Soyuz" (Union), as the pipeline was nicknamed, became a showpiece of COMECON integration. Pipes and compressors, however, were brought in from the West. In the Soviet Union, the project became controversial following Mingazprom's insistence that it be designed as a dedicated export system, without any branch lines to users located along the pipeline route. The Ukrainians, in particular, were upset by this arrangement.[15]

Another country that signed up for Soviet gas was Yugoslavia. ÖMV and ENI, the stakeholders in the Trans-Austria Pipeline, in February 1976 agreed with the Yugoslavian oil and gas companies Petrol (Ljubljana) and INA (Zagreb) to transit 1.5 bcm per year of Soviet gas to Slovenia and Croatia. Two socialist nations would thus be connected with each other by way of a capitalist transit country. The gas was to be pumped from Weitendorf in Styria to the Austrian-Yugoslavian border through a new pipeline that was to be built by ÖMV. Shipments were to commence in July 1978. In a second import arrangement, Yugoslavia contracted 1.65 bcm of Soviet gas that was to arrive by way of Hungary. This gas was to be used in Serbia, Vojvodina, and Bosnia-Hercegovina.[16]

Involving Iran

In parallel with the efforts to bring about large-scale imports of natural gas from the Soviet Union, Western Europe also started to take serious interest in imports from Iran. The Shah's vast territory was known to rest on some of the world's largest natural gas deposits, but as of the early 1970s only small volumes were actually produced. The Soviet Union, in accordance with an agreement reached in 1966, was the only country to import Iranian gas. Shipments took place through a 1,112-km pipeline known as IGAT, which entered the red empire at Astara in Azerbaijan. Some of this gas was consumed in the Soviet Union's Caucasian republics, whereas the rest was piped north to southern and central Russia. The system was ceremoniously inaugurated in October 1970 by the Shah himself, who regarded it as "a symbol of the very sober economic cooperation between Iran and the Soviet Union." Since German, British, and French engineering companies played important roles in the project, it was in practice a showpiece of cooperation with Western Europe as well.[17]

The West Europeans eagerly monitored Mingazprom's first experiences of importing Iranian gas. Although the arrangement's stability suffered from the low quality of the Soviet-made compressors that had been installed along the pipeline, deliveries had by 1974 reached the planned plateau level of 10 bcm per year, and the overall impression was that the system worked, albeit with irregularities. It was well-known that the gas price was highly favorable to the Soviets, who paid the Iranians less than half of what they charged their own European customers per cubic meter. This inspired West European gas companies to seek their own access to Iranian gas. Optimism grew in the early 1970s following the discovery of the supergiant Kangan field in southern Iran, whose reserves seemed too vast for Iran itself to absorb. The Iranian government and its national gas company, NIGC, welcomed the idea of gas exports to Western Europe, viewing it as a pathway to diversification of its energy trade away from oil.[18]

But how could Iranian gas be brought safely and economically all the way to Europe? One possibility was to construct a pipeline system from Iran through Turkey to Western Europe. Alternatively, Iranian natural gas could be imported in the form of LNG. The two variants could also be combined through pipeline transport to Iskenderun on Turkey's Mediterranean coast and from there by tanker to one or more European LNG terminals. ENI was the main supporter of a pipeline or combined pipeline-LNG arrangement, eyeing Monfalcone on the Adriatic as the ideal landing point. Belgium's Distrigaz and Sopex, for their part, emerged as supporters of a pure LNG trade. The Belgians joined forces with American companies, who similarly hoped to take part in the scramble for Iranian gas. Japanese firms were also enthusiastic.[19]

In Germany, however, a radically different variant of the envisaged Iranian trade was sketched. Ruhrgas, while taking interest in the other proposals, contemplated not primarily the creation of a brand-new export route, but suggested that the Soviet Union and its already existing pipeline network might be mobilized for the transit of Iranian gas to Europe. The arrangement could either take the form of a true transit, that is, involving physical flows of Iranian gas all the way to Germany and other West European countries, or be arranged as a "switch," implying that the Soviets would import a certain volume of Iranian gas and export a corresponding volume of its own gas to Europe. In either case, such a scheme would not only reduce the investment needs, but also place Ruhrgas in a central position vis-à-vis other West European gas companies. Ruhrgas anticipated that an LNG-based trade would turn Italy, France, and/or Belgium into central hubs for the Iranian imports, whereas Germany would be circumvented. The company also worried that southern Germany's regional gas companies—Bayerngas, GVS, and Saar-Ferngas—might use shipments of Iranian gas from the south to strengthen their independence vis-à-vis Ruhrgas and reduce its dominance on the German gas market. Ruhrgas sought to persuade both ENI and the regional companies to take part in a Ruhrgas-led consortium to further develop the Iranian-Soviet transit. Gaz de France and ÖMV were also invited to join the project.[20]

From around 1973 the attempts to get hold of Iranian gas intensified. The competition between alternative—though not necessarily mutually excluding—transit routes seemed to amount to a fierce race. Ruhrgas now received strong political backing for its Iranian visions, as Bonn spotted advantages in the project for Germany as a whole. The government agreed with Ruhrgas that a Soviet transit would ensure the Germans a dominant position that would most certainly be unattainable in the case of a more southerly import route. Besides, energy advisors Plesser and Lantzke reasoned that a Turkish transit or a pure LNG trade would be more expensive and complicated to bring about than a transit through the Soviet Union. Ruhrgas and several other large West European gas companies already had far-reaching experience of Soviet natural gas, whereas a deal involving Turkey and/or longhaul LNG shipments would be an experiment and a venture into the unknown. The only problem with Ruhrgas' proposal was that it would increase Germany's vulnerability to potential disruptions in the gas flows from the East; if the Iranian project became reality, Germany would by 1985 have around 20 percent of its gas shipped through the Soviet pipeline system. Bonn judged that such a high figure lay at the "limits of the defensible." Ruhrgas argued that large-scale LNG projects were more risky than a Soviet tranit of Iranian gas.[21]

The Iranians, seeking to encourage competition among Western importers, showed themselves highly interested in several of the proposed projects. In January 1974 the Shah stated that he imagined a double-routed export of Iranian gas to Europe: up to 40 bcm per year would preferably be shipped through Turkey and 13–21 bcm through the Soviet Union. ENI also thought the projects complementary, opting to join the Ruhrgas-led consortium with a share of 20 percent while continuing to focus on a Turkish transit as the main import route. GdF and ÖMV joined the Ruhrgas-led consortium with shares of 20 percent and 10 percent, respectively. Ruhrgas took the remaining 50 percent, whereby it was also agreed that the Germans would be responsible for actual negotiations with Iran. Apart from Ruhrgas, ÖMV was particularly enthusiastic. In addition to its own interest in adding Iranian gas to its supply portfolio, the Austrians hoped that Gaz de France would prefer to receive its share of Iranian gas by way of transit through Austria. This would ensure a sustained "key position for Austria in Europe's natural gas supply."[22]

Of the competing Iranian projects, the Soviet one soon turned out to make the best progress. The Kremlin quickly confirmed its willingness to participate, showing itself particularly interested in a "switch." In late 1973, a feasibility study was presented. After several negotiation rounds, exporter and importers in January 1974 signed a "letter of understanding". According to this first sketch, Ruhrgas, on behalf of the West European consortium, would buy the Iranian gas at Astara and make efforts to come to terms with the transiteer about the further transmission—whether physical or virtual—from Azerbaijan to a West European border-crossing point. Three alternative export volumes were to be considered: 13, 17, or 21 bcm per year. At least half would be delivered to Ruhrgas and the rest to its partners in the customer consortium. The gas price was to "orient itself to the energy price level in the Federal Republic," but NIGC was to receive an "adequate" return on capital for the necessary investments linked to the project.[23]

To facilitate the transit negotiations with the Soviet Union, the Germans decided to include the Iranian gas issue as a a point on the agenda in the high-level talks between Chancellor Brandt's skilled advisor Egon Bahr and Soviet Party leader Leonid Brezhnev that were held in spring 1974. Following these talks, Brezhnev instructed Deputy Minister of Foreign Trade Nikolai Osipov to speed up the negotiations. In early May, then, a first West European-Soviet-Iranian tripartite meeting was organized in Tehran, with the West European interests being represented by Ruhrgas. However, it turned out difficult to agree on a gas price and transit fee, and in particular on the distribution of revenues between producer and transiteer. A problem in this context was that Iran and the Soviet Union were just in the process of renegotiating their bilateral gas trade. Since the start-up of this trade a few years earlier, Iran had become increasingly dissatisfied with the low price paid by the Soviets, comparing it to the much higher profitability of Soviet sales to Western Europe. As of 1971, Moscow sold natural gas to Ruhrgas for $11.30 per 1,000 cubic meters, whereas Iran only received $5.20 for its exports to the Soviet Union.[24]

The Soviets were as always tough negotiators and initially refused to give in to any Iranian demands, a stance that the Iranians considered "illogical and unfriendly." As a result of the Iranian-Soviet conflict, the next round of talks involving the West Europeans, which was to have taken place in Essen in early July 1974, was cancelled. Plesser in Bonn thought this might have decisive consequences, since any delay would favor the other, competing attempts to bring Iranian gas to market. A further problem for Ruhrgas and the federal government was that a number of regional German gas companies—Thyssengas, Salzgitter Ferngas, and the three south German gas companies—were approaching the Iranian government independently. German ambassador Wieck in Tehran, hearing about this from Iranian government sources, strongly recommended the Ministry of Economy to ensure a joint approach from the German side. The issue was considered so important that new German chancellor Helmut Schmidt, who in May 1974 had succeeded Willy Brandt, raised it with the Soviet side on his first visit to Moscow that same summer.[25]

At about the same time, it was reported that a major accident had interrupted the gas flow through the Iranian-Soviet pipeline already in operation. A huge explosion had taken place, caused by "a weak point" in the pipeline network. Exports remained disrupted for six days. The Soviets were happy that the accident had taken place in summer and not in winter, but the suspicion arose that the "accident" was in reality a way for the Iranians to put pressure on the Soviets in the ongoing price talks. A few weeks later, a breakthrough in the Soviet-Iranian talks was actually achieved as a new export price for Iranian gas could be agreed upon. This in turn enabled the tripartite talks with Western Europe to be resumed.[26]

As for the competing projects, NIGC general director Mossadeghi in August 1974 claimed that the "prenegotiations" with a groups of companies interested in a Turkish transit and/or LNG trade had been completed. In reality, however, this variant of the project still had a long way to go, and Mossadeghi's declaration was probably more of a way to put pressure on the proponents of the competing Soviet transit. In the same vein, a much-publicized "letter of intent" was signed between NIGC and a group of companies consisting of El Paso (USA) and Distrigaz and Sopex (Belgium) about the delivery of Iranian LNG to Europe and America. The partners were to create a joint venture with the goal to develop and exploit the celebrated Kangan field, bring its gas by pipeline to a harbor on the Persian Gulf, liquefy it there, and ship it around the Cape of Good Hope. In a first phase 20 bcm per year were to be exported, a volume that would require no less than 34 LNG tankers in continuous operation. It was acknowledged, however, that several more detailed studies were needed before any precise contractual arrangement could be worked out. The project was thus, in effect, at an early and preliminary stage.[27]

The tripartite negotiations made much better progress. Thanks to the new bilateral Soviet-German and Soviet-Austrian contracts signed in autumn 1974, there was broad consensus about what the gas price would have to be at the German border, and the talks could instead concentrate on the price to be charged at the Iranian border and the Soviet transit fee. Following tough negotiations in early 1975, Ruhrgas and NIGC managed to settle the Iranian border price. After another round of "very difficult talks" held in Moscow in March, the Soviets accepted to be satisfied with the difference between the two border prices as a transit fee. Hence the price problem could be regarded as solved.[28]

This paved the way for the signing, on April 10, 1975, of a "precontract" by Ruhrgas (on behalf of the European consortium), NIGC and, on the Soviet side, the State Committee for Foreign Economic Relations and the Ministry of Foreign Trade. In addition, two bilateral agreements were signed between Ruhrgas and NIGC and between Ruhrgas and the Soviet side. Iran was to start exporting gas to Western Europe in January 1981. After a period of three years, exports were to reach a plateau level of 13 bcm. The gas was to be shipped from Kangan in southern Iran through a new, 1,440-km pipeline, for whose construction NIGC would be responsible and which would be laid in parallel with the existing IGAT pipeline. The Soviet Union would not transit Iranian gas in

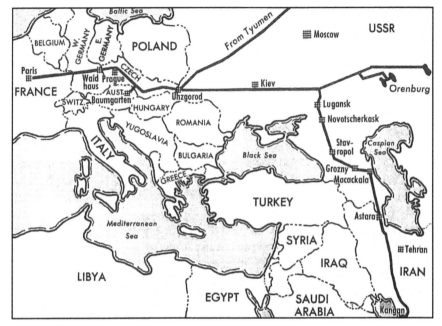

Figure 10.1 Exports of Iranian gas to Western Europe, with transit through the Soviet Union, as envisaged in the 1975 tripartite deal.

Source: Oil and Gas Journal, August 16, 1976, p. 82. Reproduced by permission.

a physical sense to Western Europe, but rather ship an equivalent volume of Siberian, Komi, and, possibly, Orenburg gas. 11 bcm of additional Soviet gas would in this way reach Western Europe, accounted for as Iranian gas; 2 bcm would be accounted for as "compression gas" along the virtual transit route. Importantly, the Soviets were obliged to deliver its gas to Western Europe only to the extent that it actually received the corresponding volumes from Iran.[29]

Ruhrgas tentatively agreed to act as a single buyer, acting on behalf of the international consortium in which Gaz de France, ÖMV, and ENI also participated. Shortly after the precontract's signing, however, the consortium was threatened by collapse as ENI, which showed itself dissatisfied with the conditions achieved by Ruhrgas on its behalf, in May 1975 decided to withdraw from it. The arrangement was saved after France and Austria agreed to take on the import volumes originally intended for Italy. On November 30, 1975, the contract in its final form could eventually be signed.[30]

The Turkish transit and the LNG project, for their part, failed to materialize. For the foreseeable future, the only way for West European gas companies to access Iranian gas would be through the tripartite switch arrangement. However, since the switch did not involve any physical shipments of Iranian gas to Europe, but rather deliveries of an equivalent volume from Soviet fields, the material outcome of Western Europe's efforts to access Iranian gas was in actual practice an increased dependence on Soviet natural gas. The Iranians, the Soviets, and the West Europeans jointly declared that they considered the deal "a perfect example of the fruits of friendship and cooperation among states with different political, economic, and social systems." Yet it remained to be seen how the arrangement would influence Western Europe's energy security.[31]

Doubts in the Kremlin

Critics to the Iranian deal argued that the project would in effect double Europe's vulnerability, as the trade would depend on smooth deliveries both from Iran to the Soviet Union and from the Soviet Union to Western Europe. German minister of economy Hans Friderichs, defending the Federal Republic's central role in the project, countered that neither Iran nor the Soviet Union would have anything to gain from letting the transnational pipeline infrastructure stand idle. Iran, in particular, which was to host—and pay for—the key components of the system, would have to sell large volumes of gas for many a year before the infrastructural investments paid off. The German government had "no doubt that Iran will correctly fulfil its contractual obligations."[32]

In neighboring Austria, it was noted that the Iranian deal would radically expand ÖMV's access to imported natural gas, boosting overall supply by no less than 42 percent. But the deal gave rise to criticism in local media, particularly as it became clear that the attempt to diversify supply through imports from Algeria, arranged as part of the larger European consortium, had collapsed. The failure meant that Austria would remain totally dependent on deliveries from beyond the Iron Curtain. The critics noted that although these deliveries had so far proved reliable, Austria was bound to become more vulnerable to problems in imports from the East as remaining domestic gas reserves, which so far had functioned as a highly effective buffer, were about to be depleted. ÖMV's general director Ludwig Bauer sought to meet this critique by pointing at gas imports as part of a wider energy import regime, referring to the fact that Austria's "entire crude oil import takes place via Trieste." The message was that the country's oil and gas imports were largely independent of each other, with positive effects for overall energy security.[33]

But the Iranian deal also gave rise to concern in the Soviet Union. Though formally a transit agreement, it effectively generated a new Soviet export commitment of roughly the same dimensions as the prestigious "Soyuz" project. To meet its new "transit" obligations, Mingazprom would have to boost exports to Western Europe by no less than 47 percent, adding 11 bcm on an annual basis to earlier bilateral export agreements. To judge from previous experience, it would not be an easy task to scale up exports so massively.

The long-term strategic question for the Soviets was whether it was really advisable to aim for further expansion of gas exports. The country's political leadership and in particular its foreign trade organizations certainly saw major economic opportunities in massive gas sales to the West. As before, they also saw political opportunities, although these remained somewhat diffuse. Not everybody, though, agreed that the opportunities outweighed the costs, risks, and troubles that already had caused Mingazprom and Minneftegazstroi so much pain in their attempts to fulfil their contractual obligations. The extreme difficulties of the early 1970s with regard to construction and operation of the export infrastructure had left deep wounds. Gas Minister Sabit Orudzhev, who, coming from the oil industry, in 1972 had succeeded Alexei Kortunov at this post, had been shocked by the chaotic nature of the gas industry's expansion, the effects of which still lingered on. Built in a haste under chaotic conditions, many pipelines worked inefficiently and were in need of more or less constant repair and maintenance. Pipeline breaks and explosions, along with constant problems of keeping compressor stations in operation, were part of everyday life for Mingazprom and its engineers. Orudzhev was furious about poor worker performance and alleged mistakes from the side of various R&D, design, and planning institutions.[34]

Moreover, the attempts to exploit Siberia's gas riches and transport them over several thousand kilometers to the European part of the country had turned out much more difficult and costly than anticipated, and those in the Soviet elite who in the 1960s had advocated a more conservative Siberian strategy, notably Gosplan Chairman Nikolai Baibakov, saw their judgments confirmed. Following the drive to shift the center of Soviet gas production from Ukraine to Siberia and boost the overall share of natural gas in primary energy supply, the average cost of producing a cubic meter of gas had grown by 120 percent in the five-year period from 1971 to 1975. In addition, enormous gas volumes were lost and wasted at production sites, along pipelines, and through ineffective use.[35]

Against this background, Orudzhev advocated a more moderate expansion for the upcoming decades. He even argued for a "long-term plan to phase out natural gas as a boiler fuel, on the grounds that it was too scarce and expensive to be consumed in nonpremium uses." Gas use should be limited to areas where it was really needed, such as in the chemical and petrochemical industries. Other advocates of a moderate expansion strategy, notably the scientific community and its powerful Academy of Sciences, argued that it would be most economic and rational to reserve Siberian gas for energy-intensive industries in Siberia itself. This would largely eliminate the immense need for pipelines to the European part of the country, while also serving to avoid premature depletion of the giant Siberian fields. From a rational scientific point of view it appeared wasteful to use such a high-quality fuel as natural gas in sectors where coal and nuclear power could do the job, notably in electricity production. The view was shared by a nascent Soviet environmental movement, whose representatives argued that "extremely valuable forests, reindeer pasturage, and trapping areas are being disturbed" by the Siberian pipeline routes.[36]

Yet putting a brake to the rapidly expanding Soviet gas system, with its burgeoning momentum, was easier said than done. Key actors such as the Tyumen branch of the Communist Party, whose earlier secretary Boris Shcherbina had advanced to become head of Minneftegazstroi and whose new secretary G. P. Bogomyakov followed in Shcherbina's footsteps, advocated continued expansion. Shcherbina and Bogomyakov joined forces with the Ministry of Foreign Trade, whose officials looked at the gas industry through different glasses but whose conclusion was the same. The big question, from their point of view, was not how Soviet energy supply could be optimized on the short and long term, but rather how the state would be able to sustain its hard-currency earnings. Soviet oil, which had been the country's most important export article since the 1950s, was no longer as abundant as it had been, and natural gas was looked upon as its logical successor. This was so despite the fact that gas exports did not offer the same degree of flexibility and freedom as oil exports. The country's oil resources were being more rapidly depleted than anticipated; the reserves had been overestimated, and no new supergiant oil field had recently been discovered. By 1977, the problems of sustaining and further expanding oil production had grown insurmountable. The overall growth rate dropped and fears were voiced both at home and abroad that Soviet oil production would peak within 2–3 years. This worrisome development in turn fueled a more general debate about the future of Soviet energy.[37]

During 1976 and 1977, when the country's energy future was most fiercely debated, no further contractual negotiations with prospective Western gas importers were initiated. The country's political leaders appeared to hesitate. By 1978, however, the Kremlin seemed to have made up its mind, judging that natural gas would have to play a crucial role both for the expansion of domestic energy supply and for the future of hard-currency earnings. A main reason was that the prospects for longhaul transmission of Siberian gas had improved. The first gas pipeline from Siberia had at last been taken into operation, proving, albeit with delay, that the efforts to exploit northwestern Tyumen's gas fields had after all not been in vain. The first of the Siberian giants, Medvezhye, already delivered large amounts of gas to the Urals as well as some volumes westward through the Northern Lights system. In 1978, then, supergiant Urengoi was scheduled for start-up. Several additional fields of almost the same size had been discovered, boosting the country's gas reserves to all but mythological levels.[38]

At the same time, thanks to new equipment import possibilities, the problems concerning access to modern compressors had largely been overcome. The successful acquisition of three large Italian compressors as agreed upon in the 1969 contracts had shown that Western export restrictions did not rule out access to advanced foreign compressor technology. After all, the extraordinary expansion of the Soviet gas industry made this country the world's biggest market for both pipes and compressors, and Western firms were eager to conquer it. Restrictions were further relaxed during the 1970s, and as it turned out the Soviets were able to purchase not only Italian, but also American and British compressor parts. General Electric and Rolls-Royce were regarded as the world's top producers, and for the new lines from Siberia and Komi Mingazprom was able to install compressors with equipment from both firms. The result was that the gas industry, after a deep crisis in the years around 1970, started to meet and surpass the planning targets set for it. Natural gas even started to be referred to as the "star performer" of the Soviet economy as a whole.[39]

Envisaging the "Yamal" Pipeline

In early 1978, the Kremlin started to openly signal an interest in concluding new export agreements with West European gas companies. The issue partly continued to be subject to internal debate, but this did not prevent concrete ideas from being developed regarding possible export arrangements. Following up on the principle deployed in the Soyuz project—in which gas from Orenburg in central Russia would be exported to communist Central Europe—it was proposed that a dedicated export pipeline be constructed that would not be used for any other purpose than shipments to the West. This meant that export flows and domestic flows would be separated from each other to a much greater extent than in previous East-West projects. The purpose was probably to convince West European importers that no export volumes would be diverted to domestic needs in case of a general Soviet gas shortage. It could also be interpreted as a way for Moscow to assure regions such as western Ukraine, Belarus, and the Baltics, which in the earlier export projects had seen domestic supplies diverted for the purpose of export, that this would not happen again.

The Soviets proposed that the new export project make use of gas from the large Yamburg field in the Yamal-Nenets Autonomous Region, and the new export infrastructure to be built accordingly started to be referred to as the Yamal pipeline. For such a line to be profitable without combining it with domestic supply, it would need to carry very large export volumes. For this reason the Soviets urged their potential customers in the West to think in terms of as large a new deal as possible. As before, the most important Soviet discussion partner was Ruhrgas. Satisfied as they were with the way in which the Iranian negotiations had been handled, the Soviets made clear that they wanted Ruhrgas to take responsibility, again, for forming a consortium of gas importers from several countries.[40]

Ruhrgas declared that it was basically positive to the new idea and set about probing the issue with other major gas companies in Western Europe. Among its "Iranian" partners, Gaz de France was the company that responded most positively. ÖMV, in contrast, made clear that although it was interested in negotiating a new Soviet contract, it would prefer to arrange it on a bilateral basis with Moscow. ENI, which had dropped out of the Iranian deal in the last minute, did not seem interested at all. On the other hand, Ruhrgas received a positive response from Belgium's Distrigaz and the Netherlands' Gasunie. This was of a certain importance not least since Ruhrgas knew that two other German gas companies, Thyssengas and BEB, were approaching the Belgians and the Dutch independently, in an attempt to form their own international consortium. In the end, although not everyone wished to join the Ruhrgas-led consortium, there clearly seemed to be a market for Yamal gas in Western Europe.[41]

But how much additional Soviet gas could a country such as West Germany import without jeopardizing supply security? Building on a tradition of constructive cooperation with the federal German government, Ruhrgas took up this issue for discussion. The energy experts at the Ministry of Economy, who by now had considerable experience of dealing with gas security issues, noted that Germany's dependence on Soviet gas was about to increase, through already concluded contracts, to 22 percent by the second half of the 1980s. The Iranian transit arrangement was included in this figure. Ruhrgas contemplated an import of an additional 8 bcm through the Yamal pipeline. This would raise the share of Soviet (and Iranian) gas to 30 percent. The government argued that this was a high level, but that "nonetheless the related risk seems acceptable." Noting that the "threshold for politically motivated interruptions lies very high," the energy experts identified three factors that reduced Germany's vulnerability:[42]

1. "Targeted actions", it was observed, were "in practice prevented through the interconnected pipeline grid Iran-Soviet Union-Federal Republic-France." In other words, the Soviet Union would not be able to disrupt deliveries to Germany without disturbing deliveries to France and the transit of Iranian gas to Western Europe. The Kremlin would have to pay a high price in terms of likely disputes with third countries in case of a major supply disruption to Germany, a circumstance that was seen to reduce the probability of such actions being attempted.
2. It was believed that unexpected interruptions lasting up to 2–3 months could be managed without noteworthy consequences for German gas users. This was because of the flexibility negotiated in import contracts with other exporters, notably the Netherlands, and a possible increase in domestic production. Gas from elsewhere would thus be able to come to rescue.
3. From the perspective of hard-currency earnings, the energy exports noted, "longer-term disruptions stand against the self-interest of the USSR." The Soviet Union had in previous years imported very large volumes of pipe and equipment from Germany, and it needed the gas export revenues to pay off the loans taken for this purpose.

It was thus seen highly improbable that the Kremlin would intentionally disrupt gas supplies to Germany. Only in the case of war could an unlimited interruption in deliveries be imagined. But even in such an extreme situation, Germany would lose only 6 percent of its primary energy needs. "Considering the fundamentally changed supply structure in such a case and Western Europe's heavy weight among other natural gas supply sources, this seems surmountable," the government's experts concluded.

Backed up by the federal government and by the prospective participants in the European customer consortium, Ruhrgas Chairman Klaus Liesen in April 1978 flew to Moscow, the purpose being to start serious elaborations on a new possible "big export project of Soviet gas to Western Europe." After a "longer talk" with Minister of Foreign Trade Nikolai Patolichev, his deputy Nikolai Osipov, and Gas Minister Sabit Orudzhev, Liesen's impression was that the idea of radically scaled-up exports to the West remained internally contested in the Soviet Union. Only by the end of 1979, Liesen was informed, would the Soviet government decide whether or not to include new export deals in their economic plans. Yet Patolichev wanted to know in advance how much Yamal gas Ruhrgas could possibly absorb. Agreeing that the project would need to be at least 20–25 bcm to be economically feasible, Liesen replied that Ruhrgas, "in cooperation with other interested West European countries," would be able to take on this volume.[43]

Back home, Liesen immediately contacted Minister of Economy Otto Graf Lambsdorff, who in 1977 had succeeded Hans Friderichs on this post. Liesen argued that the "capacity and preparedness of the Soviet Union to expand its gas export business" was not to be questioned. Echoing the earlier analysis of Lambsdorff's own advisors, he also stressed that the project would be realized in cooperation with countries further west, in such a way that large amounts of Soviet gas would be transited to France, and most probably to Belgium and the Netherlands as well. Germany would thus enjoy a central hub role, which "for reasons of supply security would constitute a major advantage."[44]

Lambsdorff was positive to the project, as it would likely have a positive influence on Germany's foreign trade balance with the Soviet Union. He also accepted the security arguments of his energy advisors, although he added that imports from the East had better not be further expanded after the envisaged conclusion of a deal for Yamal gas. Lambsdorff got an opportunity to further discuss the project in connection with a much-publicized visit by Soviet leader Leonid Brezhnev and several members of the Council of Ministers to Germany in May 1978. A meeting was organized with Minister of Foreign Trade Patolichev, who praised the Soviet-German gas-for-pipe countertrade deals that had already been concluded as exemplary, while at the same time emphasizing that nothing could yet be promised from the Soviet side regarding the Yamal project. All in all, the Soviets seemed more hesitant to the undertaking than the Germans.[45]

In the meantime Ruhrgas worked on forming its West European customer consortium. A triumph for Ruhrgas was that it managed to tie the regional gas companies in southern Germany to the consortium. In March 1978 Bavarian minister of finance Max Streibl, on a visit to Moscow, had probed the possibility of an independent Bavarian import of Yamal gas, arguing that Bavaria rather than the Ruhr would be the ideal "hub" for the new, scaled-up gas trade. The Soviets, however, appear to have taken a skeptical stance to the idea, and in the end Ruhrgas convinced Bayerngas and the other southern companies that it was "better to let Ruhrgas negotiate alone."[46]

A problem that proved more difficult to solve was seen to lie in different conceptions about the delivery begin for Yamal gas. The Soviet side, while still hesitating whether or not to actually embark on the project as such, informed its West European partners that it would prefer a start-up in 1983/1984, whereas the consortium wished to wait until 1988. The problem was unexpectedly resolved through dramatic political turns in Iran. By fall 1978 it had become obvious that an Islamic revolution was in the making in the Soviet Union's neighbor to the south, with immediate consequences for international energy relations. The country's oil industry was plagued by radical strikes, leading to a sharp decrease in production and a collapse of exports.[47]

Since Iranian exports of natural gas to the Soviet Union depended on associated gas produced at the oil fields, the revolution also brought the operation of the IGAT pipeline to a halt. Moreover, the political chaos undermined the tripartite Iranian-Soviet-West European transit project. Construction of the new export infrastructure, which so far had progressed more or less according to plan, was interrupted. The new Islamic regime, with Ayatollah Khomeini as its front figure, signaled that it considered the prices agreed upon by the Shah's government far too low. In June 1979, the new president of the National Iranian Oil Company (NIOC), Hassan Nazih, told Western journalists that the new export pipeline "with 90 percent probability will not be built." The company set out to convert already completed parts of the export infrastructure into a domestic transmission system.[48]

The new Iranian turn came as a shock both to the Soviet and the West European gas industry. In the Soviet Union, the Caucasian republics were directly affected as gas deliveries from the south were disrupted. Yet Moscow refused to give in to Iranian demands for a higher gas price. Instead, the Soviets embarked on a set of new domestic pipeline projects that would enable additional gas from southern Russia and western Turkmenistan to be brought in to the Caucasus. Western Europe, for its part, feared that the Soviet loss of 10 bcm of gas per year (the volume that had so far been shipped from Iran) would negatively impact on Mingazprom's ability to live up to its export commitments, as some gas originally destined for exports to the West might now have to be diverted to the Caucasus.[49]

The collapse of the tripartite deal, according to which 13 bcm of Iranian natural gas were to have been transferred to Western Europe starting in 1981, did not have any immediate effect on Europe's or the Soviet Union's gas supply. Yet it forced both to rethink the future of red gas exports to the West. In material terms, the Soviets had planned to arrange the Iranian-European "transit" through deliveries of Komi and Siberian gas to Western Europe, and for this purpose Minneftegazstroi had since the conclusion of the tripartite contracts in November 1975 invested heavily in expanding pipelines and compressor stations along the route from Siberia and Komi to the westernmost Soviet regions and the border-crossing point at Uzhgorod. This concerned in particular a new pipeline along the Northern Lights route. These investments now appeared pointless. At the same time, Austria, Germany, and France, which had prepared for imports of Iranian gas, faced a structural supply shortage on the mid-term, for which replacements would had to be secured very soon.[50]

From this point of view it was clearly advantageous that the Yamal project was already in the making. For the Soviets, this project offered a way to make use of its "Iranian" export infrastructure in which it had already invested so heavily. The West Europeans, for their part, were pleased to observe that "there is the possibility that the Soviet Union will jump in as a supplier of the contracted and planned deliveries from Iran," as ÖMV's general director Ludwig Bauer put it. Ruhrgas Chairman Klaus Liesen similarly expressed "confidence that the Soviets will honor their commitment to supply the gas" involved in the Iranian deal. Moreover, identifying the Yamal pipeline project as an attractive replacement for the collapsed Iranian scheme, Western importers became keen to access Yamal gas as soon as possible, rather than in the late 1980s.

All in all, the Iranian revolution strengthened the overall enthusiasm for the new Soviet export project and helped resolve the disagreements about its timing. By the same token, internal Soviet support for actually initiating negotiations with the Western gas companies about the Yamal pipeline grew stronger. Apart from the Iranian debacle, the Soviet Union's own looming oil production crisis contributed to the conviction that there was no alternative to renewed export negotiations for natural gas. Skyrocketing production costs and a decline, for the first time in many years, of Soviet oil exports from 1978 cemented the view that natural gas would have to replace oil as the country's main export commodity. Brezhnev had now also made up his mind in this respect, and in autumn 1979 the decision was eventually taken to submit formal invitations to major Western gas companies to come to Moscow for a first round of negotiations.[51]

Opposition from the United States

An interesting question in the context of the Yamal project is whether Mingazprom and the Ministry of Foreign Trade knew in advance of Brezhnev's fateful plan to invade Afghanistan. The December 1979 invasion, which allegedly took place "on request" from Kabul's new communist regime, inevitably shocked the world and sparked international protests. In mid-January 1980, the United Nations approved a resolution calling for the removal of Soviet troops from the country. The crisis made the United States and several other countries boycott the Olympic Games that were to be held in Moscow that summer, and it contributed decisively to refreezing the Cold War, fueling a new wave of anti-Soviet sentiments throughout the West.

It is not difficult to imagine that Mingazprom and the Ministry of Foreign Trade feared that the Afghan war would negatively impact Western Europe's willingness to go ahead with the Yamal project. Yet they may also have calculated that the Europeans, whatever their rhetoric, in the end would not let geopolitical developments stand in the way. If anything could be learnt from the analogous invasion of Czechoslovakia in 1968, it was that military aggression mattered little for Western Europe's preparedness to expand East-West energy relations. Rather than putting an end to the West's enthusiasm for Soviet natural gas, Moscow's brutal crushing of the Prague spring had, quite on the contrary, served as the point of departure for negotiating several pioneering pipeline deals. Eleven years later, history repeated itself. Although demands for canceling the negotiations were certainly raised in protest of the Afghan invasion, neither gas companies nor governments were in the end prepared to let political and ideological considerations jeopardize a technically and economically sound project. With only slight delay, negotiations for Yamal gas could thus be initiated as planned in spring 1980, Afghanistan notwithstanding.[52]

As in previous Soviet-Western gas negotiations, the talks were part of a countertrade scheme in which natural gas was traded for large-diameter steel pipe and key equipment such as compressor stations. The total volume of gas and equipment under discussion was now much larger than in any of the earlier projects, but the negotiating procedures and the contractual model aimed for were more or less the same. To a great extent, the negotiators also overlapped with the ones who had been responsible for the first East-West gas talks held a decade earlier. They now knew each other well and, as a result, did not have to fear any fundamental misconceptions or unexpected turnabouts rooted in factors such as differences in culture or worldview. No longer did Western gas companies have to spend long hours explaining to their Soviet counterparts how the West European gas market functioned, nor did the Soviets have to convince the companies that gas transmission over a distance of 5,000 km or more from the Soviet Arctic was feasible. Instead, the negotiators could quickly proceed to key matters such as the gas price and suitable pipeline routes. By autumn 1980, the parties had come very close to each other in their bids, and they felt confident that the Yamal pipeline would actually be built.

At about the same time, however, newly elected US president Ronald Reagan started to take interest in the project. American companies were not involved in the Yamal talks, but Reagan nevertheless turned out to have strong opinions about the project, identifying it as a threat to the security of America's West European allies and openly stating that he wished to prevent or at least delay it. This was something new. The United States had so far never intervened in or objected to Western Europe's imports of red gas. The Germans, in particular, had always asked the Americans explicitly about their stance to the deals negotiated by Ruhrgas, and had always received green light from Washington. Up to 1976, the US administration had also taken a largely positive stance to the idea of Soviet LNG exports to the United States. Reagan's predecessor Jimmy Carter had insisted that Soviet gas exports be positively evaluated from a security point of view, reckoning that "increased exports of Soviet energy would ease supply/demand pinches worldwide and lead to moderation of energy prices." This was particularly so following the second oil price shock that set in following the Iranian revolution.[53]

Reagan, however, had a radically different agenda and made clear that he would do whatever he could to prevent the project from materializing. The Kremlin was furious, interpreting the new American stance as "hysterical anticommunism." The West Europeans, too, were outraged about Washington's new stance. As exemplified by the German elaborations referred to above, European gas companies and governments had already carried out in-depth analyses of the security issue and come to the conclusion that the Yamal project did by no means pose a threat to them.[54]

Reagan and his administration spent considerable efforts trying to convince—and force—the West Europeans to abandon the project. At a summit held in Ottawa in July 1981, new US secretary of state Alexander Haig presented the West Europeans with a plan to forestall the Soviet pipeline, offering greater access to other energy sources such as North American coal. The Americans also offered Europe nuclear technology, but this was hardly popular in the aftermath of the Three Mile Island accident. Washington furthermore suggested that Europe might take increased LNG deliveries from Algeria, Nigeria, and Cameroon, indicating that the United States might step back from competing with Europe for gas from these countries. This proposition was not taken seriously by the Europeans, who did not believe that the Americans were at all interested in African LNG, nor that such deliveries would strengthen Europe's energy security.[55]

Washington was thus unable to find a way to discourage Europe from buying more Soviet gas. The main result of America's opposition to the project appears to have been that it spurred Moscow to set natural gas exports even higher on the agenda. At the Congress of the Soviet Communist Party, held in February–March 1981, Brezhnev stressed the promises of Siberia's natural gas in its international context:

> I consider it necessary to single out a rapid increase in production of Siberian gas as a task of prime economic and political importance. Western Siberia's gas deposits are unique. The biggest of these—Urengoi—contains such gigantic resources that it can for many years meet the country's domestic and export needs—including exports to capitalist countries.[56]

The only factor that seemed to reduce Western Europe's thirst for red gas was the unexpected decline in overall demand that started to become evident as the Yamal negotiations progressed. The new market instabilities following the second oil price shock had far-reaching effects on the relative competitiveness of different fuels. In particular, the interest in coal, which had not been directly affected by the oil crisis, grew markedly. The relative price of coal decreased significantly, challenging the competitiveness of natural gas especially for electricity production. It was this trend that Washington sought to exploit when suggesting that coal might replace Siberian gas. At the same time, overall energy demand decreased following economic recession.[57]

Under normal conditions, the gas industry would have regretted this negative trend. Yet in the aftermath of the Iranian failure it was most of all a relief—at least for the three companies that had been involved in the tripartite deal. The overall stagnation meant that there was not such a big hurry after all to find replacements for cancelled deliveries of Iranian gas. Since Iranian gas might have generated an oversupply, it was to a certain extent even an advantage that these deliveries did not commence as planned.[58]

Contracts for Yamal gas were eventually signed by most of the countries who had taken an interest in such deliveries, but as a result of the stagnation of gas demand the contracts became smaller than originally projected. The Germans were the first to come to agreement with the Soviet side. Ruhrgas had initially planned to import 8–10 bcm per year plus replacements for the Iranian deficits, which amounted to 5.5 bcm per year. The actual contract, signed in November 1981, was for an annual import of 10.5 bcm only, motivated by

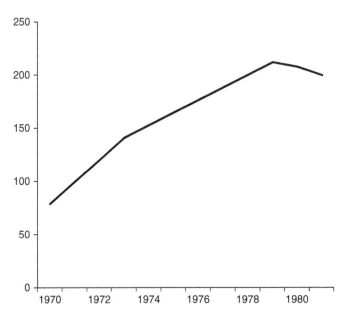

Figure 10.2 Gas consumption in OECD Europe, 1970–1981 (bcm).
Source: Estrada et al. 1988, p. 18.

revised forecasts for aggregate German gas demand. Deliveries were to start in 1984 and reach a plateau level in 1987.⁵⁹

In the case of Italy and France, the signing of Yamal contracts was delayed. The main reason was that both countries were at the time also trying to settle the terms of gas imports from Algeria. In accordance with agreements signed in 1977, Italy's ENI had just completed construction of the prestigious Trans-Mediterranean pipeline, which stretched from Algeria through Tunisia and across the Sicilian straits to the Italian mainland. It was a major—and very expensive—engineering feat. The line would enable piped Saharan gas to enter Europe for the first time. Following the second oil price shock, however, the Algerians announced that they wished to renegotiate the terms of the 1977 contract. Shocked by the dramatic increase in global fuel prices, they also demanded a price formula based on 100 percent crude-oil parity. The Italians refused to accept these demands, and in the meantime the brand-new pipeline that had been laid on the sea bottom stood empty. The French, who were already major importers of Algerian LNG but wished to scale up the trade, found themselves in a similar position, facing new Algerian price demands. GdF's failure to agree with Algeria on the terms of delivery even prompted Sonatrach to disrupt its already established LNG trade with France.⁶⁰

When the Italians and French in November 1981 indicated that they intended to sign their Yamal contracts with the Soviet Union, the Algerians approached the Soviet side with an appeal for "solidarity" regarding the demand for 100 percent crude-oil parity in gas exports. The Soviets were surprised by the move, but considered it sufficiently interesting to be taken seriously and decided to take some time to discuss it internally. In the meantime the prospective contracts remained unsigned.⁶¹

Solidarity was also the word of the day in Poland, where Lech Walesa's *Solidarnośc* movement challenged the Polish communist regime. On December 13, 1981 General Jaruzelski, fearing a full-fledged democratic revolution, introduced martial law. Thousands of people were imprisoned and many killed as the authoritarian order was restored. At the borders, Soviet military forces were ready to strike. Both Western Europe and the United States filed sharp protests, particularly regarding Moscow's obvious involvement in the Polish course of events. Once again the Yamal project, bound as it was to be interpreted as a symbol of friendship and mutual understanding between the Soviet Union and Western Europe, came under threat from dramatic geopolitical turns.

French president François Mitterand and his Socialist Party explicitly condemned the Soviet Union's involvement in Poland and opposed the Soviet gas deal that GdF was just about to sign. Yet political rhetoric was one thing, concrete action another. GdF eventually managed to resist the politicization of the Yamal project, and within a few weeks Paris had already given its gas company green light to sign. The contract was of great importance for France's gas supply, paving the way for an import of up to 8 bcm per year, in addition to the 4 bcm that were already being imported in accordance with the first Soviet-French contract from 1972. The new contract implied that the Soviet Union from the late 1980s would deliver 17.5 percent of France's natural gas.⁶²

ENI had planned to sign an analogous deal with the Soviets, aiming for an import of up to 8.5 bcm per year from the East in addition to the 6 bcm that were already being imported in accordance with the pioneering 1969 agreement. On December 30, 1981, however, the Italian government ordered a "pause for reflection," officially because of the crisis in Poland but in reality also because of the Algerian problems. It took another two years before the Italians could eventually sign its second Soviet contract. By that time the annual volume to be imported had been reduced to 6 bcm. Since world fuel prices declined substantially during the two-year "pause," ENI was able to take advantage of the delay. In the contract that eventually was signed, the gas price was seen highly favorable.[63]

In contrast to the earlier Soviet contracts, the amount of gas to be imported in accordance with the Yamal contracts was not fixed to 100 percent. Gaz de France, for example, could choose to import between 6.4 bcm and 8.0 bcm. Italy could choose between 4.8 bcm and 6.0 bcm. Moreover, the changing market situation in Western Europe promted gas companies in countries that were still negotiating with the Soviets to scale down their initially reported interests. Thus Gasunie in February 1982 indicated that it might take only half of the 4 bcm per year that it had originally considered. Distrigaz similarly contemplated reducing its 5 bcm target to 2-3 bcm only. The Dutch reduction was mainly motivated by new estimates regarding future gas demand, and Gasunie stressed that it was not to be understood as a political statement. In the Belgian case, the reduced interest in Soviet gas was linked to a discussion in which critics pointed out that the country would become dependent on Soviet gas to 38 percent if it agreed to import 5 bcm per year. The Belgian government announced that it intended to limit the share of Soviet deliveries to 25 percent.[64]

Austrian minister of trade Josef Staribacher similarly hinted that his country's interest in Yamal gas might be only half the 3-5 bcm originally envisaged. The new stance was motivated by an unexpected decline in Austria's gas consumption following price increases and recession. When ÖMV in June 1982 at last signed its Yamal contract, it covered an annual volume of 1.5 bcm only, with an option for another 1 bcm. The Belgians, in contrast, in the end never came to agreement with the Soviets, opting, instead, for increased imports from the Netherlands. The Netherlands itself also failed to come to agreement with the Soviet side. Yet if the upper limits of the signed Yamal contracts were summed up, the new pipeline from Siberia would have to designed for transmission of a staggering 27 bcm per year. This corresponded to more than a doubling of all previously contracted Soviet exports.[65]

The Compressor Embargo

Washington was not happy with the progress made in negotiating the Yamal contracts. President Reagan insisted that the project must be stopped, or that the Kremlin must at least be made to give in to certain political demands. Shortly after the finalization of Ruhrgas' deal with the Soviets, Reagan therefore launched a new strategy, imposing an embargo on technology and equipment sales to the Soviet Union. The objective was to "pressure the Soviet Union to reach a conciliation with Poland."[66] The embargo threatened to jeopardize the Yamal project due to its dependence on American and West European compressor technology.

In return for Soviet gas exports, Europe's leading compressor manufacturers had signed export contracts for no less than 125 compressor units to be installed along the Yamal pipeline, each with a capacity of 25 MW. Soviet compressor manufacturers were still plagued by chronic problems and were not yet able to produce reliably operating machines with the same capacity. The leading factory, Leningrad's Nevsky machine-building plant, was still struggling to master serial production of 16 MW units. Timely delivery of the European 25 MW units was seen critical to the Yamal pipeline's success.

Reagan's advisors had discovered that the European producers relied on American technology for some of the compressors' most critical components. This concerned in particular the tubine blades and rotors, which had been developed by General Electric. The Europeans either imported these components from the United States or produced them on licence from GE. By the time the new embargo was activated, GE had already shipped rotating components for 22 turbines to Europe. Further deliveries were now declared illegal. The Europeans responded to the new situation by boosting their own, licensed production of the GE parts.

Reagan was furious that the West Europeans did not want to cooperate with the United States in implementing the anti-Soviet technology embargo. In June 1982, Washington increased the pressure on the Europeans by extending the sanctions to "equipment produced by non-US subsidiaries and overseas licensees of US companies." It was believed that this "could delay by 1–3 years start-up of the so-called Yamal pipeline." The European companies involved in the delivery of compressors—AEG-Kanis and Mannesmann (West Germany), John Brown Engineering (Scotland), Nuovo Pignone (Italy), Creusot-Loire (France), and the French subsidiary of US-based Dresser—were all shocked.[67]

Reagan's policy was highly controversial. Washington's attempt at economic warfare was immediately attacked both by the Soviet Union and by Western Europe. In the United Kingdom, which did not intend to import Soviet gas but whose machine-building industry had major stakes in the project (amounting to around £200 million or $344 million), Prime Minister Margaret Thatcher condemned Reagan's move, threatening to "take legal steps to compel UK companies to defy US technology sanctions." The Italian and French governments similarly instructed their compressor manufacturers to "ignore US sanctions and honour their contracts with the Soviet Union."[68]

Reagan stated that he would blacklist European companies that did not adhere to the new policy. But the Soviet Union similarly threatened the companies with penalties in case of failure to meet their contractual obligations. At the same time, Moscow's decision makers stepped up their support to domestic machine-builders in their effort to produce 25 MW compressors, declaring that their turbines would be of the same or higher quality than equivalent foreign models. The "only" problem was that these units were still in the process of being developed. According to earlier plans, production of the machines was to have started in Leningrad in 1981, but by summer 1982 only three prototypes had in reality been built, and it remained unclear as to when serial production could begin. In the meantime all Soviet pipelines continued to rely on domestic compressors with a capacity of up to 10 MW in combination with more powerful foreign machines.[69]

The Commission of the European Communities calculated that compliance with the US embargo would deprive European companies of $8.5 billion in revenues—and this at a time when the region was beset by recession and high unemployment. Insisting that the Yamal pipeline did not pose a security threat, a potential decision to adhere to the embargo was pointed at as both destructive and pointless. In a joint letter submitted to the US State and Commerce departments in August 1982, The EC Council and the ten EC member states declared the US embargo against the USSR illegal. The opposition was further backed by a resolution adopted by the European Parliament.[70]

In the meantime, actual production of the huge machines had already started. Following governmental directives, the embargo was ignored and in early September 1982 three French-made compressors, manufactured by a subsidiary of the leading Texan company in the field, Dresser, left a French port en route to the Soviet Union. Six turbines produced by John Brown Engineering were similarly loaded aboard a Soviet freighter in the port of Glasgow, whereas two compressors built by Italy's Nuovo Pignone left Livorno aboard another Soviet vessel, thus violating Reagan's embargo. Turbines from AEG-Kanis in Hamburg were also being completed and made ready for shipment. The Soviets unloaded the units in Baltic Sea harbors and immediately prepared for their further transport to the respective sites along the Yamal pipeline route.

Refusing to give up its attempts to stop the project, Washington responded, as it had threatened to do, by issuing Temporary Denial Orders (TDOs), effectively blacklisting the European companies by banning all US exports to them. The consequences were absurd. In October, US customs agents in New York "seized $3 million worth of pipeline parts owned by Nuovo Pignone." However, the parts had nothing to do with the Yamal pipeline, but were destined for Algeria and its gas business. The French company Alsthom-Atlantique, for its part, was refused access to certain key technical data from the United States, which it needed for completion of a compressor that had just been contracted for export to Australia. Meanwhile the shipments of compressors from Western Europe to the Soviet Union continued as planned. The Americans were forced to recognize their inability to steer the development in the desired direction, noting that the embargo "damaged the West more than the Soviet Union." In November 1982, then, President Reagan decided to lift it. The export rules from before December 1981 were reinstated. The American attempt to interfere in the East-West gas business had failed.[71]

Europe's Contested Vulnerability

The United States opposed the Yamal project because it was feared that the Soviet Union might use (the threat of) gas supply disruptions for political blackmail against Western Europe's NATO members. Since this was a hypothetical event, however, it was difficult to assess the true extent and character of Europe's vulnerability. Reagan's advisors, referring to negative experiences from the Soviet oil trade, argued that "it is important to note that in the past the Soviet Union has used energy exports as a political lever, interrupting supplies to Yugoslavia, Israel, and China, among others," and that a politically motivated cutoff of Siberian gas to Western Europe must, therefore, be taken into account as a real possibility. But the examples of countries that, allegedly, had become victims of the Soviet energy weapon in the past did not include any West European nations, whose political relations to the Soviet Union differed markedly from those of the cited countries, nor did they refer to cases where natural gas had been involved, a fuel whose grid-based character made it unsuitable for comparisons with oil.[72]

Governments and gas companies in Western Europe tended to downplay the risk of politically motivated gas supply disruptions. The German government, back in 1978, had come to the conclusion that an unlimited interruption in supply could be imagined "only in the case of war." Chancellor Schmidt's energy advisors had noted that "longer-term disruptions stand against the self-interest of the USSR," at least in terms of hard-currency earnings. Not even the Soviet Union's two West European competitors in the field of gas exports, the Netherlands and Norway, were prepared to argue against red gas on the basis of security considerations. Gasunie President Hendrik Vonhoff believed the Soviets had nothing to gain from a cutoff in exports to the West, as it would "cause severe economic hardship in the USSR" itself. It was also argued that an interruption was highly unlikely due to the dependence of several importers on the same pipeline. The Soviet Union would thus not be able to target any single European country without hurting others, with devastating consequences for its international reputation.[73]

The probability of *unintended* supply disruptions was a different matter. Washington believed to know that "technical or seasonal difficulties—perhaps complicated by the need to divert gas from export to domestic use to make up for reduced deliveries of Iranian gas—forced the Soviets to slow some gas shipments to the West last winter and spring [1981]." For the West European gas companies, however, this was nothing new. They regularly experienced temporary interruptions due to technical problems in the East, but they had got used to such events and had learnt to handle them without consequences for users. The Soviets had proven trustworthy in their capacity to bring back supply to regular levels, and they had always compensated for lost volumes at a later point in time. The Americans argued that the "probability of further technical or seasonal interruptions may increase as the Soviets try to ship more gas from outlying and more risky Siberian provinces to western Europe." But the Europeans felt confident since both pipes and compressors to be used on the Yamal pipeline were of West European origin and could, therefore, be expected to work reliably.[74]

The Europeans also stressed the importance of seeing the allegedly risky imports from the East in relation to deliveries from elsewhere. There was agreement that intra-EEC gas reserves were limited and that, if member states wished to continue relying on natural gas to any noteworthy extent, the share of imports would have to increase substantially in the course of the next couple of decades. A turning point was expected to come in the mid-1980s, when the old Dutch export contracts would start to expire. Even if an extension of the Dutch contracts was not out of the question, the relative contribution of Dutch gas to Europe's overall supply was clearly bound to decline.[75]

Algeria was an alternative, but Western gas companies and governments could testify that Sonatrach in the past had always been a much more difficult and unreliable partner than Mingazprom. Agreements had in some cases been unilaterally cancelled before deliveries had even commenced, usually in connection with attempts from the Algerian side to force a higher sales price. Both France and the United States had experienced a number of intentional disruptions of LNG deliveries from Algeria. The Italians, for their part, had suffered from the Algerian refusal to start up deliveries of Saharan gas through the Trans-Mediterranean pipeline in 1981, as stipulated in the contract.

Norwegian gas was a more promising option. First Norwegian exports—to Emden in Germany—had come on-stream in late 1977, and during the following years exports to the United Kingdom and several continental European countries increased rapidly. The United States regarded this as a major opportunity in the context of Reagan's opposition to the Yamal pipeline. More precisely, Washington sought to push the Norwegians to offer equivalent amounts of gas to Western Europe as an alternative to increased imports of Soviet gas. Deputy US defense secretary Richard Perle visited Norway on several occasions to explore this possibility, suggesting to Norwegian oil minister Vidkun Hveding that "action to keep US companies and technology out of the Siberian project would delay the pipeline to the point where gas from the giant Troll gas field in Norwegian waters might be a viable alternative." The Norwegians disagreed. Oslo told the Americans that "we have underlined over and over again that it is just not possible...it is not technically feasible to bring any major new gas source on the Norwegian shelf on stream on the time scale we are talking about here—the middle or late 1980s."[76]

Besides, the reliability of Norwegian gas had started to be questioned. From summer 1981, just as the first Yamal contracts neared completion, Norway's Ekofisk and Statfjord fields were plagued by a series of strikes and unrest among offshore workers, who were dissatisfied with wages and working conditions. During summer, when European gas demand was relatively low, this did not cause any notable problems, but once the heating season started, the situation grew more serious. In October 1981, two major strikes in the Frigg and Ekofisk gas fields—the former for a few days, the latter for a week—led to supply disruptions both to the United Kingdom and continental Europe. The strike deprived British gas consumers of about one-third of their gas supplies. Such events, it was noted in Western media, were highly unlikely under the authoritarian conditions of the Soviet Union.[77]

Another contested issue, apart from the probability of a major cutoff in shipments of red gas, was the extent to which such a critical event would actually cause any social and economic damage. When the first Soviet export contracts were signed in the late 1960s, natural gas had been a relatively unknown energy source that was used by a minimal number of industries and households. Since then, however, the absolute volume of gas used in Western Europe had grown enormously, from 78 bcm in 1970 to 211 bcm in 1979—an increase by more than 170 percent in only nine years. A rapidly growing number of industries, households, and municipal users might thus fall victim to supply disruptions or price shocks.

The Europeans emphasized that the Yamal pipeline would deliver less than 4 percent of the EC's gas needs, and that imports from the Soviet Union actually contributed to diversifying the community's supply sources.[78] Aware of the problems with Algerian and Norwegian supplies, the Yamal pipeline was argued to have a positive rather than a negative impact on European energy security. Reagan's advisors, however, correctly noted that

> the volume of Soviet gas as a percentage of total European energy consumption is not a sufficient indicator of economic and political vulnerability... gas is a difficult fuel to replace on short notice. Unlike oil there is no spot market. It is much more expensive and technically challenging to hold large strategic stocks of gas compared with oil. Residential and commercial consumers are particularly dependent on gas. A cutoff of Soviet gas would be particularly onerous for these politically sensitive sectors.[79]

The EC's energy commissioner, Viscount Etienne Davignon, countered that the EC was "setting up a security system that would enable it to survive without Soviet gas if necessary." Following the American compressor embargo, the Commission started investigating this issue in depth. The task was to assess whether EC members would be able to "survive a hypothetical 25 percent cut in gas supplies in the winter of 1990." By then, the community's aggregate dependence on Soviet deliveries was expected to have reached 19 percent. In autumn 1982 a report was issued confirming that the Soviet Union would not "point an energy gun" at Western Europe. The Commission concluded that there would be "only some limited peak supply problems in some of the countries and in the most extreme circumstances." The EC would be able to deal with a total disruption of red gas deliveries "even in a severe winter." A key role in such a situation would be played by the Netherlands. Gasunie president Vonhoff confidently stated that his company would be able to "hike gas supplies to existing European customers by 80–100 percent in the event of a cutoff of Soviet deliveries."[80]

Such elaborations for the supply of "emergency gas" were possible thanks to the high degree of interconnectedness that by the early 1980s had come to characterize the all-European gas grid. Precisely because of the fears of possible supply disruptions from the East, gas companies such as Bayerngas and ÖMV, having secured access to Soviet gas, had worked hard to create connections to the networks of other gas companies and exporters. From 1974 the Bavarians were connected with northern Germany and, by extension, the Netherlands. From 1977 Norwegian gas could be relied on as a further backup. In the case of ÖMV, the Trans-Austria Pipeline built in 1974 for the purpose of transiting Soviet gas to Italy (and from 1978 to Yugoslavia as well), could conceivably be used for emergency shipments in the opposite direction. The inauguration of the Trans-Mediterranean Pipeline was of a certain importance to ÖMV in this context, as it, in combination with the Trans-Austria Pipeline, created a physical connection between Austria and the Saharan gas fields. Similar emergency considerations motivated ÖMV to put pressure on Germany and France to have their Soviet gas imported through Austria. The Austrians failed to come to terms with the Germans in this respect, but were more successful in the French case. An agreement was reached in October 1975, stipulating that Soviet gas destined for France would be transited through Austria. For Austria this was seen to be of "decisive importance," a major reason being that ÖMV

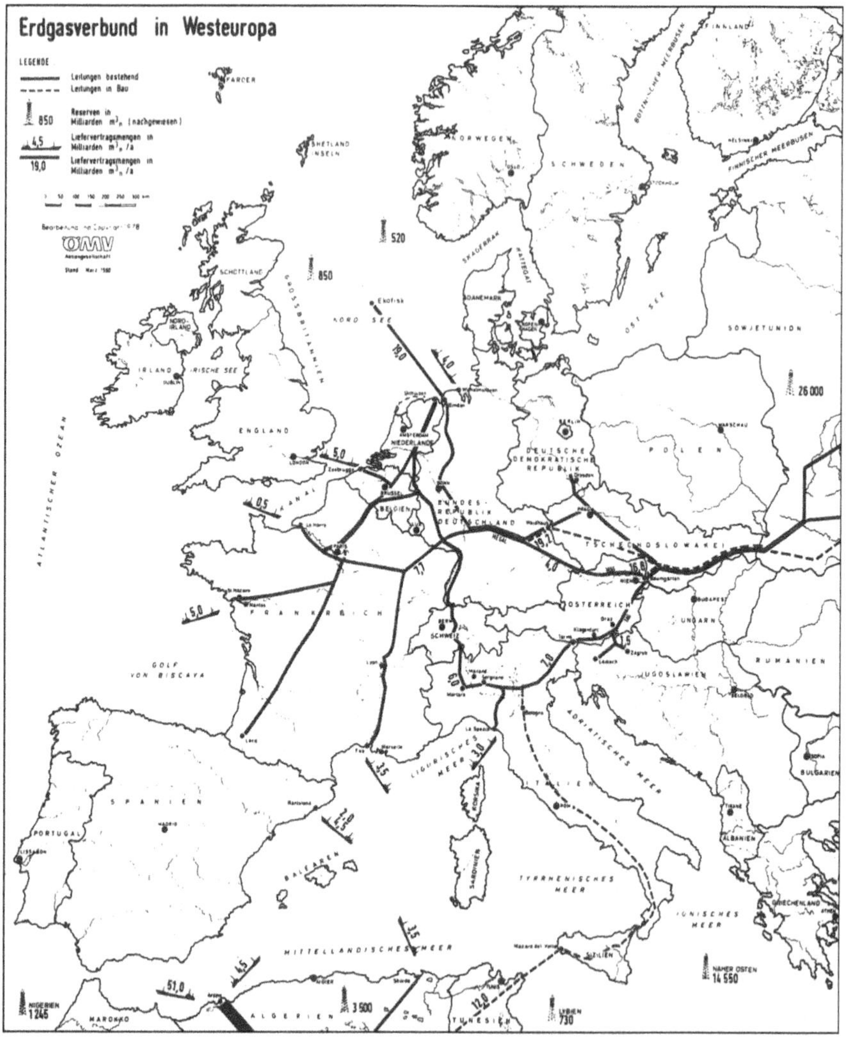

Figure 10.3 The integrated gas system of Western Europe as of 1980. This map from ÖMV makes visible the ways in which the export pipelines from the Soviet Union contributed to Western Europe's internal integration.

Source: ÖMV. Reproduced by permission.

in the negotiations with France had made sure the transit pipeline might be used for gas transmission in the opposite direction.[81]

Later on a connection was also built from the Czech-German to the German-French border, enabling France to receive red gas without Austrian involvement. It was initially conceived as a route for Iranian gas on its way to France, but following the collapse of the tripartite Iranian project it was redesigned for the purpose of Yamal gas shipments. For France the double transit system was seen to improve overall transit security.[82]

ÖMV, seeking to strengthen its hub position in the evolving European grid, also developed emergency plans that relied on large storage facilities. By 1983, the company had built up a storage capacity of 2.3 bcm, corresponding to nearly two-thirds of annual Austrian gas consumption. The main function of this capacity was not crisis fighting but regular load management, both nationally and internationally. As a partial alternative to domestic storage facilities, both Gaz de France and the Slovenian gas company Petrol signed agreements with ÖMV for the co-use of the Austrian facilities. But the capacities could clearly also be used to maintain domestic supply security in case of a major international crisis, whereby Austria would theoretically be able to survive without foreign gas for more than half a year without notable disturbances.[83]

Not all gas markets in Western Europe were yet interconnected. France, Italy, and Spain, for example, remained totally isolated from each other. The Nordic countries were also weakly connected, both with each other and with the continental system. Yet the overall European natural gas grid had by the early 1980s become a remarkably integrated structure. Although most pipelines had originally been built to connect a certain producer with a certain user region, the overall grid had more and more taken the form of an interconnected system in which natural gas from several sources were mixed with each other. These mixed flows were seen to contribute decisively to Europe's ability to deal with any shorter or longer supply disruptions, whether or not they were politically motivated. By extension, the sense of security that Europe's interconnectedness generated opened up for contemplating a further scaling up of supplies from potentially unreliable producers.

In LTS terms, this chapter has been a study of growing momentum in Soviet-West European natural gas relations. The period was characterized by political turmoil, which on several occasions made any further growth of red gas exports highly contested and uncertain. Yet the positive feedback from earlier negotiations, signed contracts, already built pipelines, and a functional system generated strong incentives for transnational system-builders to further scale up the trade. The evolving system showed itself remarkably resistant to any pressure or shocks from the international political and economic environment.

Red gas profited from the overall growing popularity of natural gas as a fuel, which in the course of the 1970s became increasingly linked with its favorable environmental characteristics. Gas from the East was also helped by the growing availability of gas from elsewhere, mainly of Algerian and Norwegian origin, which meant that the relative weight of the Soviet Union in the overall supply picture did not grow as fast as deliveries in absolute terms. Even so, the initial expectation, as often formulated in the years around 1970, that red gas would remain a "supplementary" source of fuel for the West, had by the early 1980s definitely given way to a conviction that it was a totally integral component of Europe's energy supply. And nothing, it seemed, could prevent the system from further expanding.

11
From Soviet to Russian Natural Gas

Surging Dependence

The "Yamal" pipeline was ceremoniously inaugurated in late 1983. Celebrated by Mingazprom and Minneftegazstroi as a great engineering feat and by the Kremlin as a Cold War victory against the United States, it had a major impact on the physical possibilities to transmit Siberian natural gas to Western Europe, opening up for more than a doubling of total exports. When fully equipped with compressor stations, the line would be able to handle a gas flow of no less than 40 bcm per year, which was substantially more than the volumes actually contracted with the West Europeans. As a result, Mingazprom would for the foreseeable future no longer need to worry about bottlenecks in the infrastructure.

The export pipeline was one of six new Siberian lines taken into operation during the five-year period 1981–1985. Together they paved the way for a radical geographical shift in the Soviet Union's internal gas supply structure. The earlier dominance of Ukraine and other gas-producing regions in the European part of the country gave way to northwestern Siberia as the undisputed production center. Tyumen's giant fields and the new transmission system enabled Mingazprom to boost the total flow of gas to a staggering 536 bcm in 1983. Mingazprom thereby surpassed not only the official planning target, but also, for the first time, the volume of gas produced in the United States. Khrushchev's old dream, formulated a quarter of a century earlier, of overtaking its Cold War enemy "also in this field" had at last come true.[1]

Yet the price for this apparent success was high. The new Siberian system, and in particular the new export pipeline, was built in an extreme rush that would not remain without consequence for the security of supply. The desire to show the Americans that the attempts to disturb Soviet system-building were pointless only made things worse, as everything was sacrificed for speed. In many places welding turned out to be of low quality, giving rise to numerous pipeline breaks later on. The first explosion on the Yamal pipeline was reported in October 1983, already before it had been taken into operation along its full length. The explosion caused "a fire which engulfed a bus and caused many casualties." A few years later, Soviet 1,420-mm trunk lines were reported to be experiencing on average "28 stoppages per year, each one averaging over 100 hours." The worst accident occurred in June 1989, when an electric spark from two passenger trains that passed by a pipeline in central Russia's Bashkiria republic caused the high-pressure steel pipe to explode, killing more than 600 people. It was a national disaster. Stretching from Siberia to European Russia, the 1,860-km pipeline carried natural gas liquids rather than gas proper, but it testified dramatically to the miserable material status of the overall infrastructure.[2]

Furthermore, delays in taking into operation gas treatment plants at Urengoi meant that some gas was being fed raw into pipelines, with adverse effects for sensitive users in the chemical and other industries. Construction of gas cooling stations similarly lagged far behind schedule, forcing Mingazprom to ship warm gas from Siberia to meet domestic and foreign demand. This inevitably caused the permafrost on which the pipelines rested to thaw, which in turn increased the risk that the pipe supports collapsed and thus paved the way for further destruction and malfunction. The nascent Soviet environmental movement had much to say about the ecological damage.[3]

At the same time, delays in opening up the next supergiant field, Yamburg, forced the Soviets to push Urengoi's production to barely justifiable levels. The original plans had foreseen that Yamburg, situated to the north of Urengoi in the Yamal-Nenets autonomous region, would feed the Yamal pipeline (hence its name). The failure to bring Yamburg gas on stream, however, meant that this plan had to be given up. Instead, Western Europe received its increment of red gas from Urengoi. Accordingly, the Yamal pipeline was renamed the "Urengoi-Uzhgorod export pipeline."

Western gas companies had hoped that shipments of first-class pipes, compressors, and equipment to Siberia would guarantee the technical reliability of the export system. In response to the American embargo policy, however, Moscow opted for a propaganda coup, seeking to complete the line a year ahead of schedule. This meant that the construction of compressor stations had to begin already before the arrival of the Western machines. In this situation domestically manufactured compressor units had to play a much greater role than originally anticipated. The units were pushed out from Leningrad's Nevsky factory and other plants in large series although the serious construction defects that Mingazprom had always been so upset about had by no means been eliminated. As a result, the reliability of gas transmission along the Urengoi-Uzhgorod export pipeline was not at all as high as importers had expected. As for the compressors made in Western Europe, there were also problems. The machines themselves were fine, but installing them proved hazardous. The Europeans were shocked to receive the news, in October 1983, that a British engineer assisting the Soviets with installing a John Brown compressor in northern Russia had been killed in an explosion.[4]

According to initial plans, exports through the Yamal line were to have reached a plateau level already in 1987. Plagued by slaggish demand, however, Mingazprom's West European customers opted to approach Moscow seeking slower growth and downward adjustment of the gas price agreed upon in 1981–1982. The Soviets were worried that the overall stagnation in gas demand would lead to serious underutilization of the new export infrastructure. To fill the pipe,

Moscow started offering lower prices to customers willing to take on a larger volume of gas. The Soviet Union's main competitors on the West European gas market, lacking any corresponding spare capacity in their pipelines, were less interested in boosting overall supply at the expense of lower prices, but were in effect forced to follow the Soviet Union's new, offensive pricing strategy. As a result, by early 1984 a growing number of analysts believed that in the future "the Russians will possess the power to set the price of gas in Europe."[5]

In what followed, the Soviets became increasingly active in approaching further potential customer countries, offering natural gas at prices significantly below equivalent oil prices. Switzerland had already agreed to become part of the Soviet import system, taking a share of Urengoi gas by way of transit through Ruhrgas' networks. Belgium, whose original negotiations for Yamal gas had failed, was offered a spot contract at a price reported to be 10–15 percent lower than for regular contractual shipments. Talks were also initiated with Greece and Turkey, both of which were offered Siberian supplies by way of transit through Romania and Bulgaria. Similarly, efforts were made to revive the old negotiations with Sweden, deliveries to which could

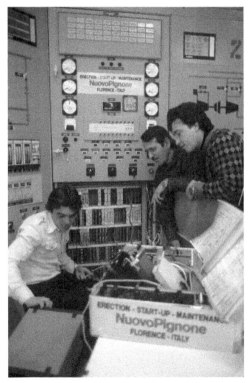

Figure 11.1 Italian technicians from Nuovo Pignone adjusting electronic equipment at one of the new compressor stations along the Yamal (Urengoi-Uzhgorod) pipeline.
Source: RIA Novosti.

preferably be combined with scaled-up exports to Finland. Spain was another target. Even in the United Kingdom, which lacked physical links with the continental European gas grid, the possibility of imports from the East started to be discussed, although Prime Minister Thatcher, on ideological grounds, was "strongly against purchasing Soviet gas."[6]

Agreement with Turkey was eventually reached in February 1986, foreseeing the construction of a pipeline that would bring Soviet gas all the way to Ankara, 800 km from the Bulgarian border. Deliveries were to begin in April 1987 and reach a plateau level of 5–6 bcm per year from 1990. Greece followed suit with an agreement signed in October 1987, according to which deliveries would start at a modest rate of 1 bcm per year in 1992 and reach a plateau level of 2.4 bcm in 2002. Greece became the fourteenth European country to sign up for red gas.[7]

While most of the countries that already were importers of Soviet gas increased their supplies in accordance with the Yamal contracts, Moscow thus also broadened its range of customers. Soviet exports to Western Europe as a whole grew from 29 bcm in 1983 to 40 bcm in 1987 and 63 bcm in 1990—an increase by 120 percent in only seven years. No longer could red gas be regarded as a "supplementary supply." Europe's dependence on the Soviet Union was surging in both absolute and relative terms.

To be sure, the scaled-up and broadened imports of red gas gave rise to concerns. In particular, the need to balance supplies from the East with gas from elsewhere was increasingly pointed at as vital for Europe's energy security. The OECD's International Energy Agency (IEA), which had so far mostly dealt with oil, became particularly active in this respect. The agency's executive director Helga Steeg, a former advisor to the federal German government, argued that "there must be certain threshold levels of import dependence on potentially insecure supplies, beyond which industry and if necessary governments must be prepared to limit commercial decisions." In other words, the gas price alone should not be allowed to determine the structure of the European gas market.[8]

Seeking to balance the Soviet Union's growing influence, gas companies started to elaborate on new pipeline projects that would connect Europe with promising gas regions other than the ones with which it was already linked. For example, the once-discussed import of Iranian gas through Turkey was revived, later on to be nicknamed "Nabucco" (a name that testified to Italy's central role in the project). Qatar, where large gas deposits had been discovered, was identified as another potential supplier from where pipelines could conceivably be built. Spain, for its part, pointed to Nigeria as a promising supplier of piped gas to Europe. For this purpose, a "Trans-Saharan" pipeline from the Nigerian fields through Algeria and Morocco to the Iberian peninsula was proposed. In addition, a number of possible LNG import projects were discussed.[9]

But these attempts to broaden Europe's supply base proved difficult to sustain in the face of excess pipeline capacity for imports of red gas, along with Moscow's new willingness to act as a price leader. From an economic point of view, Soviet gas was becoming too attractive to resist, and even though security arguments pointed to the necessity of diversification, gas from the Middle East, sub-Saharan Africa, and other distant regions could not be brought in without taking commercial realities into consideration. Some analysts predicted that

Soviet gas might outcompete gas supplies from intra-European sources as well. Norwegian gas, in particular, was since the 1970s held high as an important guarantor of Western Europe's long-term supply security—but for how long would it be able to stand up against its eastern competitor? "With lower gas price, more expensive gas fields, such as those off Norway, will not be developed, and Europe will become increasingly dependent on Soviet gas," one report predicted. In addition, since the early 1980s Norway's role in ensuring Europe's supply security had been questioned following repeated labor conflicts in the North Sea. On several occasions gas exports to the United Kingdom and the European continent were totally disrupted. The Norwegians were reportedly "irritated by reminders from Ruhrgas that the Soviet Union has been a far more reliable source of gas."[10]

Yet Ruhrgas and the other gas companies were certainly susceptible to the calls for diversification. By spring 1986, negotiations had already reached an advanced stage between Norway and a consortium of continental European gas companies, led by Ruhrgas and Gaz de France, for exploitation of Norway's two main remaining fields, Sleipner and Troll. Agreement was reached in June 1986. It was decided that deliveries from the two fields would come on stream in 1993 and 1996, respectively, and that the continental companies would increase their Norwegian supplies by 20–35 bcm per year by the early twenty-first century. The gas would be shipped by subsea pipeline to Zeebrugge on the Belgian coast, from where it would be transited to markets as far away as Austria and Spain.[11]

Sleipner and Troll served to restore what was perceived as a reasonable balance between intra-European supplies and imports from elsewhere. Precisely as a result of this sense of improved security, however, increased supplies from the Soviet Union could once again start to be seriously considered. The temptation to do so became particularly strong when overall European gas demand, after a period of stagnation in the early 1980s, started to grow again in the second half of the decade. Several new trends contributed to this growth. One was the debate on global warming, which took off in earnest in the late 1980s. Since combustion of natural gas emitted only half as much carbon dioxide as the combustion of coal, the introduction of new carbon taxes and other policy measures tended to make natural gas commercially more attractive. From now on natural gas was considered environment-friendly in relation to coal and oil not only because of its lower sulfur and nitrogen contents, but also because of its potential to come to rescue in the context of climate change. Of course, renewable energy sources were argued to be an even better option, but to the extent that natural gas replaced old coal-fired capacities, it was interpreted as part of the solution rather than as part of the problem.

Second, the Chernobyl disaster in April 1986 put an abrupt end to the fast expansion of nuclear power in Europe, and large gas-fired power plants were identified as suitable for replacing old, unsafe nuclear power plants or planned new ones that, in the aftermath of Chernobyl, were not built. This logic became particularly important in the discussions about Sweden's energy future, where nuclear power had so far generated up to half of total electricity while natural gas still played a negligible role. In 1988, after years of debate following a referendum held in 1980, the Swedish parliament decided to embark on a total phase-out of the country's 12 reactors. Both Norway and the Soviet Union lobbied the Swedes to have natural gas-fired capacities take nuclear's place, and previously failed negotiations with both countries were resumed. By 1989, the Swedish-Soviet negotiations, in which Finland was also involved, were reported to have reached an advanced stage.[12]

Third, the neoliberal shift in European and world politics tended to favor an increased use of natural gas, particularly in electricity production. In Britain, whose conservative government in 1989 pushed through a major reform aimed at deregulation of both electricity and gas, a variety of new actors appeared on the market, eager to exploit the opportunities of combined-cycle gas turbine technology for electricity production. Britain had sizeable domestic reserves in its sector of the North Sea, but since the producers of this gas were unable or unwilling to meet the Soviet price pressure, imports from the East started to be seriously considered. Trade Minister Lord Trefgarne and Energy Secretary Cecil Parkinson showed themselves very interested in the opportunity. Apart from a possible transit through continental Western Europe, the idea of a new subsea pipeline from the Soviet Union through the Baltic and North Seas—or through Sweden, the supply of which could conceivably be integrated into a prospective Soviet-British system—started to be sketched. Parkinson argued that it was simply "up to the Russians to see if they can match the competition through a pipeline to Britain." An Anglo-American consortium became the first to initiate negotiations with the Soviets. Its business idea was to feed red gas into a new gas-fired power plant to be built at Richborough on the southern English coast. British Gas also took an interest in possible imports from the East.[13]

Everything thus seemed to point in the direction of an increased role for natural gas in general, and of Soviet natural gas in particular, on European energy markets. As of 1988, Moscow reasoned that Europe's thirst for gas was bound to "create a demand so large by the end of the century that it can be met only by greater dependence on the Soviet Union." By 1990, Western analysts tended to agree that "before too long Europe will need new trunk lines to move Russian gas." Thanks to the 1986 deal with the Norwegians, along with rapidly growing imports of Algerian gas through Italy's Trans-Mediterranean Pipeline—both of which were seen to contribute to a diversified supply structure—Western governments and gas companies were able to argue that such new lines were acceptable from a security point of view. The stage thus seemed set for further expansion.[14]

The Biggest Geopolitical Disaster of the Twentieth Century?

Through the rapid expansion of gas exports during the second half of the 1980s, the Soviet Union had embarked on a transition from oil to natural gas as its main export commodity. Oil was still king, but gas was quickly approaching, its share in overall hydrocarbon exports doubling from 12.8 percent in 1978 to 24.7 percent in 1989. On the other hand, due to falling oil and gas prices, the share of hydrocarbons in overall Soviet export earnings decreased from 52 percent in 1984 to 36 percent in 1989. Overall revenues from both oil and gas exports actually fell in absolute terms during a period of several years in the late 1980s, reaching a minimum in 1988—despite record sales volume.[15]

From a historical point of view, an interesting question is to what extent higher world energy prices might have prevented the fateful downturn of the Soviet Union's national economy in the late 1980s, often cited as one of the main causes behind the red empire's collapse and dissolution in 1991. Falling income from oil and gas exports certainly contributed to the Soviet Union's rapidly growing indebtedness during these last years, and although any analysis of the country's collapse necessarily needs to take a range of other factors into account as well, it cannot be excluded that greater oil and gas export revenues could have prolonged the empire's death dance by months or even years. This was so not least because the export earnings were crucial for Soviet access to advanced Western technology, on which the Soviet economy was so dependent.

The actual demise of the Eastern Bloc started in Central Europe rather than in the Soviet Union itself. In June 1989, the Hungarian government decided to open up its borders to neighboring Austria, creating a first hole in the Iron Curtain. The decision launched a chain of events that culminated with the fall of the Berlin wall only five months later. In late 1991, then, the Soviet Union was itself dissolved, giving way to 15 independent nations. Czechoslovakia, meanwhile, was split into a Czech and a Slovak Republic in January 1993. The division of Yugoslavia became a much more painful and bloody experience that would find its end-point only with Kosovo's declaration of independence in 2008.

In his 2005 "State of the Nation" speech, Russian president Vladimir Putin argued that the collapse of the Soviet Union constituted the "biggest geopolitical disaster of the [twentieth] century." For the gas industry, a key question was whether the East-West natural gas trade would prove robust enough to survive this political apocalypse. Following the pioneering Soviet-Austrian gas contract signed in 1968, West Europeans had imported Soviet natural gas for 23 years without much deviation from contractual obligations. But to what extent would this cooperation be able to survive the new political and economic turmoil? The gas trade was firmly rooted in decisions taken and traditions established during the Cold War. It was the result of West European gas companies' and governments' trustful interaction with communist governments and enterprises, and of the totalitarian system's ability to force planners, engineers, and workers to live up to the tasks imposed on them. The collapse of communism and the dissolution of the Soviet Union suddenly removed much of this social and institutional foundation, and no one knew what would take its place.

Several of the countries involved in the gas trade had not only gone through political revolution, but had even ceased to exist. Germany had imported its red gas in cooperation with the Soviet Union and Czechoslovakia, but Ruhrgas now found that both of these partner states had disappeared from the European map, and that the pipelines through which it received its Siberian gas now crossed the territories of no fewer than five "new" countries: Russia, Belarus, Ukraine, Slovakia, and the Czech Republic. At the same time the Germans had to cope with the integration of the former GDR, another dissolved country, into the West German transmission system.

Another reason for worry stemmed from the fact that the main system-builder on the Soviet side, Mingazprom, was divided up into several national entities.

The Russian part of the ex-Soviet ministry was transformed into a state-owned company, RAO Gazprom. It was by far the largest of Mingazprom's successor agencies. It was immediately obvious, however, that the unified transmission grid could not simply be cut into pieces. "Our gas distribution system is shaped in a way that makes it simply impossible to divide it according to the borders of CIS member nations," ex-Mingazprom officials noted. "If we tried to carve up the system, even Russia, with its enormous gas reserves, would simply not be able to meet its own gas needs."[16] This was because, for example, parts of southern Russia were supplied by pipelines that passed through eastern Ukraine. Ukraine's impressive system of gas storage facilities was also of great importance to other ex-Soviet republics. Clearly, the Soviet Union's successor states would have to find a new way to cooperate with each other. In particular, contractual arrangements would have to be made for gas sales between the republics, and for gas transit to markets in Central and Western Europe.

The economic collapse in the former communist countries added to the perceived uncertainty about the future. The severe economic downturn throughout Central and Eastern Europe meant that financial resources were unlikely to be sufficiently available for maintenance and investment in the physical infrastructure. It also increased the risk that the transit countries might not be able to pay for their own imports of Russian gas. This was particularly so due to the Russian ambition to raise formerly subsidized Soviet prices to world market levels. No one knew what the long-term consequences of "nonpayment" would be for supply security in the transit countries and for customers further downstream.

Intentional Disruptions

During the Cold War period, the Soviet Union had never disrupted its exports of natural gas for political reasons. Unintended disruptions of a technical nature had plagued the East-West gas trade ever since its inception in the 1960s, but the fears that red gas would be used for political blackmail had never materialized. This track record had contributed decisively to the perception of red gas as secure and to the willingness of Western importers to scale up their imports from the East.

The post-Soviet chaos changed everything. The creation of a new export-import regime among the former Soviet republics turned out to be anything but straightforward. The result was a series of disputes over contractual terms, and when the conflicts could not be resolved, Gazprom opted to cut supply. In some cases these intentional disruptions were interpreted as political. Gas exports to Western Europe were bound to be indirectly affected, since the disputes often involved major transit countries.

Lithuania was among the first to be hit. When the republic in March 1990 declared itself independent, Moscow protested by threatening to disrupt energy deliveries if the declaration was not withdrawn. Vilnius refused to give in, and as a result Gazprom was instructed to reduce deliveries of both oil and gas to the republic. Gas supplies were reduced by no less than 80 percent. Whereas the Lithuanian government immediately set out to negotiate oil replacements with Western countries, the lack of gas pipelines across the Baltic Sea or LNG terminals on the coast meant that no similar replacements could be expected in the case of natural gas. The only potential rescuer was the neighboring Latvian SSR, with which Lithuania was connected by pipeline and which possessed one of the Soviet Union's largest gas storage facilities. But Gorbachev explicitly forbade both Latvia and Estonia, at the time still union republics, to assist the Lithuanian separatists. The Latvians and Estonians were shocked by the way their Lithuanian neighbors were treated, an experience that may well have contributed to the decisions in Latvia and Estonia to pursue a more gradual path toward political freedom.[17]

The frequency of intentional disruptions increased radically after the demise of the Soviet Union in late 1991. Winter gas supply had always been problematic in many Soviet republics, but the first post-Soviet winter became more horrible than anyone had ever experienced during the Soviet era. Nothing seemed to work as it should, leaving millions of households, municipal institutions, and industrial enterprises without gas.

Most shortages resulted from unintended failures to manage and coordinate gas supply in the huge ex-Soviet system. However, a number of political disputes between former union republics also produced cutoffs. The newly independent Caucasian nations—Armenia, Azerbaijan, and Georgia—were probably the ones most severely affected. These had been connected to each other through a heroic system-building project initiated in the 1960s, the Azerbaijan-centered "Friendship of the Peoples" pipeline grid (see chapter 2). By the early 1990s not much of this alleged friendship remained. Armenia and Azerbaijan headed into an armed conflict over the Nagorno-Karabakh exclave, and in winter 1991–1992 the former Armenian SSR was cut off from its Azeri gas supplies after Azerbaijan disrupted the flow along the main pipeline. Only thanks to emergency deliveries of Russian gas, arriving through Georgia, could some volumes be shipped. These deliveries were also suspended, however, after Azeri-friendly terrorists operating in North Ossetia and Chechnya, through which the gas was transited, blew up the pipeline and rendered transmission to Georgia physically impossible. The pipeline was soon renovated, only to be blown up again.[18]

Another conflict raged in newly independent Moldova, which, similar to the Baltic states, hosted sizeable ethnic Russian and Ukrainian minorities. Gas deliveries from Russia, reaching the tiny republic through Ukraine, were disrupted, but only to the westernmost part of the country, where most ethnic Moldovans lived. The government interpreted the cutoff as "part of an economic blockade to achieve political objectives." Meanwhile Ukraine was also hit by cutoffs.[19]

East Germany, which in October 1990 had ceased to exist as an independent country, was the westernmost territory to be shaken by cutoff threats in winter 1991–1992. In this case the background was economic rather than political. Since natural gas was seen to offer great environmental opportunities, with the potential to replace the ex-GDR's outdated lignite-fired power stations, eastern Germany was considered an extremely promising market for Western gas companies. Ruhrgas had in 1990 managed to acquire a large ownership stake in the GDR's transmission operator VNG and in this way extended its influence from western to eastern Germany. Ruhrgas' competitor Wintershall, however, also sought access to the new East German market. VNG had so far enjoyed a monopoly on gas sales in East Germany, but Wintershall forged a new partnership with Gazprom and challenged VNG and Ruhrgas. Gazprom had come to understand that it could potentially boost its sales by dealing directly with its German customers rather than first selling its gas to VNG or other German transmission companies. In 1991 Gazprom and Wintershall, operating through a joint venture known as Wingas, announced a sharp increase in the price of gas sold to VNG. VNG refused to accept the increase and simply continued to pay the pre-1991 price. Gazprom threatened with a total cutoff of supplies to eastern Germany from January 1, 1992, should VNG continue to refuse the new gas price. It also announced, however, that Wingas was willing to take over the supply to VNG's customers.[20]

VNG turned to the federal German government with a plea to activate emergency measures, but since eastern Germany still lacked pipeline connections with the western part of the country, there was no possibility for gas from elsewhere to come to rescue. Gazprom and Wintershall did their best to exploit this physical isolation in their price negotiations with VNG. In the end, a temporary one-year agreement was reached between the parties involved and the feared supply disruption was avoided. When the temporary contract was to be renegotiated a year later, both Ruhrgas and Wintershall had built new pipelines linking eastern Germany with the West, and Gazprom did no longer have the same leverage. From now on, the former GDR was physically able to import not only Russian, but also Dutch and Norwegian gas, making the east Germans less vulnerable to future supply disruptions from the East.[21]

Most of the former communist countries were in a similar situation as the ex-GDR in terms of limited interconnectedness with the West European gas grid. Intense efforts were launched nearly everywhere to improve the situation through the construction of new pipelines and negotiations with suppliers other than Gazprom. Hungary agreed with Austria's ÖMV on the construction of a connecting line to Baumgarten, Austria's powerful transit hub for East-West gas flows. Poland sought access to Norwegian gas and contemplated the construction of a pipeline for this purpose through the Baltic Sea to the Polish coast. Latvia similarly suggested that a Baltic Sea pipeline be built, though without Polish involvement. The Latvian idea was to interconnect the three Baltic states with Scandinavia, whose gas companies would thereby gain access to the large gas storage facility outside Riga while the Balts themselves would profit from access to Norwegian and Danish supplies. A problem, however, was that Sweden, the big white spot on the European gas map, still hesitated to invest in a nation-wide pipeline grid.[22]

Apart from seeking access to supplies from the West, several CIS member states revived the Soviet Union's old interest in imports from Iran. The old IGAT-1 export pipeline through which the Soviet Union had imported 10 bcm per year of Iranian gas during the 1970s stood empty since more than a decade as a result of the Iranian revolution, and IGAT-2—through which large volumes of Iranian gas were to have been transited to Western Europe—remained half-built. Talks aiming to reinstate the trade had been held during the second half of the 1980s, only to be interrupted by the political chaos of 1989–1991. It did not take long, however, before Azerbaijan, through which Iranian gas had earlier been imported, approached the Iranians showing itself interested in reviving the old import regime. The Azeris joined forces with Ukraine, which similarly sought a way out of its dependence on Russia. Although Iranian gas would have to be transited both through the politically unstable Caucasian region and parts of southern Russia, the Ukrainians regarded possible supplies from Iran as a promising option, not least because the pipeline infrastructure necessary for the transit was already well developed. Azerbaijan's military enemy, Armenia, whose gas supply had been severely hit by targeted political cutoffs, approached Iran with a proposal for an alternative import pipeline.[23]

The potential benefit of a better integration with the international gas system was convincingly demonstrated by Czechoslovakia. By virtue of its central role in the transit of Soviet natural gas to Western Europe, Czechoslovakia was the only former communist country that was already well integrated with the rest of Europe. The Czechoslovaks proved able to exploit this position for both security and economic purposes. The country was a major gas user, with an aggregate consumption of 22 bcm per year at the time of the 1989 velvet revolution. Nearly all of this gas was red. In 1991, the new democratic government in Prague and the national Czech gas company set out to diversify its supply by negotiating an import of 4 bcm of Norwegian gas, to be shipped through Germany along the pipelines built in the 1970s and 1980s for transit in the opposite direction, and 2 bcm of Algerian gas by way of the Trans-Mediterranean and Trans-Austrian Pipelines. Prague declared that it intended to "cut in half" its 97 percent dependence on the former Soviet Union by 2005. In physical terms, however, neither Norwegian nor Algerian gas would reach Czechoslovakia, since a much more powerful stream of Russian gas traveled in the opposite direction through the same system.[24]

A key circumstance that enabled both Western and Central Europe to avoid the most severe effects of the post-Soviet turmoil was the existence of long-term supply contracts. Russia, taking over most of the Soviet gas contracts, saw no reason to abandon or change the arrangements made with its Western customers years and decades earlier, and the overall institutional regime thus remained stable. In contrast, deliveries across previously internal Soviet borders had never been contractually regulated and there was no established tradition on which to build a stable institutional framework. Union republics had received their gas in accordance with annual and five-year plans negotiated with and coordinated by Gosplan and Gossnab in Moscow. There had been no contractual relations of the kind that regulated export flows to Central and Western Europe and no formulae for calculating the gas price. After the Soviet Union's collapse, everything thus had to be negotiated from scratch. This was so despite the fact that the pipeline infrastructure already existed and gas had flown through it for several decades already.

Difficulties to agree on gas prices and transit fees and, once agreed upon, to make distributors and customers pay them became recurring themes throughout the former Soviet Union. The problems were exacerbated by political tensions between former union republics, so that in times of crisis it became difficult to discern whether gas exports were disrupted for economic or political reasons. In June 1993, for example, Estonia's new national parliament adopted a controversial citizenship law that seemed to seriously discriminate the large Russian-speaking minority living in the country. The Kremlin was furious, regarding the new Estonian law as a form of "apartheid." Russian premier Andrei Kozyrev explicitly stated that Russia might turn to economic boycotts of Estonia in protest of the new citizenship law, emphasizing that natural gas, for which Estonia was totally dependent on Moscow, would function as a main instrument.[25]

A few days later, the gas supply to Estonia was actually disrupted. But what was the real motive? Kozyrev's statement indicates that politics certainly played a part. This was also a reasonable interpretation in view of a report issued in 1992, written by a group of top Russian decision makers, including two deputy ministers—for energy and for defense—in which it was clearly stated that Moscow was going to use both oil and gas exports for putting troublesome neighbors under economic and political pressure. For Gazprom, however, the main motivation to disrupt supplies had to do with the failure from the side of the new Estonian gas company to pay for earlier deliveries. The accumulated debt, which was not questioned by the Estonians, amounted to $8 million.[26]

Indeed, gas supplies to Estonia were resumed after four days following the conclusion of an agreement between the two gas companies on how the debt was to be regulated. The overall impression was that Gazprom managed to use the political crisis between Russia and Estonia for speeding speed up the process of solving the debt issue. The ambiguity of the crisis increased when, two days after the Estonian cutoff, Russia stopped supplying Lithuania as well. This country had not adopted any controversial citizenship law, but its gas debt was five times as large as Estonia's. Gazprom also threatened Latvia and Belarus with cutoffs on similar grounds. All in all, the Estonian crisis was clearly part of a broader campaign from Russia's side to make clear to its customers that nonpayment would under no circumstances be accepted in the post-Soviet transnational gas trade.[27]

The most serious post-Soviet dispute was between Russia and Ukraine, the main new transit country for Russian gas on its way to Western Europe. Framed both as a political and an economic conflict, this dispute was to plague Europe for decades. Following the dissolution of the Soviet Union, the Ukrainian economy collapsed and many gas customers found themselves unable to pay their bills, which in turn caused problems for Mingazprom's Ukrainian successor agency, Naftogaz, to regulate its payments with its external suppliers. A huge debt accumulated, and in the same vein as in the Baltics, Gazprom repeatedly threatened to—and often did—disrupt supplies. Since nearly all Russian gas destined for Western Europe was transited through Ukraine, West European gas companies were strongly affected.

The first major Russian-Ukrainian crisis began in October 1992. Serious supply disruptions were reported both by Naftogaz and by Central and West European gas companies further downstream. At first, the problems were interpreted as the unfortunate result of extreme weather conditions. As such it did not seem to deviate in any notable way from experiences made during the Soviet era. A few weeks later, however, it was revealed that the harsh weather had not been the only cause of the cutoff. Instead, Russia had reduced its shipments to Ukraine "because of unresolved problems regarding prices and volumes."

The Ukrainians judged that they, in this situation, had the right to take some of the gas destined for export and use it for domestic purposes, since existing agreements between Ukraine and Russia allegedly stipulated that "any curtailment of gas deliveries by Russia's Gazprom through pipelines crossing Ukraine's territory must be applied proportionately among all customers," whether in Ukraine or abroad. As a result, German, Austrian, Italian, and Swiss gas importers experienced cutoffs as large as 50–75 percent of normal supplies during a period of two weeks. Ruhrgas was shocked by this experience, which was the first time ever that Germany was hurt by an intentional cutoff in its supplies from the East. The company had always defended large-scale imports from Russia as more reliable than deliveries from other suppliers, Norway included. This view was now challenged.[28]

The price and payment problems in Russian-Ukrainian gas relations worsened after Ukraine introduced its own national currency. Implemented in 1993, this key reform meant that Ukraine gave up its monetary union with Russia, and as such the move was not well seen in Moscow. The Kremlin, without explicitly referring to Ukraine, responded by ordering Gazprom to "charge world prices for oil, gas, and other natural resources to former Soviet republics that no longer use the ruble." Moscow would no longer "subsidize the economies of neighboring states" by selling them gas at below-market prices.[29]

In what followed, Ukraine's debt problems were dramatically scaled up. The Russian government made an attempt to solve the payment crisis by linking it to political issues that had little to do with natural gas. It proposed, in particular, that Ukraine's gas debt be cancelled under the condition that the country "returned control of the Black Sea Fleet to Russia and returned all remaining nuclear warheads." Kiev did not accept this offer, nor did it give in to Gazprom's attempts to acquire large stakes in the Ukrainian transmission system and its gas storage facilities. As a result, the crisis lingered on and the debt continued to increase. By early 1994, the total combined debt of Ukraine and neighboring Belarus already amounted to a staggering $1.14 billion—a huge sum for two impoverished ex-Soviet republics. In March, the Russians made a new attempt to force Ukraine to pay, by reducing gas deliveries to a minimum. The Ukrainians also had accumulated a large debt to Turkmenistan, the main successor state of the Soviet Union in the Central Asian gas business, which similarly suspended supplies. Again, the cutoffs had severe implications for gas supply to Western Europe, where ÖMV and ENI, for example, reported a 50 percent drop in pressure on the Trans-Austrian Pipeline. Turkey was another importer that was affected.[30]

Gazprom president Rem Vyakhirev complained that failure to collect revenues from the company's gas deliveries to Ukraine and other former Soviet republics, as well as from domestic customers, created "considerable difficulties in financing new gas production and transportation projects as well as reconstruction of current operations." Pipeline construction came to nearly a complete halt, whereas maintenance and upgrading of existing lines, compressor stations, and other vital components was left to suffer. As a result, the overall vulnerability of the system increased—in a way that was bound to have repercussions on gas export security.[31]

Managing Dependence

The European gas industry followed the early post-Soviet trends with unease. As for the long-term prospects of Russian gas, however, Western gas companies were not necessarily that pessimistic. The end of the Cold War and the transition to market economy throughout Central and Eastern Europe increased the possibilities for Western technology to strengthen the former Soviet pipeline grid, and foreign investment seemed to offer a partial solution to the lack of capital. The overall expectation was that the price disputes and nonpayment dilemma in the intra-CIS gas trade would be resolved within a few years, that the technical reliability of supply would rise considerably following refurbishment of the infrastructure and the establishment of a stable contractual framework, that Gazprom would undergo transformation into a "normal" gas company, guided by market economy principles and sound business logics, and that Russia would eventually emerge as a worthy successor of the Soviet Union in terms of export security.

Given this optimism, *reducing* Europe's dependence on Russian natural gas was for most importers not an issue. Rather, the challenge was to *manage* dependence. During the Cold War period, countering import vulnerability had mainly taken the form of initiatives to strengthen—materially and institutionally—the intra-European system. Improving the interconnectedness between different regional pipeline networks and creating underground storage facilities had been key challenges. In the post-Soviet period, Western gas companies sought to expand these initiatives into Russia itself and the transit countries. Post-Soviet gas companies were approached with bids for closer cooperation, investments, and joint ventures.

As early as 1992, Gaz de France, which at the time imported 31 percent of its gas from Russia, was reported to have set up a "coordinating committee" with Gazprom "to deal with security of the gas transmission network, develop underground storage projects, and save energy all along the gas chain." It also set up offices in Prague and Bratislava as a basis for joint ventures with the most important Central European transit country, Czechoslovakia. Ruhrgas, for its part, tried to take an active role in the future of Russian natural gas by participating in the privatization of Gazprom. By 2004 the Germans had acquired 6.4 percent of the former gas ministry's shares. Supported by the federal government, Ruhrgas also spent considerable effort trying to bring about an "international consortium" of gas companies, consisting of Gazprom, Naftogaz, and Ruhrgas, with the possible inclusion of further Western importers. Its task would be to fund and organize the refurbishment and further development of Ukrainian transit objects.[32]

Ruhrgas, GdF, ENI, and ÖMV all dreamed of acquiring stakes in the transmission companies that were in charge of transiting Russian gas to the West. Actual success in this field, however, was limited. One reason was that the companies faced competition from Gazprom, which similarly aimed to increase its influence over the transit system. In the end the Russian company emerged as the main foreign partner in several Central European countries. But Gazprom did not make halt at the former Iron Curtain; it continued its expansion into Western Europe. In Germany it was already well-established through its joint venture with Wintershall, and similar initiatives were on the way in other importing countries. In addition, the Russians became active in negotiating access to underground gas storage capacities in the Netherlands. The Dutch gas industry, once Western Europe's most important source of natural gas, was in the post–Cold War period about to be transformed from a major exporter into a provider of "hub services." Many of its gas fields were no longer used for production purposes, but for load management in the wider West European system. Through its stakes in these storage facilities, Gazprom strengthened the reliability of its exports, since Russian gas stored in the Netherlands could now come to rescue in case of problems along the troubled Ukrainian transit route.[33]

The activities of the gas industry itself was accompanied by political efforts to manage dependence. National governments in the importing nations were keen to take up problems of the gas trade for discussion at high-level political meetings with Russia and the transit countries. At the international level, an early political attempt to improve the security of East-West gas flows was the EC Commission's proposal, originally presented in 1990, to construct a "charter of principles governing long-term energy cooperation between the EC and the Soviet Union." The initiative eventually led to the so-called Energy Charter Treaty, which was signed in 1994. Although Russia ultimately did not ratify this document, Moscow considered the Charter the best forum for negotiating gas transits at the international level.[34]

The Energy Charter was later followed up by additional efforts, mainly by the EU, to improve supply security. Brussels' main goal was to push through a liberalization of the European gas market, but these attempts became closely related to Commission chairman Romano Prodi's initiative, launched at the EU-Russia Summit in Paris in October 2000, to embark on an "EU-Russia Energy Dialogue." This dialogue contributed significantly to shaping the EU's much-debated gas liberalization policy, which in its early version had been severely criticized by Gazprom for doing "serious damage to European energy security." At stake was mainly the tradition of long-term contracts and of prohibiting gas importing companies from freely reexporting or reselling Russian gas. The Commission eventually showed itself willing to give in to some of Gazprom's demands, particularly the ones concerning long-term contracts, and, as a result, relations improved.[35]

Overall, activities from the side of both governments and companies in the turbulent post-Soviet period testified to a persistent conviction that the problems encountered in the East-West gas trade could and would be solved through sound and mutually beneficial cooperation. Dependence, it was believed, could be effectively managed in such a way as to keep Europe's vulnerability at a minimum. Whereas opposition parties in several importing countries, along with journalists and a general public that was alarmed by the rise of intentional cutoffs of Russian gas, increasingly argued that dependence on Russia must be reduced, this idea appears to have been virtually nonexistent as far as governments and gas companies were concerned.

An important trend that influenced this stance was that forecasts of Europe's gas demand were adjusted upward. This circumstance, combined with the fact that Dutch natural gas production had peaked, while the future of Norway's offshore fields looked uncertain, made Russian gas appear extremely attractive. Only the Soviet Union's former satellites in Central Europe, many of which were already dependent on Russian gas to 100 percent, at times vigorously opposed a scaling up of the system. Whereas West European imports of Russian gas grew by a staggering 70 percent from 63 bcm in 1991 to 107 bcm in 2004, Central European imports remained constant at 42 bcm.[36]

The "Molotov-Ribbentrop" Pipeline

An important aspect of—and argument in favor of—continued expansion of Russian exports to Europe was that it might be combined with a diversification of the transit routes. During the Cold War period, nearly all Russian gas destined for Western Europe had been transited through a narrow Ukrainian-Czechoslovak corridor. This corridor had originally been created in 1967 for the Soviet-Czech "Bratstvo" pipeline, and all subsequent export projects—except the lines to Finland, Greece, and Turkey—had been built in parallel with this line. Such a heavy reliance on a single route was considered inconvenient. Both exporter and importers agreed that any new export pipelines had better take alternative routes. Diversification of the transit routes was considered important both for security reasons and because it would enable actors to exploit economic competition between the transit countries.

The principal alternative to the Ukrainian transit was the route through Belarus and Poland to northern Germany. Apart from a pipeline built in 1986 for the purpose of exports to Poland, it had remained unexploited during the Cold War period. One reason was the GDR's unclear status and its difficult relations with West Germany, another the political unrest in Poland in the early 1980s. In the post–Cold War era, however, nothing seemed to stand in the way for realizing the project. It was highly attractive not only for Russia and Germany, but also for the Poles, who saw it as an excellent opportunity to turn their country into an important hub for Central Europan gas flows.

If realized, the new transit system promised to improve Poland's access to Russian natural gas while also enhancing its connectivity with the West European gas grid. This would be beneficial both for security reasons and because it, at least in principle, opened up for competition between Russian and Western suppliers on the Polish gas market. The Poles hoped to further strengthen their position by combining the envisaged East-West system with a North-South pipeline. The latter, it was imagined, would be used for importing and transiting Norwegian gas, which would arrive by way of the Baltic Sea and traverse Poland on its way to Czechoslovakia and other former communist countries.

Gazprom's German partner in the Polish project was Ruhrgas' competitor Wintershall. The project gained momentum following a deal in which the latter's owner, the chemical giant BASF, agreed to build a huge chemical complex in Western Siberia in return for exclusive rights for Wingas to market the gas that entered Germany through the new pipeline. Shortly afterward, in August 1993, the Russian and Polish governments under ceremonial forms—with Presidents Boris Yeltsin and Lech Walesa attending—signed an intergovernmental agreement on "construction of a transit gas pipeline system for transportation of Russian gas through the territory of the Republic of Poland and on deliveries of Russian gas to the Republic of Poland." The agreement stated that the transmission capacity of the system would eventually reach 67 bcm per year. (By comparison, Ukraine's transit capacity amounted to around 100 bcm.) Poland itself would take 14 bcm for domestic use.[37]

In 1994 the Polish pipeline was formally included as a priority project in the EU's Trans-European Networks (TEN) program, and started to be referred to as the "Europol" link. Thanks to strong support from both national governments and Brussels, actual construction could soon start. The first section to be built was a 107-km long part stretching from Lwówek near Poznań to Górzyca on the German border. It was completed in November 1996. At Górzyca, this link was connected with a short pipeline section built by Wingas, which crossed the Oder River and continued into the ex-GDR. This link was merely 11 km long, but it was of historical importance, since it together with the Polish section of the Europol system and the already existing, older Belarusian-Polish link for the first time enabled Russian gas to be exported to Germany without involving Ukraine.[38]

Two more sections were successfully completed and taken into operation in September 1999, considerably improving transmission capacity. The route for Russian gas exports through Belarus and Poland seemed to be on good way to emerge as a serious competitor to the dominant Ukrainian route. In what followed, however, system-building stagnated. To enable transmission of the volumes mentioned in the Russian-Polish agreement, two large parallel pipelines would have to be built. In reality, it now seemed highly uncertain as to whether the second pipe would actually materialize. By 2005, the amount of gas shipped along the first pipeline amounted to 20 bcm, which was certainly not a negligible volume but still a far cry from the 67 bcm target.[39]

A main reason for this stagnation was that Russian-Ukrainian gas relations, from around 2000, seemed to be improving. This coincided with the election of Vladimir Putin to the Russian presidency and the appointment of a new management board for Gazprom. In what followed Gazprom launched a major effort to improve its Ukrainian relations. Winter 1999–2000 had seen renewed disturbances in the gas flow through Ukraine to Central and Western Europe. According to Gazprom the delivery failures occurred because Ukraine diverted some of Gazprom's export gas to domestic users. The dispute was not at all the first of its kind, but Gazprom now responded by starting concrete preparations for a radical new project that would make Russian exports independent of Ukraine for a large portion of its exports to the West. The project took the form of a pipeline through Belarus and from there into Poland's southeastern corner. In contrast to the Europol system it would not continue westward to Germany, but head south into Slovakia, where it would link up with the larger East-West transit system already in place. The project was radical because it did not intend to *complement* the Ukrainian transit, but *replace* a large part of it.[40]

The Ukrainians interpreted Gazprom's new plan as a real threat to Ukraine's profitable transit business, and the result was that Kiev and Naftogaz started to take greater interest in improving their relations with Gazprom. By 2001, new agreements on exports to Ukraine and transit to Europe had been signed. Although Ukraine's debt still needed to be regulated, it was, as Jonathan Stern notes, a "carefully structured attempt to break with a lawless past." Relations seemed to be normalizing, and Gazprom's replacement project was dropped. At the same time, ideas were developed to expand Ukraine's transit role through the construction of additional pipelines.[41]

From 2003–2004, however, both the Ukrainian and Belarusian transit again faced serious problems. Increasing Belarusian gas debts made Gazprom threaten with delivery suspensions. In February 2004, following several failed attempts to regulate the situation, supplies were disrupted. Importers further downstream the new Belarusian-Polish transit system protested, particularly Poland itself. This crisis was followed by political turmoil in Ukraine, where the "Orange Revolution" paved the way for stronger political and economic ties with the West. The revolution first seemed to improve the prospects for strengthening transit security, as the chances for West European companies to take part in the refurbishment of the Ukrainian infrastructure increased. Yet this development was soon overshadowed by mounting political tensions between Kiev and Moscow. The relative stability that the 2001 agreements had helped establish quickly gave way to renewed chaos, culmating in a series of major supply crises.[42]

The growing turmoil and unpredictability of Russian-Ukrainian relations strengthened both Russia's and Western Europe's interest in alternative transit routes. With both Ukraine and Belarus being regarded as highly problematic partners, the main emphasis was now placed on a Russian–West European route that did not include any transit country at all: through the Baltic Sea. Such a route had been discussed as an interesting possibility already during Soviet times, though without much concrete action taken. The original plan in this context had been to extend the already existing Soviet-Finnish export system to Sweden and, possibly, onward to Britain and/or northern Germany. The project had at times seemed close to a breakthrough, but had in the end always failed to take off, a main reason being Sweden's hesitation.

Following the collapse of communism, the Baltic Sea visions gained momentum again, whereby several alternatives were considered. Apart from a pipeline through Finland and Sweden, these included a line through Finland with a subsea extension to Germany, a subsea pipeline from the Russian Kaliningrad exclave across the Baltic Sea and onward through Denmark to Britain, and, finally, a direct subsea line from the St. Petersburg area to continental Western Europe and, ideally, onward to Britain. A study released in February 1992 found the Danish-British project "technically feasible and economically viable." Throughout the 1990s, however, the Finnish-Swedish version of the new export route was at the forefront in the discussions. In 1997, Gazprom and Finland's Neste set up a joint venture, North Transgas, to study its feasibility. The line would be dimensioned for transmission of 20–35 bcm of Russian gas per year, on par with Western Europe's imports of Norwegian Troll gas.[43]

As before, however, Sweden hesitated, and as a result the focus shifted from a Scandinavian transit to other arrangements. Joining forces with Ruhrgas, North Transgas first proposed a pipeline that would traverse Finland and continue from there on the bottom of the Baltic Sea to Germany. In the end, however, Gazprom and Ruhrgas opted for a route that would bypass Finland as well, from Vyborg near St. Petersburg to Greifswald on the former GDR's Baltic coast. Greifswald was considered an excellent landing point since it was home to a vast Soviet-designed nuclear complex that had just been decommissioned and for which gas-fired capacities were eyed as a possible replacement. The new subsea pipeline would be an ideal way to guarantee such a power plant's fuel supply.[44]

Apart from Germany, Britain was expected to become a partner in the project. Already during the Soviet era, Gazprom had dreamt of expanding its sales to across the Channel, and the Baltic Sea pipeline was regarded as a key link for large-scale deliveries to the United Kingdom. In 1999, Gazprom for the first time entered Britain on a spot market basis, but the company had much higher ambitions. It imagined an extension of the new export system from Greifswald westward through northern Germany and the Netherlands, and from there in the form of another subsea line to the English coast. This ambitious extension of the Baltic project would, in addition, improve Gazprom's access to Dutch storage capacity, a key factor both for security of Russian gas supply in Western Europe and for Gazprom's participation in the increasingly liberalized EU gas market. In 2003 the United Kingdom and Russia signed a "bilateral energy pact" that incorporated a tentative agreement to jointly create a new export system based on this vision.[45]

During the first few years of the twenty-first century, the seeming stabilization of gas relations between Russia, Ukraine, and Belarus made the Baltic Sea project look less urgent, and the main emphasis was on expanding the Belarusian-Polish transit by adding a second pipeline along this route. Following the 2004 gas dispute between Russia and Belarus and the subsequent political revolution in Ukraine, however, the Baltic project started to gain momentum. In September 2005, a "basic agreement" was reached for the construction of the subsea line, paving the way for the North European Gas Pipeline Co. (NEGP) to be established. In its initial version, this was a Russian-German joint venture in which Gazprom held 51 percent, Ruhrgas (which at this time had become part of the larger E.ON group) 24.5 percent, and BASF (the owner of Wintershall) 24.5 percent. NEGP announced that the pipeline would be 1,200 km long and that the aim was to have it in operation by 2010. The ultimate transmission capacity would be 27.5 bcm per year, but preliminary plans were already drawn up for a doubling of this capacity through addition of a second, parallel line.[46] In 2007, by which time NEGP had been renamed Nord Stream AG, an agreement between the three original partners allowed for the Netherlands' Gasunie to take over 9 percent of the shares from the two German companies.[47]

The need for the Baltic Sea pipeline seemed dramatically confirmed soon after the signing of the 2005 agreement, as Russian gas exports through Ukraine were threatened by a new, serious conflict. On January 1, 2006, gas supplies to Ukraine were disrupted, and again several importers further downstream in Central and Western Europe were hit. The recent Orange Revolution and the election of Viktor Yushchenko, a decidedly pro-Western political figure, to the Ukrainian presidency added a strong political dimension to the crisis. The crisis was temporarily solved through negotiations on debts and prices, but similar gas crises became a recurring headache for both Eastern and Western Europe during the following years. The Russian-Ukrainian conflict culminated in a much-publicized, even more serious cutoff in January 2009, which affected nearly all European countries in one way or the other. Never before had Western Europe suffered in such a way from Russian gas supply insecurity.

Meanwhile the planned Baltic Sea pipeline met with opposition, particularly from actors in countries that the line would bypass. In Sweden and Estonia, the Baltic pipeline project was criticized mainly for its possibly adverse effects on the Baltic Sea's sensitive ecosystem. But the project was also debated from a geopolitical and even military perspective. The most severe criticism came from Poland, where it was noted that Nord Stream was likely to kill the prospects for construction of the second Europol transit line. Polish foreign minister Radosław Sikorski acidly dubbed the project "the Molotov-Ribbentrop Pipeline," since it in his—and many others'—view aroused unpleasant associations to the infamous Soviet-German pact of 1939.[48]

Through this provocative interpretation of the new transit project, East-West natural gas relations had in a sense returned to where it had all started. It was precisely the Molotov-Ribbentrop Pact that had paved the way for Soviet annexation, in 1944, of Galicia, at the time one of Europe's most important oil- and gas-producing regions. A system of long-distance gas pipelines had already started to take form in Galicia, and the new Molotov-Ribbentrop borders turned this system into an international one, allowing the Soviet Union, for the first time, to export natural gas. Gas from the same Galician sources started to be exported to Czechoslovakia in 1967 and to Austria in 1968. In view of the fact the Soviet gas geography of the late 1960s made it impossible to export natural gas from other gas-producing regions, these pioneering deals would probably not have come about, had Galicia, following the Molotov-Ribbentrop division of Europe, not been annexed.

12
Conclusion

"Will Europe Come to Depend on Russian Natural Gas?" read a headline in the *Oil and Gas Journal*'s August 28, 1961 issue. It was a radical suggestion at the time, and few analysts—to the extent that they took notice of it at all—believed that imports of red gas, let alone any dependence on it, would ever come about. A number of factors seemed to speak against it.

Most fundamentally, natural gas was for most Europeans still a largely unknown energy source that did not play more than a negligible role in the energy debate. This made it far-fetched to think of it as a dependence-generating fuel. To the extent that energy dependence was an issue in the early 1960s, it centered on oil imports, and it was precisely for the purpose of countering oil dependence that some European countries had taken an interest in natural gas in the first place.

Moreover, to the extent that gas imports had started to be considered, Algeria was eyed as the logical supplier. Regarding the Soviet Union, it was far from certain that it would have any gas to export and, if so, that it would be willing to do so. Siberia's vast gas Eldorado still remained to be discovered, and Ukraine's gas was piped in eastern rather than western directions.

Furthermore, factors of a more political nature spoke against a future European dependence on Soviet gas. Through the erection of the Berlin wall in mid-August 1961, the Cold War had reached a new height. A year later, the Cuban missile crisis followed, bringing the world to the brink of nuclear war. Such geopolitical tensions, in the eyes of most observers, made it highly improbable that Western Europe would choose to make itself dependent on piped energy from the East. The recent discovery of vast natural gas resources in the Netherlands, large enough to cover West European gas demand for the foreseeable future, added to this perceived improbability and nonnecessity of gas imports from the Soviet Union.

Yet a few years later, Europe had chosen precisely that path, opting for imports of Soviet natural gas large enough to make several Western countries highly dependent on it. In the period from 1968, when small volumes of Soviet gas for the first time started flowing into Austria, through 2011, when the controversial Nord Stream pipeline was inaugurated, Russian natural gas rose to become one of Western Europe's most important energy sources. In absolute terms, imports grew from a modest 1.5 bcm per year in the early 1970s to 29 bcm at the time of the "Yamal" pipeline controversy in the early 1980s, to 63 bcm when the Soviet Union was dissolved in 1991, and 107 bcm in 2004. Currently, apart from most ex-Soviet republics, 15 West European and eight Central European countries depend on Russian natural gas via regular, long-term contracts. A few additional countries such as Belgium and the United Kingdom import Russian natural gas on a spot market basis. Through the planned addition of new pipeline capacity, the gas flow from the East is scheduled for continued growth.

This book set out to explain why Western Europe, in the midst of the Cold War, chose to make itself dependent on Russian natural gas, and why it was prepared to successively scale up these imports so massively. It also sought to understand how the involved actors, during the Cold War period and beyond, learnt to live with—and profit from—its dependence on piped energy from the East. This chapter summarizes and synthesizes the findings.

Dependence in Retrospect: Four Phases

The making of the East-West gas system can be divided into several more or less distinct phases. The *first phase* started in the late 1950s, at which time it was still uncertain whether natural gas, though already widely used in North America, would ever become a fuel of any significance in Europe. This was because the known intra-European gas resources were very limited. It was precisely this internal scarcity that inspired a number of visionary actors to consider the idea of longhaul imports. The main interest was in French Algeria's large gas fields. The possibility of Soviet gas imports was largely formulated in analogy with the Algerian visions. For the time being, however, none of the prospective system-builders managed to mobilize sufficiently strong support for any of the proposed projects. In the early 1960s it then gradually became clear that the Netherlands rested on huge natural gas reserves. On the short term, this tended to lower the interest in possible gas imports from far away. On the long term, an important effect of the Dutch finds was that they convinced Europeans of natural gas' potential as a fuel of the future. As a result, overall demand for natural gas started growing in earnest and long-distance pipeline construction took off.

By 1964 the Soviet Union had made up its mind to become a gas exporter, aiming for market shares in both Central and Western Europe. A first export agreement with Czechoslovakia inspired several governments and gas companies in Western Europe to seriously consider the possibility of linking up with the Soviet grid. Concrete negotiations were initiated in 1966 and 1967 with Italy, France, Austria, Finland, and Sweden, as well as with Japan. But the actual outcome was meager, being limited to a contract with Austria, signed in June 1968. Three months later, Soviet natural gas was for the first time piped across the Iron Curtain.

The *second phase* in the making of Europe's dependence started with the Warsaw Pact's invasion of Czechoslovakia in August 1968, just as Austria was preparing for its first imports. As it turned out, this critical event did not disturb the emerging gas trade, but rather helped to boost it. This was because most Western governments reasoned that Soviet-style totalitarianism could be more effectively managed—and ultimately eliminated—"through rapprochment," as Willy Brandt and Egon Bahr put it, than through confrontation and isolation. Natural gas was in this context identified as an important vehicle for strengthening West European–Soviet relations, and as a result a new wave of gas negotiations were initiated in early 1969. The prospective importing countries were the same as in 1966–1967, with the important addition of West Germany. Pioneering contracts were eventually signed with Italy in December 1969, with Germany in February 1970, with Finland in December 1971, and with France in July 1972, whereas analogous talks with Sweden failed.

The new contracts firmly established the Soviet Union as a major contractual partner for several of Europe's largest gas companies. Whereas the pioneering imports to Austria had been realized through a minor extension of an already existing pipeline from the Soviet Union to Czechoslovakia, the new contracts, foreseeing a ten times larger gas flow, necessitated construction of a much more complex and costly transmission system. The Soviet Union faced enormous difficulties to complete its part of the system on schedule, but from winter and spring 1973–1974 the new pipelines could eventually be taken into use. Germany, Italy, and Finland now received their first Soviet gas deliveries, followed by France in 1976.

In the *third phase*, those countries that had already negotiated Soviet gas contracts sought access to additional volumes of red gas. In parallel, several new prospective importers signaled their interest in linking up with the emerging East-West system. Expensive pipelines had now already been built, transgressing the Iron Curtain, and red gas already reached customers in many parts of Western Europe. The challenge eyed was to further scale up the system. Western Europe was shocked by the 1973/1974 Arab oil embargo and the radical increase—and growing unpredictability—of world oil prices. This stimulated attempts to diversify energy supply both in terms of fuel and geography. Environmental concerns also contributed to boosting the popularity of natural gas. Fears were raised that the Soviet Union might turn out an equally unreliable partner as the OPEC, but most actors judged that Soviet natural gas was part of the solution rather than part of the problem.

For those countries that already had experience of negotiating gas import contracts with the Soviets, and of importing red gas in practice, the talks were now much easier to conduct than in the 1960s. The trust that had been established through earlier cooperation turned out to be a valuable asset in the new, more turbulent energy era. New prospective importers—notably Sweden, Switzerland, Belgium, and Spain, but also the Netherlands and the United States—failed to come to agreement with Moscow, a strong interest in Soviet gas notwithstanding. Apart from imports of red gas proper, supplies from Iran, to be transited through the Soviet Union, were hailed as a major opportunity for Western Europe. Germany, Austria, and France in 1975 agreed with Moscow and Tehran on a major tripartite "switch," in which Iran was to pipe large volumes of gas to the Soviet Union, which in turn would transmit a corresponding volume to Western Europe. This was followed by the much-debated Yamal negotiations, which started in 1978 and resulted in a series of bilateral contracts, most of which were signed in 1981–1982. Whereas implementation of the tripartite deal was ultimately jeopardized by the Iranian revolution, the Yamal project materialized as planned, paving the way for more than a doubling of previously contracted imports. However, the resurgence of Cold War in the early 1980s made the Yamal deals highly controversial. The United States, which up to then had not objected to European imports from the East, vigorously opposed the ambitious project and actively sought to prevent or at least delay it—though in vain.

In the *fourth and last phase*, European gas importers had to cope with the radical fact that East European communism had collapsed and the Soviet Union ceased to exist. A period of organizational and institutional chaos followed. The sudden emergence of new transit countries—Ukraine, Belarus, and Moldova—generated new vulnerabilities in the overall system. Rather than trying to *reduce* its dependence on Russian gas, however, Western Europe opted to do what it could to *manage* it. At the same time proposals for a further scaling up of imports were developed. The West Europeans believed that the post-Soviet political and economic crisis was of a temporary nature. In reality it turned out to be chronic. Importers sought to actively contribute to the crisis' solution, but in vain. Failure to establish a stable contractual regime for the gas trade between former union republics was a key factor behind a series of intentional Russian supply cutoffs to the new transit countries. Western Europe was indirectly affected by these events. Seeking a way around the problem, Russia and its Western customers increasingly channeled their efforts to create new transit routes, of which the most controversial was a direct pipeline from Russia to Germany, laid on the bottom of the Baltic Sea.

Remarkably, neither the resurgence of Cold War in the 1980s nor the collapse of the Soviet Union in 1991 had any notable impact, quantitatively speaking, on the overall system's continued expansion and growth. The Iranian revolution, the second oil price shock, the Soviet Union's invasion of Afghanistan, the Polish crisis of 1981–1982, and Reagan's new confrontative policy toward Moscow—none of these international crises were able to stop the West Europeans from radically scaling up their imports of red gas in the 1980s. Similarly, the fall of the Berlin Wall, the division of the Soviet Union into 15 independent states, and the extreme institutional and economic disorder in the former red empire were unable to bring about any slowdown in East-West system-building. Imports continued to grow and additional pipelines were constructed. This sustained expansion is especially remarkable if seen in relation to mounting opposition in the West, starting in the 1980s, to dependence on the Soviet Union and Russia as voiced by powerful actors in Europe and elsewhere. In LTS terms, the East-West gas system had by the 1980s acquired considerable "momentum."

From where, then, did this momentum come? Why did Soviet and West European stakeholders in the first place decide to establish a transnational gas system? What compelled them to take the first steps into this unknown and obviously risky realm? Why did they not follow in the footsteps of, for example, Europe's electricity system-builders, who for political and military reasons had created separate East and West European grids, with only minimal interaction across the Iron Curtain? Based on the stories told in this book, the following sections explore this intriguing issue first from a Soviet, then from a West European perspective.

Energy Weapons: Real and Imagined

Whether Russia, through its natural gas exports, possesses an "energy weapon" and the extent to which it, if so, makes use of it in actual practice, has been subject to fierce debate in the wake of repeated post-Soviet gas crises. The wide range of conclusions arrived at by different analysts is striking, suggesting that energy weapons do not (always) objectively exist, but that we are more likely to grasp their nature if we treat them as social constructs. In other words, it may be useful to think of an energy weapon as existing only to the extent that it is believed to exist, and to place perception rather than objective reality at the center. Whether or not Soviet/Russian natural gas has "actually" constituted an energy weapon, the involved actors have been forced to relate themselves to such a weapon's perceived existence. Importantly, Russia's energy weapon, whether or not "real" or "imagined," has thereby had a very concrete influence on system-building activities such as the dimensioning of Western Europe's underground gas storage facilities, its efforts to build interconnecting pipelines with alternative gas suppliers, and its overall ambitions to diversify supply.

Moreover, the material presented in this book suggests that the usual understanding of Russian gas exports to Western Europe as a "threat" needs to be broadened. The debate so far has focused almost exclusively on the fear of *politically motivated supply disruptions* and, accordingly, Russia's alleged ability to influence West European politics through the mere potentiality of such cut-offs. The available material does not support the thesis that Moscow, during the Cold War, contemplated the use of natural gas exports for this purpose. Particularly in the early phase of the trade, the Soviets were well aware that it would be pointless to try and use it for political blackmail, simply because the imported volumes were much too insignificant in relation to total European energy use. Later on, the Kremlin was able to observe how the importing companies developed advanced mechanisms that would enable them to respond effectively to any supply crisis, irrespective of the causes of such an event. Importantly, the Soviets were not disappointed by such developments, but, on the contrary, felt encouraged, since these measures indicated that the Europeans were interested in a further scaling up of imports from the East.

There are at least three alternative ways in which Soviet natural gas exports may be—and have been—viewed as a potential weapon. First, the perceived danger of importing Soviet gas was historically discussed in terms of *potential dumping of Soviet gas on West European markets*. In 1967, for example, as the Germans started discussing possible imports of Soviet natural gas, their main fear was not so much the possibility that the Soviets might deliberately disrupt supplies, but rather that the Kremlin might seek to disturb the Ruhr's politically sensitive coal industry, which at the time was facing severe difficulties, by flooding the Federal Republic with cheap red gas. This interpretation of the Soviet energy weapon was rooted in a previous debate that had taken off in the 1950s following an aggressive Soviet oil export strategy. These fears may have been warranted, yet the available sources, while showing that this was a major concern in the West, do not support the idea that it was actually a major purpose of the Soviet Union's attempts to sell natural gas to West Germany, nor to any other country. On the contrary, in the actual negotiations the Soviets quickly earned a reputation for being extremely tough concerning their demands for a high gas price, and several prospective deals failed precisely because of reluctance from the Soviet side to lower its bids.

Second, there was the possibility of using gas exports as a more *indirect* weapon in the Cold War struggle. A main Soviet foreign policy strategy was for a long time to encourage cooperation with some countries though not all. The purpose was *to divide the capitalist world and the NATO*. The available material suggests that the Soviet gas export strategy indeed was molded to fit this overarching foreign policy goal. The Soviet-Austrian gas negotiations of 1966–1968, for example, must clearly be analyzed in relation to Austria's attempts at the time to associate itself more closely with the European Economic Community (EEC). The Soviets vehemently opposed any such association, and natural gas became a brick in the wider political struggle about Austria's future. Moreover, while the Soviets in 1966 and 1967 initiated gas negotiations with Austria, Italy, France, Finland, and Sweden, they deliberately refused to do the same with West Germany. This was well in line with the Kremlin's dominant policy at the time of politically isolating the Federal Republic. When the Yamal pipeline was negotiated in 1980–1982, the Soviets similarly used the East-West gas trade to exploit and encourage intra-Western (and more specifically transatlantic) political tensions.

Third, natural gas exports were used in the Cold War ideological struggle between capitalism and communism for the purpose of strengthening the Soviet Union's *international prestige*. This was emphasized by Gas Minister Alexei Kortunov in his attempts to persuade the Soviet leadership of the importance of bringing about an East-West gas trade in the first place. Soviet media also referred repeatedly to the country's growing natural gas exports with great pride, pointing to the fact that the West Europeans had turned to the communist world's reliable suppliers to solve their energy problems.

Broadening the interpretation of what may constitute an "energy weapon" to the possible ways in which a country may use energy to deliberately hurt or weaken another country or region, directly or indirectly, it may thus be argued that the Soviet Union did possess an energy weapon and that it did make use of it. It was clearly used by the Kremlin as a tool for dividing the capitalist world and for increasing the Soviet Union's international prestige. Yet it was not used for political blackmail, nor for deliberately disturbing West European fuel markets.

In general, the evidence presented in this book indicates that many analysts have exaggerated the role of Soviet and Russian natural gas as an energy weapon. Importantly, even the "softer" political functions of Soviet natural gas exports—that is, its potential to divide the capitalist world and boost the Soviet Union's prestige—were only secondary motives for the Soviets to embark on the gas trade. Regarding the primary motives, these were interpreted differently by different actors. For the Soviet Union's political leadership, the main purpose of gas exports was clearly to generate hard currency, which in turn could be used to cover trade deficits that would otherwise loom large vis-à-vis the capitalist world. This motive grew increasingly important as global fuel prices increased, a trend that was much discussed around the world precisely at the time when the first East-West gas contracts started to be negotiated. The economic dimension of gas exports grew even more important from the mid-1970s, when the Kremlin was forced to acknowledge that it would not be able to continue increasing its oil exports to the West. Scaled-up exports of natural gas, which at the time still accounted for only a small share of overall export income, was identified as the only way forward.

Mingazprom, for its part, had a different primary motive for advocating exports of natural gas. It was Mingazprom, under Alexei Kortunov, that played the main role in bringing up the idea of transnational system-building as a real possibility. Yet the ministry was not first and foremost interested in strengthening the Soviet Union's foreign trade balance, but rather in boosting the role of natural gas in relation to other branches of the Soviet energy and fuel complex. Kortunov had three major motives: first, to use gas exports to the West as a way of mobilizing resources and political support for his ministry's ambitious but controversial plan for a radically expanded domestic pipeline system. Second, to strengthen the domestic legitimacy and prestige of the Soviet gas industry, at the time a young branch of the Soviet economy whose future was still contested. Third, to combine gas exports with the import of high-quality Western steel pipe and advanced equipment, access to which was considered crucial in view of chronic technical problems and coordination failures in the domestic pipe and equipment industries.

Notably, some Soviet actors also argued *against* exporting gas to the West. This was because they considered such a trade riskier than the export of crude and refined oil products. Natural gas demanded a dedicated export infrastructure that would have little alternative use, should the West suddenly decide to withdraw from the project. Gas exports were bound to be inflexible: they would have to take place on a long-term basis, making it difficult to quickly increase or decrease the flow so as to optimize the foreign trade balance—a characteristic Soviet technique in the case of oil exports.

The ways in which the Soviet Union has and has not used or intended to use its gas exports as an "energy weapon" does not necessarily tell us whether Russia might do so in the future. It is one of the salient characteristics of large technical systems that they may be built for one purpose and later on be exploited for another. For the time being (hard currency) revenues clearly remains the primary motive for both the Kremlin and the gas industry. Notably, Russian export earnings are nowadays often interpreted as a political weapon in their own right. In an age of huge state budget deficits and accelerating indebtedness in large parts of the Western world, natural gas export revenues contribute decisively to Russia's overbalanced state budget, and thereby strengthens Russia's power and independence on the international arena. Moreover, the contribution of gas exports to Russia's international prestige remains an important factor. Having lost its previous status as one of the world's two superpowers, Russia's global leadership in gas exports constitutes a more important source of national pride than ever before.

If we turn to Russia's relations with other former Soviet republics, the situation looks different and new motives have appeared since the collapse of communism. Russia's gas exports to the Baltics, Belarus, Ukraine, Moldova, and the Caucasian republics have been plagued by repeated intentional supply disruptions and contractual controversies. For Gazprom, these cutoffs have as a rule followed as a result of nonpayment from the side of the importers. For the government, however, potential and actual supply disruptions have come to form an integral component of the Kremlin's foreign policy. This was apparent already under Yeltsin in the 1990s, and it has become an even more salient feature under Putin and Medvedev. The ambiguous nature of intra-CIS supply disruptions, which are likely to continue plaguing post-Soviet gas supply for the foreseeable future, is at the heart of the overall debate about Russia's present-day reliability as a supplier. Even so, the CIS experience is of little value for understanding Western Europe's evolving dependence, which is of a very different nature.

Understanding Europe's Enthusiasm

Western Europe feared, from the very outset, possible negative consequences of the East-West gas trade. Actors discerned a number of risks, including the abuse of natural gas as a political and economic "weapon," but also technical and organizational risks related to unwanted disturbances in the pipeline system and Soviet failure to make the contracted gas volumes available. Why, then, did West European countries, in the midst of the Cold War, eventually accept these risks? Why did they voluntarily choose to embark on a journey that was bound to make them highly dependent on their main ideological and military enemy?

The stories told in this book provide several complementary answers. First of all it should be emphasized that West European actors, while aware of the risks, identified far-reaching *opportunities* in gas imports from the Soviet Union. In the most fundamental sense, Soviet gas offered a major *supply opportunity*. It was considered a way of getting access to a fuel that was not (sufficiently) available domestically and, as in the case of Finland and Austria, might not have been available from any other supplier on acceptable terms. This is to say that without imports from the Soviet Union, natural gas might have had to play a much less prominent role in Western Europe's primary energy supply. The Soviet contracts were typically negotiated at a time when additional gas supplies were direly needed, usually because domestic gas resources were about to be depleted while demand continued growing exponentially. This was obvious in the case of Austria and Bavaria, which were, therefore, the ones most in a hurry to negotiate Soviet contracts, whereas it was a less salient feature for Ruhrgas and Gaz de France. ÖMV, in particular, was extremely relieved having concluded its first Soviet contract, emphasizing that it prevented the domestic situation from becoming critical.

But there were also *economic opportunities* in linking up with the East. Clearly, countries such as Germany, Italy, and France would have been able to cover their gas demand for the foreseeable future through imports from elsewhere. However, they might have had to accept substantially higher prices. Importantly, this does not mean that Soviet gas was (much) cheaper than Dutch, Libyan, Algerian, or Norwegian gas. Although this was sometimes perceived to be the case—notably for Austria and Bavaria, both of which were located near the Iron Curtain and at considerable distance from other suppliers—the Soviets closely monitored the prices West European gas companies paid for gas from elsewhere, and skilled negotiators such as Nikolai Osipov earned a reputation for being extremely tough in demanding prices at roughly the same level. Failure to agree on the gas price appears to have been the main reason for the initial failure of the Soviet-Italian negotiations in 1966–1968, and the harsh price dispute in Ruhrgas' pioneering negotiations in 1969 was close to killing Germany's prospects for imports from the East. ÖMV, for its part, was severely criticized domestically for having agreed on a too high price for its Soviet supplies. The Soviets changed their pricing strategy only in the 1980s, lowering their bids in order to make maximal use of the newly constructed Urengoi-Uzhgorod export pipeline. By that time, however, the overall East-West gas regime had already been firmly established. In the preceding, formative period of the 1960s and 1970s the Soviets were *not* prepared to go significantly below the price of competing exporters.

To a much greater extent, the economic opportunity lay in the potential of Soviet natural gas to *stimulate the overall competitive dynamics* of the West European gas market. Importers wanted Soviet gas because it would offset the economically unfavorable reliance on what were perceived to be monopolistic suppliers, notably the Esso-Shell group, which was in control of Dutch gas exports, and Algeria's state-owned oil and gas company Sonatrach. In practice, importers such as ÖMV, ENI, Ruhrgas, and GdF tried to make use of this competitive dynamics by negotiating in parallel with several prospective exporters. ENI thus negotiated at the same time with the Soviet Union and the Netherlands, playing them off against each other. Ruhrgas negotiated its first Soviet contracts while simultaneously renegotiating the terms of its earlier Dutch supplies and seeking access to Algerian gas. GdF, for its part, negotiated in parallel with the Soviet Union and Algeria. This behavior often had the desired effect, and there is no doubt about the fact that the Soviet Union played an important role in increasing the overall competitiveness of natural gas vis-à-vis other fuels in Western Europe, and thus in boosting the overall popularity of natural gas in this part of the world. But the strategy of parallel negotiations was also perceived as risky in the sensitive formative phase of East-West system-building, as it threatened to destroy the positive atmosphere that was being built up between the negotiating parties. Moscow initially suspected that the Western companies were not truly interested in imports of red gas, but merely used the East-West talks as a lever in their negotiations with other prospective suppliers. In the German case, the federal government had to intervene in Ruhrgas' negotiations with the Soviet side, reassuring the Soviet delegation of the sincere German interest in actually coming to agreement.

Another reason for Western Europe's interest in Soviet gas was that it offered *environmental opportunities*. This aspect was significant already from the outset, but it grew more important with time. Soviet gas was used for replacing coal and oil on environmental grounds, and it also played a role in replacing nuclear power. Neither Austria, which in 1978 decided to abandon its nuclear program at a time when its first reactor had just been completed, nor Italy, which in the aftermath of the Chernobyl disaster decided to quickly decommission its reactors, might have been able to do so, had Soviet gas not been available in large quantities. After the 2011 Fukushima disaster, Russian natural gas can be seen to play a similar antinuclear role in Germany. All in all, gas imports from the East have been perceived as a way of solving environmental problems throughout Western Europe.

Finally, actors saw significant *political opportunities* in the prospective import of Soviet natural gas. Austria hoped that by linking up with the Soviet gas system the Kremlin would be less irritated at the country's attempts to associate itself more closely with the EEC. Italy's communists favored an East-West gas deal hoping that it would strengthen the country's relations, in a general sense, with the world's leading communist power. For Willy Brandt and Egon Bahr in West Germany, Soviet gas was used as an instrument in implementing the Social Democrats' New Eastern Policy. Brandt and Bahr deliberately aimed to make Germany dependent on the Soviet Union, anticipating that this would convince the Kremlin of Bonn's sincere intentions to embark on a new, reconciliatory political path in Soviet-German relations. France also appears to have regarded natural gas pipelines as an effective way of strengthening overall French-Soviet ties. Needless to say, speeches by government representatives pointing to the benefits of Soviet gas for international understanding were given ample room at the inauguration ceremonies of practically all new East-West pipelines.

Because different actors supported (and opposed) imports for different reasons, it is not possible to specify any primary or most important Western purpose of importing Soviet gas. Actors had different agendas and different opportunities in mind, although they were typically eager to rhetorically make use of all the above opportunities when seeking to convince others and mobilize actor networks. Otto Schedl, for example, was primarily interested in supply and economic opportunities (in the regional Bavarian context). Egon Bahr in Bonn became a supporter of Schedl's ambitions, but for completely different reasons, stressing the political opportunities (for the Federal Republic, not for Bavaria). Supporters of the project referred to each other's arguments, and Bahr prepared a long list of seemingly unrelated advantages that a Soviet deal would offer and which could then be used for convincing others. This made it possible for Bahr to enroll supporters from throughout the political spectrum as well as from the business elite, while the number of opponents were reduced. In this way a strong coalition of actors could be built.

A Gradual Learning Process

Prospective system-builders in Western Europe needed to convince themselves and others that the risks linked to red gas were manageable and, therefore, worth taking. A first necessary step in this context was to *learn to trust the intentions of their Soviet partners*. This was easier in some cases than in others. A factor that influenced the degree of difficulty turned out to be whether or not the prospective importing country already had well-developed relations with the Soviet Union, that is, in general political and economic terms. Austria, Italy, and France—as well as Finland and Sweden—here turned out to be in favorable positions. Despite their location on the Western side of the Iron Curtain, they had opted for an overall path of cooperation rather than confrontation with Moscow. Although this did not mean that relations were friction-free, a certain level of basic trust was already in place at the time when possible gas imports started to be discussed. From this perspective, it is hardly surprising that the above five countries became the first to initiate serious negotiations.

In a much more difficult position were those actors in Germany who wished to import Soviet gas. At the time when Bavaria started working for a gas import from the East, general Soviet-German relations were characterized by fear, hostility, and suspicion. The former wartime enemies sought to mutually isolate each other on the foreign policy arena. Bonn was extremely suspicious about the true intentions of the Soviets, and this became a major reason for Bavaria's initial failure to gain support from the federal government for its plans—an absolutely necessary condition for Schedl's vision to have any chance of being realized.

An important facilitating factor in learning to trust Moscow's intentions took the form of experience of importing other Soviet energy sources, notably oil. The East-West oil trade had a long history and for the actors involved the gas trade was regarded as its logical extension. This became obvious especially when the main oil and gas actors in a prospective importing country coincided. Both ÖMV and ENI—the Austrian and Italian state-owned oil and gas companies, respectively—had already developed close links with Soyuznefteexport and the Soviet Ministry of Foreign Trade in connection with large-scale oil imports. The prospective partners of the gas trade thus already knew each other, and above all they knew that contracts signed for imports of red oil had been fulfilled. To judge from the oil trade, there was no reason to distrust the Soviets regarding their sincere intentions to live up to a contract once it had been signed. In the case of Germany, in contrast, imports of red oil had mainly been handled by a private Hamburg-based company with no connections to the gas industry. This possibly contributed to the late start of the German-Soviet gas negotiations.

But it was also necessary to learn to *trust the Soviet Union's technical and organizational ability* to carry out gas exports in practice. This was arguably more difficult. Exporting natural gas was technically and organizationally much more challenging than exporting oil, and no one had ever attempted to build a large-scale pipeline infrastructure that transgressed the Iron Curtain. No one knew if it would work in a technical sense, nor to what extent unanticipated and perhaps unsolvable problems would appear on the way. Against this background, it was not surprising that Austria became the first capitalist country to import Soviet gas—five years before Germany and six years before Italy and Finland. Soviet exports to Austria were crucially facilitated by the fact that already existing pipelines offered a highly convenient interconnection possibility. Only five kilometers of new pipelines had to be built to interconnect the existing national systems of Austria, Czechoslovakia, and the Soviet Union! All other West European importers would have to commit themselves to much more far-reaching investments in transit pipelines and domestic infrastructure. Austria's import of Soviet gas was further facilitated by this country's internal gas geography, as its main gas fields were located next to the Czechoslovak border. This made it possible, from an Austrian point of view, to treat imports from the east as just another gas field that fed natural gas into the domestic pipeline grid, while in case of supply disruptions domestic production could easily be accelerated so as to compensate for the loss. No expensive additional investments were needed, at least not on the short term.

The key role of Austria's imports of Soviet gas as a test-case for Europe as a whole can hardly be exaggerated, and it is far from certain that Western Europe would have come to import Soviet gas at all, had Austria not opted to do so. The Soviet Union did its utmost to live up to its contractual obligations vis-à-vis ÖMV during the initial delivery years, to the point that domestic gas needs were sacrificed and Ukrainian, Belarusian, Lithuanian, and Latvian consumers were left to freeze in the middle of the winter. When Mingazprom still failed to live up to what it had promised, so that ÖMV received much less gas than promised and at a much more irregular pace, the Soviets tried to hide its failure through innovative bookkeeping. With hindsight it is easy to understand why the Soviets were so keen to make a good impression in terms of its reliability as an exporter: a good Austrian track record was a prerequisite for other, larger European customers to join in during the second phase.

Apart from some unconfirmed reports hinting at the chaos of Soviet system-building, Western gas companies and governments in this early period remained unaware of the Soviet difficulties to bring about gas exports in practice. Despite the technical problems, Austria's imports were interpreted as functioning satisfactorily, and its allegedly positive experience became an argument for other countries to downplay the technical risks involved. This in turn paved the way for export contracts to be signed with Italy in 1969, with Finland in 1971, and with France in 1972. Even the West German Ministry of Economy, which had earlier been extremely suspicious about the Soviet Union, changed its mind regarding the red empire's trustworthiness as a gas exporter, championing a first German contract that could eventually be signed in 1970 and a second one in 1972. The second contract was notable because it was finalized before the exports that had been agreed upon in the first contract had even commenced! It is difficult to imagine that this would have been possible, had Austria not offered a seemingly positive example.

The negotiations themselves were also important arenas for building trust and creating "resonance." Soviet negotiators got the opportunity to explain in detail how their gas system functioned, what the main problems and challenges were, and how the new export flows to Western Europe would be brought about. When held in the Soviet Union, the talks often included field trips to major gas fields, pipeline construction sites, research institutes, and the like. Even so, Western gas industry representatives failed to get a realistic view of Mingazprom's undertakings, which were in a more or less constant state of crisis. Ruhrgas' top managers, in particular, repeatedly testified to the Soviet gas ministry's organizational skills and high technological level. They failed to grasp the harsh Soviet realities.

At the same time, Western gas companies, in their negotiations with the Soviet side, were forced to spend much effort trying to explain how the gas markets in their own countries worked. While the technical aspects of the West European gas system were easily explained, this was not the case regarding key market phenomena such as competition and pricing. The Soviets treated the gas system as a technical construct and were suspicious about Ruhrgas', ÖMV's, and others' insistence that it was absolutely decisive that the gas price was set in such a way that it would be competitive vis-à-vis gas from elsewhere and in relation to other primary fuels such as oil. Failure to agree on this repeatedly threatened to stall the talks. Eventually, however, the Soviets learned the trade and accepted the West European market price as a point of departure. Moreover, when contracts were extended and renegotiated from around 1971, and in particular after the first oil price shock in 1973/1974, they skillfully exploited the logic of the market, paving the way for a dramatic increase in the profitability of gas exports.

The perceived trustworthiness of the Soviet Union, both regarding its intentions and its technical ability, increased gradually through positive feedback: the more the Soviet Union proved able to export, the more convinced were Western governments and gas companies that additional imports from the East would be safe. By 1969, in the case of Germany, an import of red gas corresponding to up to 10 percent of total German demand was considered acceptable from a security perspective. Three years later, the perceived vulnerability had decreased so that a level of 14 percent was not considered problematic. By 1975, a 22 percent dependence on the Soviet Union was seen acceptable, and in connection with the Yamal negotiations a few years later the share had increased to 30 percent. By the early twenty-first century, Russian gas covered around 35 percent of total German gas demand.[1] Positive feedback over a period of several decades thus constitutes an important explanatory factor behind Europe's current dependence on Russian natural gas.

The Evolution of a Transnational System

The perception of opportunities and risks in the East-West gas trade explains why Soviet and West European actors opted to engage in dependence-generating system-building with the Cold War enemy in the first place. However, to fully grasp the internal dynamics and long-term evolution of the East-West gas system, including its ability to resist radical shocks from the geopolitical environment, we need to scrutinize the system's complexity and intricate sociotechnical character.

To the system's key technical or material components belonged transnational pipelines, underground gas storage facilities, compressor stations, control technology, and a wide range of additional equipment, whereas the social part of the system centered on national and regional governments, gas transmission and distribution enterprises as well as pipe and equipment manufacturers. International organizations such as NATO, the International Gas Union (IGU), and the European Union (and its forerunners) were also part of the system, though not at all to the same extent as national and regional actors. Apart from the organizations and individuals involved, the exports depended on innovative contractual arrangements and a variety of informal institutions for enabling communication and cooperation across military and ideological divides.

The system as a whole could come into existence and grow only when all components—technical and social—were in place and were allowed to interact in a meaningful way, mutually supporting and reinforcing each other. Enabling and managing this interaction was a prime task for system-builders. Given the system's transnational extent and, in particular, the need to overcome the Iron Curtain, no one could tell whether or not they would succeed. System-building organizations such as Mingazprom, Ruhrgas, ÖMV, and ENI had earlier been in charge of national or subnational gas grids over which they had far-reaching control, and they were used to operate in fairly homogeneous regional and national settings. When embarking on transnational projects, they faced the very different challenge of extending their pipeline networks to territories where others were in charge, and of linking up with systems whose character and style differed radically from that of "their" system.

In particular, systems in East and West differed from each other in terms of the reverse salients with which system-builders had to cope. From the perspective of transnational interlinking, this was not necessarily a problem—on the contrary, much of the dynamics of East-West integration in natural gas stemmed from the successful exploitation of Soviet and West European reverse salients that were largely complementary and could be resolved precisely through integrative efforts. In the Soviet Union, it was pipelaying that for a long time lagged behind. Throughout the first and second phases outlined in the beginning of this chapter, Mingazprom identified the shortage and low quality of domestically produced steel pipes as its overarching critical problem. Domestic manufacturers were not able to keep pace with Mingazprom's rapidly growing need for more and ever wider steel pipes, and this became one of Mingazprom's key motivations for probing the possibilities of cooperation with the West.

By contrast, Western Europe's main problem was a structural lack of gas resources. Access to high-quality steel pipes was not a problem. Not surprisingly, then, the first major East-West contractual arrangements took the form of countertrade deals in which Soviet natural gas was traded for West European steel pipe. In this way the most critical problems of the Soviet and West European gas systems, respectively, could largely be solved. Complementary reverse salients turned into drivers of transnational expansion.

Another problem, particularly evident in the first but also in the second phase of East-West system-building, was uncertainty on the Western side as to who would be the main system-builders. At one point international organizations such as the United Nations' Economic Committee for Europe (UNECE) aspired to a coordinating role in interlinking Western and Eastern Europe's natural gas systems. Regional actors also aimed to take a leading role, whereby they identified transnationalization as a tool in the domestic struggle against more dominant actors on the national scene. This was the case, for example, with Austria Ferngas, a joint venture formed by three regional gas companies, in its attempt to outmaneuver Austria's state-owned oil and gas company ÖMV, and with Bavaria, whose regional government under Otto Schedl's lead joined forces with the regional gas company Bayerngas in a struggle against Ruhrgas' dominance on the German gas market. In the end, however, UNECE proved too weak for the task, and the regional actors were found unsuitable to handle relations with Soviet system-builders. "Resonance," in social systems terms, was more easily established between the Soviet Union's powerful state agencies and Western Europe's state-owned gas companies—such as ÖMV, ENI, and GdF—and the German de facto national monopolist Ruhrgas.

In the third phase, organizational responsibilities had been defined, and the system-building process became more stable. The character of reverse salients and critical problems identified by Soviet and West European actors now changed. In the importing countries, the arrival of red gas shifted the focus from dealing with structural gas shortages and building actor networks to guaranteeing short- and mid-term supply security. Western Europe's gas companies approached this challenge by developing plans for new domestic or intra-West European pipelines that would enable emergency supplies in case of crisis, for conditioning facilities that would assure harmonization of emergency gas with Soviet gas, and for the construction of strategic gas storage facilities that for shorter or longer periods of time would be able to come to rescue in case of disrupted supplies from the East.

On the Soviet side, the main critical problem shifted from pipelaying to construction of powerful compressor stations. In the countertrade deals concluded from the mid-1970s onward, it were, therefore, compressors rather than pipes that were at focus on the equipment side. From the mid-1980s, then, the main reverse salient shifted again. With ample access to both pipes and compressors from Western manufacturers and the existence of several high-capacity export pipelines, the new challenge was to raise the load factor in the East-West system. Lagging Western demand in the wake of growing world energy prices and slow economic development was increasingly perceived of as a major obstacle for continued expansion of the export regime. The new Urengoi-Uzhgorod export pipeline, completed in 1983, proved difficult to fill. The Soviets identified the gas price as the critical problem and responded by lowering the price.

The result was that the Soviet Union for the first time emerged as the clear price leader on the West European gas market and that exports continued growing. Several new countries were now added to Moscow's list of customers. However, the growing imports created a new problem: lack of sufficient gas from elsewhere. Only through such deliveries, which were deemed necessary in order to diversify and balance overall supply, was a further scaling-up of the East-West system considered acceptable. Back in the 1960s and 1970s, this had not been as pressing a problem as it became in the 1980s, because most importers had still been in the possession of domestic gas reserves that were fairly large in relation to the level of imports from the East. Although these domestic reserves had not necessarily been depleted by the 1980s, the vast expansion of imports from the East had made them much more insignificant. In this situation, imports from elsewhere was seen as the only way to balance Soviet supplies. In what followed, West European customers of red gas set out to negotiate very large imports of gas from Norway, Algeria, and elsewhere.

One important aspect of the emerging East-West infrastructure was that the new transnational links were not added onto an already existing West European gas system. The first Soviet export pipelines were built at a time when Western Europe was not yet internally integrated. Indeed, intra-European connections were to a great extent created precisely for the purpose of handling growing imports from the East. Austria and Germany were linked up with each other thanks to the transit of Soviet gas along the Danube, and Germany and France were similarly interconnected as a result of the construction of transit pipelines for Soviet gas destined for France. Italy and Yugoslavia became linked to Austria through completion of the Trans-Austria Pipeline designed for transit of Soviet gas. Most strikingly, Czechoslovakia, thanks to its central role in the transit system for red gas, came to host more transnational gas connections than any other European country. In other words, internal Western and Central European integration was largely a product of East-West system-building. Red gas further contributed to West European integration through the perceived need to construct a unified EEC gas grid, "so that the balancing of our energy supply with neighbor states and allied, which is especially necessary in the case of crisis, can be carried out," as one leading German expert put it.[2] It was thus seen possible to reduce vulnerability through deeper integration among West European countries themselves.

It is from this perspective that we must understand the fact that the East-West gas trade survived the radical political and economic turmoil of the 1980s and 1990s. Since system-builders had designed the West European system in such a way as to be able to handle large-scale imports of red gas—including both real and imagined problems linked to this trade—it appeared irrational to scale down or phase out imports from the East. A mounting momentum pushed system-builders to identify and respond to reverse salients rather than to aim for a dismantling of the existing system or a reduction in the share of Soviet imports.

In the fourth phase of Europe's dependence, the Soviet Union had collapsed and new reverse salients were identified. The main one was the lack of a stable institutional regime for transporting gas between the former Soviet republics. The most pressing critical problem identified was the need to agree on gas prices and transit fees in the intra-CIS trade. The transit infrastructure, based as it was on a single Ukrainian-Czechoslovak pipeline route, was also identified as a problem, the solution of which was seen to lie in the creation of alternative routes. When the problems of price and nonpayment turned out to be a chronic phenomenon that could not be easily resolved, the emphasis turned increasingly to finding routes that would render transit negotiations unnecessary. The Nord Stream Pipeline, stretching directly from Russia to Germany through the Baltic Sea, made this dream come true.

The Soviet Union as a Victim

Western Europe's fear of falling victim to intentional supply disruptions from the East did not materialize during the Cold War. As for unintended delivery failures, the Soviets had difficulties living up to annual export targets during the start-up phase, but from 1974–1975 the contracts were "precisely fulfilled" and the Soviet Union earned a reputation for being a trustworthy partner—particularly in comparison to alternative suppliers such as Algeria and Libya, and from the 1980s even Norway. Short-term disturbances and irregularities continued to occur, but they were always compensated for at a later point and were not regarded as particularly troublesome. Moreover, worries among the population in the importing countries and pressure from state bureaucracies—both at the national and the EU level—forced gas companies to implement effective protection mechanisms for countering potential disturbances. These added to the perceived security of imports.

Much more vulnerable to the Soviet Union's export business were, paradoxically, gas users in the Soviet Union itself. Ukraine, Belarus, Lithuania, and Latvia were all hard hit. This was because users in these republics competed directly with Western importers for scarce Soviet gas. Failure to expand gas production fast enough and build necessary pipelines for distributing the fuel often meant that there was simply not enough gas available for everyone. In this situation, Mingazprom and the Kremlin faced the delicate choice of either breaking their export commitments or sacrificing domestic needs. Judging that exports, particularly to Western Europe, must under no circumstances be disrupted, decision makers opted to disrupt internal supplies.

The result was devastating both for Soviet industry and the general public. Families found themselves living in ice-cold houses without cooking possibilities. Schools and municipal institutions had to close down. Industrial production was forced to a stand-still. The crisis was worsened by the fact that large gas users, for whom reserve fuels in the form of coal and oil had been allocated, was often unavailable or insufficient. Factory managers and ordinary citizens used their local communist party organizations to ventilate their anger. Desperate letters were sent to Moscow, begging the country's leaders to resolve the supply crisis. Gosplan, the powerful planning organization, in cooperation with Mingazprom responded by working out detailed lists that prescribed how much gas a certain factory or municipal distribution network might use in case of gas shortages. But the instructions were rarely followed and users located at the far end of pipelines, notably in Latvia, became defenseless victims, despite repeated attempts from Moscow to prevent upstream users from using more gas than they were entitled to.

The completion of several new, powerful pipelines from Siberia improved the situation in a structural sense. From now on, Mingazprom had to deal with the problem of too much rather than too little transmission capacity. Still, the situation was far from harmonious. The legacy of the extreme hurry in which the export system had been created in the first place lingered on in the form of low welding quality, unreliable compressors, and the like. Moreover, in the stagnating Soviet economy, investments in and maintenance of the export pipelines and compressor stations were often neglected. The results were frequent accidents, explosions, and temporary interruptions of a "technical" nature. West European countries, with their strategically diversified supplies and expensive emergency systems, were well protected against these breakdowns. In the East, however, where gas storage facilities and other emergency arrangements were often missing, industries and households were directly affected. In the post-Soviet era, the legacy of this Cold War experience has continued to play a major role in shaping Europe's vulnerability geography. The postcommunist countries of Central and Eastern Europe, and in particular the former Soviet republics, thus continue to be the most vulnerable to supply disruptions.

A Long Duration

Europe's uneven vulnerability geography, as pointed at above, can be taken as evidence of an East-West divide in the long-term evolution of Europe's natural gas system. At the same time, however, the emergence of the East-West natural gas system also constitutes a remarkable case of integration between Cold War Europe's main enemy camps. Europe's gas system-builders managed to put an infrastructure in place that spanned the continent, seemingly without regard to any "Iron Curtain," and on which industries, power plants, municipal institutions, and households in both East and West became highly dependent for their daily activities. Remarkably, several countries and regions in capitalist Western Europe—notably Austria, Bavaria, Finland, northern Italy, and Greece, as well as western Turkey—became part of the Soviet-based natural gas system *before* linking up with any other foreign supplier, including intra-West European sources, and they became more dependent on Soviet than on Dutch and Norwegian gas.

The choice to import red gas was controversial to the extent that it challenged simplistic ideological and military conceptions of postwar Europe as neatly divided into an eastern and a western half. Yet in a longer historical perspective, Europe's hidden integration in gas does not necessarily come as a surprise. After all, natural gas was but the latest among the natural resources and agricultural products that Western Europe had long imported from the East in return for advanced industrial goods. In particular, gas system-builders could build on a century-long tradition of importing Russian oil. As we have seen, the main actors involved in the gas trade were in many instances even the same as in the oil trade.

The attempts from the side of the United States to prevent Western Europe from cooperating with the communist bloc became a major hallmark of the Cold War. During most of the Cold War period, Washington preferred a divided Europe and sought, instead, to favor a tightly integrated mini-Europe in the West, with strong links to North America. West European countries themselves were less inclined to give up their traditional Eastern relations for the sake of ideological and military considerations. To judge from the material presented in this book, most Europeans regarded a much more open Europe, with large-scale flows of energy and technology between East and West, as the natural and historically justified path.

If historical legacies of East-West interaction, in the above sense, inspired West European system-builders to form coalitions with their Soviet counterparts and create a vast East-West system for natural gas, it is also clear that this system, once in place, has had a major influence on Soviet-European and Russian-European relations. At the present time, there is hardly any aspect of Russia's relations with the EU or its member states, dependent as most of them are on Siberia's blue gold, that can be dealt with without (directly or indirectly) taking into account natural gas. This is because exporters, transiteers, and importers are all much too dependent on the system's continued operation for its demise or abandonment to be conceivable. Whereas many of the countries whose governments and gas companies were originally responsible for creating the system—the Soviet Union, Czechoslovakia, East Germany, and Yugoslavia—have ceased to exist, the system itself lives on, forcing today's actors to deal with it in one way or the other, regardless of how the geopolitical environment happens to look like at any particular moment.

To borrow a term from French historian Fernand Braudel, the East-West natural gas system can arguably be said to define a "long duration," spanning a period that may well be longer than the lifetime of countries, empires, and other political conjectures. To the extent that it is difficult to radically alter the system—that is, with predictable and acceptable consequences—the pipeline grid that crisscrosses Europe can be said to have more in common with ecology than economy. It has almost become part of Europe's nature, superimposed on an existing European geography of seas, rivers, forests, and mountains. Like this natural geography—itself more often than not a human construct—the infrastructured geography of natural gas can certainly be changed, though only with huge effort and at enormous cost.

Acknowledgments

The idea to write a book about the historical origins of Russia's natural gas exports and Europe's energy dependence was born in connection with the much-publicized Russian-Ukrainian gas crisis in January 2006. My colleague Arne Kaijser and I had just completed a book on the internationalization of electricity markets, and were in the early phase of another research project addressing the globalization of the nuclear fuel cycle. Western Europe's highly internationalized gas supply seemed to have the potential to become an exciting addition to our research on international energy relations in historical perspective.

Shortly afterward, two important calls for research applications opened. The first was launched by the Swedish Energy Agency and was for general energy systems studies. It helped us secure funding for a project on the ways in which the Swedish energy system has developed links to foreign ones. A major part of it took the form of a PhD project on the history of Sweden's natural gas supply, carried out by Anna Åberg. Her thesis is currently being finalized and can be regarded as an interesting complement to *Red Gas*. The second call, "Inventing Europe," was launched by the European Science Foundation (ESF) in cooperation with several national research councils, and was for collaborative research projects with a transnational scope in the history of technology. Our group in Stockholm joined forces with seven other universities in six European countries, which resulted in a major grant for a project addressing the emergence and governance of Europe's critical infrastructures, EUROCRIT, with Arne Kaijser as project leader. Funding for the EUROCRIT project from ESF and the Swedish Research Council (VR) made the research underlying this book possible.

I am indebted to all EUROCRIT participants for fruitful and stimulating discussions on Europe's evolving natural gas dependence as well as on the dynamics of Europe's critical infrastructures more broadly. From 2007 to 2010, our international team gathered for a series of intense workshops held in Rotterdam, Utrecht, Stockholm, Lisbon, Helsinki, Athens, and Sofia. Without the insights gained through this unique European cooperation, this book would hardly have seen the light of the day. In a similar vein, *Red Gas* has profited from countless discussions with my fellow historians of technology in the "Tensions of Europe" network.

In Stockholm, the Division of History of Science, Technology, and Environment at the Royal Institute of Technology (KTH) has been a great base for writing a book of this kind. I would like to thank both my students and my colleagues and, in particular, the participants in the Swedish Energy Agency project and the Stockholm node in EUROCRIT. Apart from Arne, Anna, and myself, this group initially included Björn Berglund. The two projects produced an important spin-off in the form of a new graduate course at KTH, "Energy and Geopolitics," which from 2009 offered excellent opportunities to test new ideas and discuss preliminary research results linked to the East-West gas trade.

In the final phase of writing, Arne Kaijser, Vincent Lagendijk, Paul Josephson, and an anonymous reviewer provided very useful comments and suggestions.

I would like to thank the helpful personnel at the Russian State Archive of the Economy (Moscow), the Central State Archive of the Highest Organs of Government and Administration of Ukraine (Kiev), the Austrian State Archives (Vienna), the Federal German Archives (Koblenz and Berlin), the Political Archive of the German Foreign Office (Berlin), the Bavarian State Archives (Munich), and the Archives of the United Nations Office at Geneva. Likewise, smooth access to old issues of several key specialist journals at the libraries of KTH and Chalmers University of Technology, much cited in this book, saved me a great of pain in the research process. Many thanks also to the *Oil and Gas Journal*, *Süddeutsche Zeitung*, Oldenbourg Industrieverlag, and OMV for permission to reproduce a number of maps and photographs. Tommy Westergren at KTHB helped me process a large part of the illustrations.

I dedicate the book to my wife, Liu Yuanyuan.

Per Högselius
Stockholm, September 2012

Notes

1 Introduction

1. For example, Umbach 2009; Webb and Barnett 2006.
2. For example, Goldthau 2008; Solanko and Sutela 2009; Fernandez 2009.
3. IEA 2011, pp. 159–165; Söderbergh 2010.
4. Misa and Schot 2005.
5. Stent 1981; Von Dannenberg 2007; Rudolph 2004.
6. The main source of inspiration for these studies has been Thomas P. Hughes, who in his book *Networks of Power: Electrification in the Western World 1880–1930* (1983) used the case of electricity to develop a historical theory of what he called Large Technical Systems (LTS).
7. Hughes 1983.
8. See, for example, Misa and Schot 2005; Van der Vleuten and Kaijser 2006; Badenoch and Fickers 2010.
9. Luhmann 1995, p. 187.
10. For a theoretically interesting discussion of formative periods in the evolution of energy systems, see Thue 1995.

2 Before Siberia: The Rise of the Soviet Natural Gas Industry

1. Bokserman to Sedin, April 24, 1944, Russian State Archive of the Economy (RGAE) 8627-9-272.
2. Akt Pravitelstvennoi komissii po priemke v ekspluatatsiyu magistralnogo gazoprovoda Dashava-Kiev, December 21, 1948, Central State Archive of the Highest Organs of Government and Administration of Ukraine (TsDAVO Ukrainy) 2-7-800; V. S. Chernovol, "10 let raboty gazoprovoda Dashava-Kiev," *Gazovaya promyshlennost*, January 1959, p. 35.
3. Derzhavnyi komitet naftovoi, gazovoi ta naftopererobnoi promyslovosti Ukrainy 1997, p. 147; Dinkov to Medvedkov, January 20, 1971, RGAE 458-1-2526.
4. Bokserman 1958; *Gazovaya promyshlennost*, September 1960, p. 2.
5. N. Talyzin, "Gazifikatsiya gorodov k 40-i godovshchine sovetskoi vlasti," *Gazovaya promyshlennost*, November 1957, p. 4. The (mis)quote appears again in an article in the same journal's October 1967 issue, p. 18.
6. Pravda quoted in *Oil and Gas Journal*, June 6, 1955, p. 89; Runov and Sedykh 1999, p. 46; Mawdsley and White 2000, p. 145f.; *Oil of Russia*, International Quarterly Edition, no. 2, 2006; *Oil and Gas Journal*, June 6, 1955, p. 89.
7. A. K. Kortunov, "Gazovaya promyshlennost k 40-letiyu velikogo oktyabrya," *Gazovaya promyshlennost*, November 1957, p. 2.
8. Runov and Sedykh 1999, p. 35.
9. Ibid., p. 37.
10. Kortunov, "Gazovaya promyshlennost k 40-letiyu velikogo oktyabrya," p. 2.

11. Bokserman 1958; Kortunov, "Novyi etap v razvitii gazovoi promyshlennosti," *Gazovaya promyshlennost*, April 1958.
12. Khrushchev cited in *Gazovvaya promyshlennost*, January 1960, p. 2.
13. *Gazovaya promyshlennost*, September 1960, p. 3.
14. Kortunov, "Gazovaya promyshlennost nakanune XXII syezda KPSS," *Gazovaya promyshlennost*, October 1961, p. 5.
15. *Gazovaya promyshlennost*, July 1964, p. 1.
16. *Gazovaya promyshlennost*, November 1961, p. 2.
17. See, for example, Slavkina 2005, pp. 148–151.
18. Yu. I. Bokserman, "Razvitie transporta gazov po magistralnym gazoprovodam v SSSR," *Gazovaya promyshlennost*, November 1957, p. 16; For a detailed study of Estonia's shale gas, including "exports" to Leningrad, see Holmberg 2008.
19. For details on the construction of the Dashava-Minsk-Vilnius-Riga system, see A. P. Yanchenko, "Gazifikatsiya Belorusskoi SSR za 1959–1961 gg.," *Gazovaya promyshlennost*, October 1961, pp. 37–39; K. K. Smirnov, "Gazoprovod Dashava-Riga," *Gazovaya promyshlennost*, October 1962, pp. 1–3.
20. Diordina and Bodyul to Kortunov, October 14, 1966, RGAE 458-1-103.
21. A. M. Abdullaev and A. A. Narimanov, "Razvitie gazovogo khozyaistva Azerbaydzhanskoi SSR za 1959–1961 gg.," *Gazovaya promyshlennost*, October 1961, pp. 29–31; Z. Piriashvili, "Gazifikatsiya Gruzinskoi SSR za 1959–1961 gg.," *Gazovaya promyshlennost*, October 1961, p. 32; I. G. Papiev, "Gazifikatsiya Armyanskoi SSR za 1959–1961 gg.," *Gazovaya promyshlennost*, October 1961, p. 33; A. Kochinyan (secretary of the Central Committee of the Communist Party of Armenia) and B. Muradyan (chairman of the Council of Ministers of the Armenian SSR) to the Soviet Council of Ministers, April 29, 1966, RGAE 458-1-101.
22. *Gazovaya promyshlennost*, January 1960, p. 2; "Russians Push Work on Giant Gas Lines," *Oil and Gas Journal*, May 29, 1961, pp. 118f.
23. *Gazovaya promyshlennost*, January 1960, p. 3, and September 1960, p. 3.
24. For example, *Gazovaya promyshlennost*, January 1965, p. 3.
25. Bokserman, "Tekhnicheskiy progress—osnova uspeshnogo razvitiya gazovoi promyshlennosti," *Gazovaya promyshlennost*, January 1960, p. 9.
26. Ibid., p. 10; Bokserman 1958, p. 90f.
27. Rudolph 2004, pp. 157ff. and 197; Stent 1981, p. 101.
28. Stent 1981, p. 101. The imported amounts of 1020 mm pipes were very large if seen in relation to the total pipe demand for the Seven-Year Plan launched in 1958, which amounted to 1.7 million tons of 1,020 mm gas pipes for the period 1959–1965.
29. Bokserman 1958, p. 97. The Gorky compressors were seen to be justified for pipelines with an annual transmission capacity of around 2–3 bcm only.
30. Ibid., p. 98f.
31. Kortunov, "Gazovaya promyshlennost v 1960 godu," *Gazovaya promyshlennost*, January 1960, p. 4.
32. *Gazovaya promyshlennost*, June 1958, pp. 34f.; Sidorenko to A. M. Lalayants (deputy chairman of Gosplan), March 5, 1966, RGAE 458-1-112.
33. *Gazovaya promyshlennost*, December 1960, p. 54, July 1964, p. 8; November 1965, p. 2, and October 1969, p. 2f.
34. Bokserman, "Tekhnicheskiy progress—osnova uspeshnogo razvitiya gazovoj promyshlennosti," p. 10; *Gazovaya promyshlennost*, February 1963, p. 53.
35. *Gazovaya promyshlennost*, August 1960, p. 1; *Oil and Gas Journal*, August 28, 1961, p. 60.
36. *Gazovaya promyshlennost*, August 1960, p. 1.
37. William R. Connole, "Russian Pipelining Is More than a Little Different," *Oil and Gas Journal*, August 14, 1961, p. 86, William R. Connole, "Soviet Gas Pattern Is

Much Like Ours," *Oil and Gas Journal*, August 21, 1961, p. 64; William R. Connole, "Will Europe Come to Depend on Russian Natural Gas?" *Oil and Gas Journal*, August 28, 1961, p. 60.
38. As referred to by Connole, "Will Europe Come to Depend on Russian Natural Gas?" p. 59.
39. Ibid., p. 58f.
40. Kortunov cited in *Oil and Gas Journal*, June 1962.

3 Toward an Export Strategy

1. A. I. Sorokin and F. A. Trebin, "Razvitie gazosnabzhenie v SSSR" (Tezisy doklada na VIII Mezhdunarodnom gazovom kongresse), published in *Gazovaya promyshlennost*, June 1961, p. 7; Connole, "Where the Russian Gas Industry's Headed," August 7, 1961.
2. *Gazovaya promyshlennost*, January 1962, p. 4.
3. Cf. *Gazovaya promyshlennost*, May 1963.
4. Ibid., p. 54, and November 1963, p. 1.
5. *Gazovaya promyshlennost*, March 1966, p. 23.
6. Slavkina 2002, pp. 35ff. and 56; *Gazovaya promyshlennost*, January 1965, p. 9.
7. Slavkina 2002, p. 152.
8. For detailed discussions of the pipe embargo, see Stent 1981, Chapter 5 and Rudolph 2004, Chapter 7.
9. Slavkina 2002, pp. 35ff. and 56; *Gazovaya promyshlennost*, January 1965, p. 9.
10. *Oil and Gas Journal*, April 30, 1962, p. 57, September 23, 1963, p. 138, December 23, 1963, p. 42, and January 20, 1964, p. 58.
11. Stent 1981, p. 122.
12. Starting at 270 mcm in 1967, the exports were to increase gradually and reach a plateau level of 1 bcm in 1970. The contractual details are outlined in a letter from Kortunov to the Central Committee of the Communist Party, June 6, 1968, RGAE 458–1–967. See also "Czechs Plan New Pipeline," *Financial Times*, January 16, 1964, and "Soviets Have Big Pipeline Plans for '65," *Oil and Gas Journal*, December 28, 1964. Further details on the Bratstvo agreement are found in RGAE 458–1–1115.
13. Runov and Sedykh 1999; Kortunov and Sidorenko (Soviet minister of geology) to the Central Committee of the Communist Party, February 8, 1967, RGAE 458–1–505. See also references in chapter 4.
14. Stent 1981, p. 137; Kortunov to Baibakov, June 7, 1966, RGAE 458–1–114.
15. Brezhnev 1966, p. 19 and 46.
16. Chadwick et al. 1987, p. 69; Gustafson 1989, p. 267.
17. Kortunov to Baibakov, June 7, 1966, RGAE 458–1–114; The latest Soviet-Italian cooperative deal, signed in June 1964, had included the import of a large Italian gas processing plant with an annual production capacity of some 500 mcm of associated gas. Kamarov (deputy minister of foreign trade) to Tikhonov (deputy chairman of the Soviet Council of Ministers), June 7, 1966, RGAE 458–1–100; see also *Gazovaya promyshlennost*, May 1964, p. 50.
18. "Fueling Italy's fires—Russian style," *Oil and Gas Journal*, November 7, 1966; *gwf*, vol. 107, no. 47, November 25, 1966, p. 1350; Schlieker to the German Foreign Office, December 12, 1966, Bundesarchiv (BArch) B102–152193.
19. Sorokin to Gusev (head of Gosplan's Division of Foreign Trade) and Komarov (deputy minister of foreign trade), May 5, 1966, RGAE 458–1–113; Protokol no. 33 Zasedaniya Kollegii Mingazproma, November 12, 1966, RGAE 458–1–77.
20. Bokserman to Ryabenko, March 9, 1966, RGAE 458–1–112; Kortunov to the Council of Ministers, December 29, 1966, RGAE 458–1–110; see also Rudolph 2004 and documents in RGAE 458–1–495.

21. Baibakov and Kortunov to the Council of Ministers, February 28, 1966, RGAE 458-1-108; Suloev to Novikov, May 13, 1966, RGAE 458-1-113; Protokol soveshchaniya v otdele neftyanoi i gazovoi promyshlennosti Gosplana SSSR, May 23, 1966, RGAE 458-1-114; see also *Gazovaya promyshlennost*, March 1966.
22. *Gazovaya promyshlennost*, June 1965, p. 1 and March 1966, p. 12–14.
23. *Gazovaya promyshlennost*, January 1966, p. 1.
24. Ryabenko (deputy chairman of Gosplan) to Kortunov, December 31, 1965. RGAE 458-1-6.
25. Ibid.
26. Brezhnev 1966, p. 79; Kosygin 1966, p. 196; Smirnov to the Council of Ministers, July 19, 1966, RGAE 458-1-101.
27. Kortunov to Baibakov, June 7, 1966, RGAE 458-1-114; Protokol soveshchaniya v otdele neftyanoi i gazovoi promyshlennosti Gosplana SSSR, May 23, 1966, RGAE 458-1-114; Bokserman to Yudin (deputy head of the division of Oil and Gas Industry of Gosplan), July 15, 1966, RGAE 458-1-114.
28. Ibid.; Kortunov to Baibakov, June 7, 1966, RGAE 458-1-114. The cited price may be compared with the 5 roubles paid by Poland for its Soviet imports, see Sidorenko to Ryabenko, March 25, 1966, RGAE 458-1-112
29. Kortunov and Sidorenko to Central Committee of the Communist Party, February 8, 1967, RGAE 458-1-505.
30. This decision is extensively quoted by Kortunov in a letter to the Council of Ministers, October 5, 1966, RGAE 458-1-109.
31. Kortunov to Baibakov, June 7, 1966, RGAE 458-1-114.

4 Austria: The Pioneer

1. Rambousek 1977, p. 24; "20 Jahre ÖMV: 2 Jahrzehnte im Dienste Österreichs," *ÖMV-Zeitschrift* 2/1976.
2. Rambousek 1977; Ludwig Bauer, "Erdgas—Konkurrenz oder notwendige Ergänzung des Energieangebotes?" *ÖMV-Zeitschrift* 2/1975, p. 2.
3. This concerned in particular the demands from the side of Western oil companies with prewar activities in Austria to be compensated for Nazi-era expropriation.
4. Rambousek 1977, pp. 53ff. In August 1955, Austria produced 67,000 barrels of oil per day. Second largest producer in Western Europe at the time was West Germany with 63,200 barrels per day. See *Oil and Gas Journal*, November 7, 1955, p. 83.
5. ÖMV, Bericht über das Geschäftsjahr 1957, p. 18; Rambousek 1977, pp. 68 and 78f. After losses in the Soviet occupation times as high as 72 percent in 1950 and 40 percent in 1955, ÖMV impressively reduced the figure to 1.3 percent.
6. ÖMV, Bericht über das Geschäftsjahr 1957, p. 19. The basic arrangements that paved the way for the formation of the regional companies and the gas supply from ÖMV's fields were agreed upon in 1956 and 1957.
7. ÖMV, Bericht über das Geschäftsjahr 1968, p. 22; OÖ Ferngas 2007, p. 18f. In 1967, 80 percent of ÖMV's sales was to NIOGAS and Wiener Stadtwerke.
8. Rambousek 1977, p. 79f. and 215, Anhang IV.
9. Ibid., p. 212f., from 1955 referring to ÖMV Annual Reports.
10. ÖMV, *Bericht für den Aufsichtsrat über das Jahr 1967*, Österreichisches Staatsarchiv (OeStA), ÖIAG-Archiv, Box 135.
11. OÖ Ferngas 2007, p. 22; Davis 1984, p. 179; "Austria plans gas imports," *Financial Times*, November 16, 1962.
12. OÖ Ferngas 2007, p. 22; *gwf*, May 14, 1965, p. 529.
13. Ibid.
14. "Austria to buy Algerian natural gas?" *Financial Times*, February 3, 1966; *Oil and Gas Journal*, March 7, 1966. Lukesch also reported on this project to VÖEST's supervisory

board on its meeting on October 24, 1966. It was concluded that the project was in a very advanced stage, although its financing was still to be clarified. The World Bank was expected to play a part, although it had its hesitations to give credits to two communist countries (Yugoslavia and Czechoslovakia).
15. ÖMV, *Bericht über das Geschäftsjahr 1967*, p. 18.
16. Semjonow 1973.
17. See, for example, "Soviet natural gas for Austria," *Financial Times*, December 16, 1964; "Soviet have big pipeline plans for '65," *Oil and Gas Journal*, December 28, 1964; *gwf*, May 20, 1966, p. 531.
18. VÖEST, Niederschrift über die 2. ordentliche Aufsichtsratssitzung, February 17, 1967, OeStA ÖIAG-Archiv, Box 325.
19. Rudolph 2004.
20. Schlieker to the German Foreign Office, December 12, 1966, BArch B102–152193; Kortunov to Novikov, "Spravka k voprosu o vozmozhnoi podache prirodnogo gaza iz SSSR v Vengriyu," December 20, 1966, RGAE 458-1-110; "ENI-Soviet disagreement on gas pipe diameter," *Financial Times*, January 11, 1967; *Oil and Gas Journal*, May 16, 1966, p. 168; *gwf*, February 16, 1968, p. 187. The imports from Romania amounted to 0.2 bcm per year. Italy's reserves grew from 33 bcm to 100 bcm.
21. Kortunov to the Soviet Council of Ministers, October 5, 1966, RGAE 458-1-109; VÖEST, Niederschrift über die 3. ordentliche Aufsichtsratssitzung, October 24, 1966, OeStA ÖIAG-Archiv, Box 325. See also *Der Spiegel*, October 17, 1966.
22. This interpretation was made by Schlieker in a letter to the German Foreign Office, December 12, 1966, BArch B102–152193.
23. Protokoll über das sowjetisch-österreichische Arbeitsgespräch, November 15, 1966, OeStA II-Pol, UdSSR 1966.
24. Joint Soviet-Italian communiqué, ca. February 1, 1967, OeStA II-Pol, UdSSR 1967; Kortunov to Lalayants, December 30, 1966; Kortunov to Gosplan, December 17, 1966, RGAE 458-1-116; "France may buy Russian natural gas," *Financial Times*, December 1, 1966.
25. *Financial Times*, January 20, February 22, and March 15, 1967.
26. Schlieker to the German Foreign Office, December 29, 1966; Plesser, March 2, 1967, BArch B102–152193; "Austria seeks Soviet order," *Financial Times*, December 8, 1966.
27. Wortlaut des gemeinsamen Schlusskommuniqués über den Staatsbesuch des Bundeskanzlers, March 22, 1967, OeStA II-Pol, UdSSR 1967.
28. Mommsen to Engelmann and Schedl, April 4, 1967, BArch B102–152193.
29. *Financial Times*, April 24, 1966, May 19, 1967, and May 20, 1967; Hufnagel to Schiller, June 15, 1967, BArch B102–152193.
30. Yergin 1991, pp. 554–560.
31. Protokoll der Pressekonferenz des Stellvertretenden sowjetischen Ministers für die Gasindustrie, Sorokin, June 9, 1967, BArch B102–152193. See also *gwf*, August 18, 1967, p. 939. For the Finnish talks, see *Gazovaya promyshlennost*, February 1967, pp. 46f. The Soviets seemed to worry that Finland's energy was increasingly supplied by West European and American companies.
32. *WID Energiewirtschaft*, July 13, 1967.
33. Hufnagel to Schiller, June 15, 1967, BArch B102–152193; Guinot to Dufrasne, October 23, 1967, Note sur "La situation du gaz en France en 1966 et ses perspectives," UNOG GX.11/13/11, Jacket 3; cf. Wilfried Czerniejewicz, "Frankreichs Gaswirtschaft im Strukturwandel," *gwf*, November 21, 1969, p. 1307; *Oil and Gas Journal*, January 23, 1967, p. 60; *WID Energiewirtschaft*, July 13, 1967.
34. Kortunov to Council of Ministers, June 9, 1967, RGAE 458-1-496; VÖEST, Niederschrift über die 2. ordentliche Aufsichtsratssitzung, June 29, 1967, OeStA ÖIAG-Archiv, Box 325.

35. VÖEST, 2. ordentliche Aufsichtsratssitzung, June 29, 1967, OeStA ÖIAG-Archiv, Box 325.
36. Verhandlungen mit der Delegation der UdSSR über den Import von Erdgas nach Österreich, September 2–16, 1967, BArch B102–152193.
37. Ibid.
38. Ibid.
39. Ibid.
40. ÖMV, Bericht für den Aufsichtsrat über das 1. Quartal 1968, OeStA ÖIAG-Archiv, Box 135; Verhandlungen mit der Delegation der UdSSR über den Import von Erdgas nach Österreich, September 2–16, 1967, B102–152193. See also *Handelsblatt*, September 7, 1967.
41. *gwf*, May 26, 1967, p. 593.
42. Ibid., p. 594; *gwf*, October 27, 1967, pp. 1241f.; "Interruption of Italo-Russian oil talks," *Financial Times*, August 31, 1967.
43. Verhandlungen mit der Delegation der UdSSR über den Import von Erdgas nach Österreich, September 2–16, 1967, BArch B102–152193.
44. Ibid.
45. Ibid.; OeStA II-Pol, UdSSR 1967.
46. ÖMV, Bericht über das Geschäftsjahr 1967, p. 18; ÖMV, Sitzung des Aufsichtsrates, February 27, 1968, OeStA ÖIAG-Archiv, Box 135; "10 Jahre Erdgasimport," *ÖMV-Zeitschrift* 3/1978. The figures are cited in a memorandum dated December 12, 1968, which was signed in connection with the later cooperation between Austria, Czechoslovakia, and the Soviet Union. RGAE 458–1–1115, and in ÖMV, Bericht für den Aufsichtsrat über das 1. Quartal 1968, OeStA ÖIAG-Archiv, Box 135.
47. Verhandlungen mit der Delegation der UdSSR über den Import von Erdgas nach Österreich, September 2–16, 1967, BArch B102–152193.
48. Wodak to Tončić-Sorinj, December 7, 1967, OeStA II-Pol, UdSSR 1967; ÖMV, Sitzung des Aufsichtsrates, February 27, 1968, OeStA ÖIAG-Archiv, Box 135; see also Kortunov to the Central Committee of the Communist Party, June 6, 1968, RGAE 458–1–967.
49. The price offered was $14.10 per 1,000 cubic meters for gas measured at 20 degrees centigrade, which was the Soviet standard for measuring gas. This translated into $15.13 per 1000 cubic meters for 0 degrees, which was the West European standard.
50. ÖMV, Sitzung des Aufsichtsrates, February 27, 1968, OeStA ÖIAG-Archiv, Box 135.
51. "Verhandlungen über Erdgasprojekt in Wien," *Frankfurter Allgemeine Zeitung*, August 29, 1967; "Soviet trade mission visits Rome," *Financial Times*, March 14, 1968.
52. It was only from 1969, when the oil price started to rise markedly, that ÖMV's general director Ludwig Bauer could note that the deal had become "recognized as favorable" by the international community. ÖMV, Sitzung des Arbeitsausschusses des Aufsichtsrates, June 18, 1969, OeStA ÖIAG-Archiv, Box 136.
53. Protokoll über das Gespräch mit Ministerpräsident Kossygin, March 19, 1968, OeStA II-Pol, UdSSR 1968.
54. Copies of the general framework contract and the detailed gas export contract are in OeStA ÖIAG-Archiv, Box 135.
55. Since the contract followed Soviet measurement standards, according to which the heat content of the gas was measured at 20 degrees centigrade, rather than 0 degrees as in the West, the interval as formulated in the contract was actually 8,100–8,400 kcal/m^3.
56. ÖMV, Sitzung des Aufsichtsrates, September 18, 1968, OeStA ÖIAG-Archiv, Box 135.
57. A copy of the agreement is in OeStA ÖIAG-Archiv, Box 135.
58. ÖMV to Soyuznefteexport, June 5, 1968, OeStA ÖIAG-Archiv, Box 135.

59. ÖMV, Sitzung des Aufsichtsrates, June 18, 1968, OeStA ÖIAG-Archiv, Box 135; VÖEST, Niederschrift über die 2. ordentliche Aufsichtsratssitzung, May 8, 1968, and 3. ordentliche Aufsichtsratssitzung, July 11, 1968, OeStA ÖIAG-Archiv, Box 325; Klarenaar, May 5, 1969, PA AA B63–435.

5 Bavaria's Quest for Energy Independence

1. The total volume of Soviet-German trade remained below the 1962 level throughout the period 1963–1966. Stent 1981, p. 137.
2. Brezhnev 1966, p. 51f.
3. Ibid., p. 46.
4. Stent 1981; Rudolph 2004, p. 202.
5. Dahl 1997.
6. Birkenfeld 1964.
7. Bayerisches Hauptstaatsarchiv (BayHStA), NL Schedl.
8. *gwf,* June 10, 1966, p. 653; Mattei to Schedl, March 10, 1962, BayHStA NL Schedl, file 199. Schedl met personally with ENI President Enrico Mattei on several occasions.
9. Göpner, Vermerk, July 10, 1969, BArch B102–152194.
10. A. Volk, "Technisch-wirtschaftliche Erfahrungen beim Bau des nordbayerischen Ferngasnetzes," *gwf,* March 4, 1966, p. 229, Bild 3.
11. Göpner, July 10, 1969, BArch B102–152194; *gwf,* March 4, 1966, p. 229, and March 3, 1967, p. 236.
12. Schedl to the Bavarian Parliament, March 1, 1965, BayHStA StK, file 18791.
13. Ibid.
14. *gwf,* August 20, 1965, p. 929.
15. See, for example, Laurien 1974, p. 8.
16. See map in BayHStA NL Schedl.
17. *gwf,* November 25, 1966, p. 1351; Wedekind to Woratz, February 24, 1967, BArch B102–152183.
18. Schlieker to the Foreign Office, December 12, 1966, BArch B102–152193.
19. Wedekind to Woratz, June 23, 1966, BArch B102–152183.
20. Wedekind, November 11, 1966, BArch B102–152183.
21. Rechenberg to Heitzer, October 21, 1966, and Heitzer, November 17, 1966, BayHStA MWi, file 27219. Chaired by Bayerngas, the consortium held its founding meeting in Munich on October 17, 1966.
22. Plesser, December 2, 1966, BArch B102–152183.
23. Plesser, November 15, 1966, BArch B102–152183.
24. See chapter 4 and Kirchhoff, March 13, 1967, BArch B102–152193.
25. Von Dannenberg 2007, p. 27.
26. Schedl, September 20, 1967, BayHStA NL Schedl, file 188.
27. Schlieker to the Foreign Office, December 12, 1966, BArch B102–152193.
28. Steidle, January 3, 1967; Grimm to Referat V C 5, January 23, 1967, BArch B102, file 152193.
29. *gwf,* July 21, 1967.
30. Grimm to Referat V C 5, January 23, 1967, BArch B102–152193.
31. Plesser to Referat V C 5, January 27, 1967, BArch B102–152193.
32. Ibid.
33. Plesser, November 15, 1966, BArch B102–152183.
34. Kirchhoff, March 13, 1967, B102–152193; Blumenfeld, March 9, 1967, Politisches Archiv des Auswärtigen Amtes (PA AA), B41–51. For Tyrol's interest in importing Soviet gas, see *gwf,* June 9, 1967, p. 653.
35. Plesser, March 2, 1967; Woratz to Schiller, March 7, 1967, BArch B102–152193.
36. Plesser, March 2, 1967, BArch B102–152193.

37. Kaiser 1968; Von Dannenberg 2007, p. 29.
38. Plesser, March 2, 1967, BArch B102-152193.
39. Ibid; cf. Plesser to Woratz, March 1, 1967, BArch B102-152193.
40. Lantzke to Schiller, April 6, 1967, BArch B102-152193.
41. Ibid.
42. Neef to Schiller, April 14, 1967, BArch B102-152193.
43. Mommsen to Engelmann and Schedl, April 4, 1967, BArch B102-152193.
44. Neef to Schiller, April 14, 1967, B102-152193.
45. Ibid.; Eichborn to Brandt, July 22, 1969, BArch B102-152194; *Süddeutsche Zeitung*, April 22, 1967.
46. VGW, April 26, 1967, BArch B102-152193. *gwf*, August 2, 1968, p. 862.
47. Ibid.
48. Neef to Schiller, May 22, 1967 and Schiller to Mommsen, June 14, 1967, BArch B102-152193.
49. Kortunov to Efremov, May 5, 1967, RGAE 458-1-506.
50. Van Beveren, June 14, 1967, BArch B102-152193.
51. Plesser, July 26, 1967, BArch B102-152193.
52. "Erdgas aus der Sowjetunion? Russische Gasfachleute in Stuttgart," *Stuttgarter Nachrichten*, June 16, 1967.
53. Eichborn to Brandt, July 22, 1969, BArch B102-152194.
54. Klarenaar to Schiller, August 31, 1967, BArch B102-152193.
55. Bahr to Brandt, June 28, 1967, in *Akten zur Auswärtigen Politik der Bundesrepublik Deutschland 1967*, vol. 1, p. 236-238; Referentenbesprechung bei Herrn Botschafter Emmel, September 22, 1967, PA AA B41-51.
56. Ibid.
57. Rudolph 2004.
58. Bahr to Brandt, June 28, 1967, in *Akten zur Auswärtigen Politik der Bundesrepublik Deutschland 1967*, vol. 1, pp. 236-238.
59. The struggle for dominance of the emerging south German gas markets escalated during 1967 and early 1968 to what was referred to as a "natural gas war." See *gwf*, February 16, 1968.
60. Volze to the Foreign Office, October 25, 1967, BArch B102-152193.
61. Ibid.
62. Von Vaoano, September 6, 1967, PA AA B41-51.
63. Plesser, September 8, 1967, BArch B102-152193.
64. Sackmann to Goppel, November 15, 1967, BayHStA StK, file 18790.
65. Engelmann and Plesser to Referat V C 5, September 26, 1967, BArch B102-152193.
66. Ibid.; Referat II A 4, Referentenbesprechung bei Herrn Botschafter Emmel, September 22, 1967, PA AA B41-51.
67. Von Dannenberg 2007, p. 37f.
68. Becke to Referat V C 5, September 5, 1967, BArch B102-152193; "Tauziehen um den süddeutschen Erdgasmarkt," *Handelsblatt*, November 20, 1967; *gwf*, February 16, 1968.
69. Wedekind to Woratz, October 6 and 23, 1967, BArch B102-152183; *gwf*, July 7, 1967, p. 770, February 16, 1968, and August 2, 1968, p. 861; "Algerien bietet sein Erdgas wohlfeil an. Staatsgesellschaft Sonatrach sucht deutsche Partner und Abnehmer," *Die Welt*, November 25, 1967.
70. Wedekind to Woratz, October 6, 1967, BArch B102-152183. The French had agreed to pay 0.66 Pf/Mcal for the Algerian gas at Marseille.

6 From Contract to Flow: The Soviet-Austrian Experience

1. Truboprovidnyi transport URSR, TsDAVO Ukrainy 337-15-397.
2. Kortunov to the Central Committee of the Communist Party, June 6, 1968, RGAE 458-1-967; *gwf*, November 21, 1969, p. 1309.

3. ÖMV, Bericht für den Aufsichtsrat über das 1. Quartal 1968, OeStA ÖIAG-Archiv, Box 135; "10 Jahre Erdgasimport," *ÖMV-Zeitschrift* 3/1978, p. 1.
4. ÖMV, Bericht für den Aufsichtsrat über das erste Halbjahr 1968; ÖMV, Sitzung des Arbeitsausschusses des Aufsichtsrates, February 27, 1968; ÖMV, Antrag auf Vergabe der Errichtung einer Kompressorenstation in Baumgarten, March 18, 1968; ÖMV, Antrag auf Vergabe von 3 Gasmaschinenverdichter Clark TLA 6 für die Errichtung der Gasstation Baumgarten, March 18, 1968; ÖMV, Memo, June 5, 1968, OeStA ÖIAG-Archiv, Box 135.
5. ÖMV, Bericht für den Aufsichtsrat über das erste Halbjahr 1968 and ÖMV, Bericht für den Aufsichtsrat über die ersten drei Quartale 1968, OeStA ÖIAG-Archiv, Box 135.
6. ÖMV, Sitzung des Aufsichtsrates, September 18, 1968, OeStA ÖIAG-Archiv, Box 135; ÖMV, Berichit über das Geschäftsjahr 1968, p. 18; Tass, September 6, 1968, quoted in *Oil and Gas Journal*, October 14, 1968, p. 74.
7. ÖMV, Bericht für den Aufsichtsrat über das Jahr 1968, OeStA ÖIAG-Archiv, Box 136.
8. ÖMV, Bericht für den Aufsichtsrat über das 1. Quartal 1969; Sitzung des Arbeitsausschusses des Aufsichtsrates der ÖMV AG, September 17, 1969, OeStA ÖIAG-Archiv, Box 136.
9. Glavgazdobycha, Soyuznefteexport, Metalimex, CSSR Gazovye Predpriyatiya, and ÖMV, December 12, 1968, RGAE 458-1-1115.
10. ÖMV, Bericht für den Aufsichtsrat über das Jahr 1969, OeStA ÖIAG-Archiv, Box 137; Lantzke and Plesser to von Dohnanyi, October 14, 1969, Barch B102-152195.
11. ÖMV, Sitzung des Aufsichtsrates, September 21, 1970; ÖMV, Bericht für den Aufsichtsrat über das Jahr 1970, OeStA ÖIAG-Archiv, Box 137.
12. ÖMV, Bericht für den Aufsichtsrat über das Jahr 1970; ÖMV, Bericht für den Aufsichtsrat über das 1. Quartal 1971, OeStA ÖIAG-Archiv, Box 137; *ÖMV-Zeitschrift* 2/1975, p. 2.
13. ÖMV, Bericht für den Aufsichtsrat über das 1. Halbjahr 1971; ÖMV, Bericht für den Aufsichtsrat über das Jahr 1971; ÖMV, Bericht für den Aufsichtsrat über das 1. Quartal 1972, OeStA ÖIAG-Archiv, Box 138.
14. A+B reserves in Soviet gas terminology. *Gazovaya promyshlennost*, May 1965, p. 6.
15. Ukrainian Council of Ministers to Baibakov and Kortunov (undated), TsDAVO Ukrainy 337-33-7; Kortunov to the Central Committee, June 6, 1968, RGAE 458-1-967. The volume transited to the other republics was about twice the volume consumed regionally in western Ukraine, see Kutsevol (secretary of the Lvov Regional Committee of the Ukrainian Communist Party) to the Soviet Council of Ministers, January 18, 1967, RGAE 458-1-494.
16. Kutsevol (secretary of the Lvov Regional Committee of the Ukrainian Communist Party) and Stefanik (chairman of the Executive Committee of the Regional Council of Worker Deputies) to Kortunov, November 28, 1966, RGAE 458-1-104; Kutsevol to the Soviet Council of Ministers, January 18, 1967, RGAE 458-1-494; Tikhonov to Dymshits (Gossnab), Kortunov (Mingazprom) and Neporozhnii (Minenergo), February 4, 1967, RGAE 458-1-493.
17. Khalatin (chief engineer, Glavgazoprovodov), January 19, 1967, RGAE 458-1-105.
18. Ruben to the Soviet Council of Ministers, April 11, 1967, RGAE 458-1-495.
19. Kutsevol to the Soviet Council of Ministers, January 18, 1967, RGAE 458-1-494.
20. Ukrainian Council of Ministers to Baibakov (Gosplan) and Kortunov (Mingazprom) (undated), TsDAVO Ukrainy 337-33-7.
21. Kortunov to the Council of Ministers, March 14, 1967, RGAE 458-1-494; Sidorenko to Galonskii (Gosplan), August 27, 1968, RGAE 458-1-975; Sidorenko to Gosplan, December 16, 1968, RGAE 458-1-975; Dinkov to Medvedkov (deputy manager of Gosplan's division for Development of Economic Cooperation

between the USSR and Socialist Countries), January 20, 1971, RGAE 458-1-2526. At Efremovka new gas deposits had recently been discovered.
22. Bokserman to Khorkov, director of Giprospetsgaz and Ruben, January 15, 1968, RGAE 458-1-502; Sidorenko to Galonskii, August 27, 1968; Bokserman to Galonskii, September 11, 1968; Galonskii to Sidorenko, July 17, 1968, RGAE 458-1-975.
23. Kortunov to the Central Committee of the Communist Party, June 6, 1968, RGAE 458-1-967; Kortunov to Shelest (secretary of the Central Committee of the Ukrainian Communist Party) and Shcherbitskii (chairman of the Ukrainian Council of Ministers), May 1968, TsDAVO Ukrainy 337-3-20.
24. Kortunov to the Central Committee of the Communist Party, June 6, 1968, RGAE 458-1-967.
25. Spravka o perspektive gazosnabzheniya g. Kieva v zimnii period 1968–1969 g.g., September 3, 1968, TsDAVO Ukrainy 337-3-20.
26. Smirnov to Gosplan and Stroibank SSSR, January 7, 1969, RGAE 458-1-1508; Ukrainian Council of Ministers to Baibakov and Kortunov, October 7, 1968, TsDAVO Ukrainy 337-3-20; Ryabenko to Kortunov, October 8, 1968, RGAE 458-1-1508.
27. Bokserman to Tikhonov, January 7, 1969, RGAE 458-1-975; Sidorenko to Gosplan, December 16, 1968, RGAE 458-1-975. It was primiarly the Ministry of Chemical Industry that complained about shortages.
28. Kortunov to Efremov, February 27, 1970, RGAE 458-1-2048.
29. Ibid.
30. Dinkov to Galonskii, February 22, 1971, RGAE 458-1-2528; Dinkov to Galonskii, April 1, 1971, RGAE 458-1-2530.
31. Ukrainiain Council of Ministers to Kortunov, March 24, 1972, TsDAVO Ukrainy 2-13-6972; Burmistrov (deputy chairman of the Ukrainian Council of Ministers) to Dymshits, April 14, 1972, TsDAVO Ukrainy 2-13-6972; Burmistrov to Lalayants, June 7, 1973, TsDAVO Ukrainy 2-13-7344.
32. Perechen predpryatii i elektrostantsii, gazosnabzhenie kotorykh podlezhit regulirovaniyu putem chastichno ili polnogo perevoda ikh na rezervnoe toplivo, August 15, 1972, TsDAVO Ukrainy 2-13-6972; Zherechov (deputy minister of heavy, energetic, and transport machine-building), September 28, 1972, to Rozenko (Ukrainian Council of Ministers) and Dinkov (Mingazprom), TsDAVO Ukrainy 2-13-6972.
33. Ploshchenko and Maksimov to the Ukrainian Council of Ministers, November 30, 1972, TsDAVO Ukrainy 2-13-6972; Dinkov to Belozerov (deputy minister of the food industry), November 13, 1972, TsDAVO Ukrainy 2-13-7344.

7 Willy Brandt: Natural Gas as Ostpolitik

1. See, for example, "Czech blitz craters Soviet gas deals," *Oil and Gas Journal*, October 14, 1968, pp. 74–75.
2. Lantzke to Arndt, March 18, 1969, BArch B102-152194; Limbourg, August 26, 1969, BArch B102-240343.
3. "The Budapest appeal: message from Warsaw Pact states to all European countries," quoted from the English translation of the full text as reprinted in Remington 1971, pp. 225–228. The original Russian text appeared in *Pravda*, March 18, 1969.
4. Von Dannenberg 2007.
5. Rudolph 2004, p. 287.
6. Woratz's speech was published in *gwf*, October 1968, p. 93f.
7. *gwf*, January 3, 1969, p. 20; Wedekind, May 16, 1969. For example, DIW carried out a market analysis indicating that total German gas demand for 1975 would be 28 bcm rather than 25 bcm.
8. Brandt, February 11, 1969, PA AA B150. See also Gespräch des Staatssekretärs Duckwitz mit dem sowjetischen Botschafter Zarapkin, April 8, 1969, in *Akten*

zur Auswärtigen Politik der Bundesrepublik Deutschland 1969, vol. 1, p. 454f. The Copenhagen meeting is referred to by Kaun, April 30, 1969, BArch B102–152194.
9. Kaun, April 30, 1969, BArch BArch B102–152194.
10. Lantzke to Arndt, March 18, 1969, BArch B102–100003; Gespräch zwischen Brandt und Zarapkin, April 4, 1969, cited in Akten zur Auswärtigen Politik der Bundesrepublik Deutschland 1969, vol. 1, p. 443f.
11. Schiller to Kiesinger, April 29, 1969, BArch B102–152194. The same document is found in PA AA B63–439.
12. Ibid.; Klarenaar, April 30, 1969, PA AA B63–439.
13. Wedekind to Lantzke, May 16, 1969; Lantzke to von Dohnanyi, May 20, 1969, BArch B102–152194.
14. Wedekind, May 16, 1969, BArch B102–152194; Lantzke cited by Herbst, June 27, 1969, in Akten zur Auswärtigen Politik der Bundesrepublik Deutschland 1969, vol. 1, pp. 740ff.
15. Wedekind, May 16, 1969, BArch B102–152194.
16. Von Dohnanyi to Schiller, June 18, 1969, BArch B102–152194.
17. Lantzke to von Dohnanyi, May 20, 1969; Von Dohnanyi to Schiller, May 30, 1969, BArch B102–152194.
18. Lantzke to von Dohnanyi, May 20, 1969; Wedekind, May 30, 1969; Schelberger to Schedl, June 10, 1969, BArch B102–152194.
19. Lantzke, May 29, 1969, BArch B102–152194.
20. Allardt to the Foreign Office, May 26, 1969, in Akten zur Auswärtigen Politik der Bundesrepublik Deutschland 1969, vol. 1, p. 641; Bericht über die Reise von Staatssekretär Dr. von Dohnanyi nach Moskau (undated), PA AA B63–439.
21. Lantzke, May 29, 1969, BArch B102–152194; Bericht über die Reise von Staatssekretär Dr. von Dohnanyi nach Moskau (undated), PA AA B63–439; Allardt to the Foreign Office, May 26, 1969, in Akten zur Auswärtigen Politik der Bundesrepublik Deutschland 1969, vol. 1, p. 641; Herbst to Brandt, July 3, 1969, BArch B102–152194.
22. Heitzer (Bavarian Ministry of Economy) to van Beveren (Mannesmann Export GmbH), Hermann Linde (Linde AG), Bogner (Deutsche Bank AG), and Presuhn (Bayerische Ferngas GmbH), June 16, 1969; Von Dohnanyi to Schiller, June 18, 1969 and July 7, 1969, BArch B102–152194.
23. Lantzke to von Dohnanyi, June 24, 1969, BArch B102–152194.
24. Ibid.
25. Schelberger to von Dohnanyi, June 24, 1969, BArch B102–152194.
26. Von Dohnanyi to Schiller, July 7, 1969, BArch B102–152194; Plesser, July 8, 1969, BArch B102–152194.
27. Plesser, July 10, 1969, BArch B102–152194; Klarenaar, July 10, 1969, PA AA B63–435.
28. Plesser, July 11 and 14, 1969; Plesser to Lantzke, June 24, 1969, BArch B102–152194.
29. Plesser, August 7, 1969, BArch B102–152194; Klarenaar, August 8, 1969, PA AA B63–435.
30. Plesser, September 1, 1969; Lantzke to von Dohnanyi, September 2, 1969, BArch B102–152194.
31. Brandt to Schiller, July 10, 1969, BArch B102–152194; Gespräch zwischen Kiesinger und Nixon, August 7, 1969, in Akten zur Auswärtigen Politik der Bunesrepublik Deutschland 1969, vol. 2, p. 896.
32. Plesser, July 25, 1969, BArch B102–152194.
33. Ibid.
34. Schroeter (Referat I C 2) to Referat III B 3, July 22, 1969; Plesser, August 20 and 21, 1969, BArch B102–152194.

35. Plesser, August 20, 1969, BArch B102–152194.
36. Bahr, July 25, 1969, in *Akten zur Auswärtigen Politik der Bundesrepublik Deutschland 1969*, vol. 2, p. 857.
37. Lantzke and Plesser to von Dohnanyi, October 14, 1969, BArch B102–152195.
38. The revaluation of the D-Mark was finally carried out on October 29, 1969, and raised its parity by no less than 9.3 percent from 4.00 to 3.66 DM per USD.
39. Allardt, October 30, 1969; Lantzke to Arndt, November 4, 1969, BArch B102–152195.
40. Lantzke, January 29, 1970; Plesser, December 3, 1969; Lantzke to Arndt, December 3, 1969, BArch B102–152196; cf. *Die Welt*, November 29, 1969.
41. Plesser, November 28, 1969, Referat III C, January 27, 1970; Lantzke, January 29, 1970; BArch B102–152196; Mannesmann Export, press-release, February 1, 1970; *VWD Montan*, February 2, 1970.
42. Henze, January 28, 1970, BArch B102–240343. The bank consortium was led by Deutsche Bank. The other involved banks were Dresdner Bank, Commerzbank, and Westdeutsche Landesbank Girozentrale; Plesser, September 24, 1973, BArch B102–257471.
43. Von Dohnanyi to Grund, July 23, 1969, BArch B102–152194; Lantzke to von Dohnanyi, June 3, 1969; Plesser, June 19, 1969, BArch B102–152194.
44. Plesser, June 19, 1969, BArch B102–152194.
45. Plesser, July 10, 1969, BArch B102–152194.
46. Wedekind, July 8, 1969; von Dohnanyi to Grund, July 23, 1969, BArch B102–152194.
47. Telle and Lony (Wirtschaftsverband Erdölgewinnung e.V.) to Plesser, July 30, 1969; Stellungnahme der in der Bundesrepublik tätigen und im Wirtschaftsverband Erdölgewinnung (WEG) zusammengeschlossenen Erdgasproduzenten zur Frage der Einfuhr von russischem Erdgas, August 19, 1969, BArch B102–152194.
48. Plesser to Lantzke, August 19, 1969, BArch B102–152194.
49. Wedekind to Lantzke, August 27, 1969, BArch B102–152195.
50. Ibid.
51. Van Well (West German Embassy in Paris), April 18, 1969; Klarenaar, April 30, 1969. PA AA B63–439; Lantzke to Arndt March 18, 1969, BArch B102–100003.
52. See, for example, the article by the president of Gaz de France in the Soviet gas industry journal, *Gazovaya promyshlennost*, September 1970, p. 45.
53. Von Dohnanyi to Schiller, July 7, 1969, BArch B102–152194.
54. Von Dohnanyi to Schiller, May 30, 1969; von Dohnanyi to Schiller, July 7, 1969, BArch B102–152194.
55. Limbourg, August 26, 1969, BArch B102–240343; Herbst, July 21, 1969; Lantzke to Herbst, June 16, 1969, PA AA B63–435. The Soviet gas issue was also discussed by a French-German special working group on energy in June 1969, see von Bismarck-Osten, July 17, 1969, PA AA B63–435 and Referat III B 3, August 25, 1969, BArch B102–152195.
56. Wedekind, July 14, 1969, BArch B102–152194.
57. Klarenaar to the West German Embassy in Paris, August 19, 1969; Lantzke, September 4, 1969, BArch B102–240343; Referat III B 3, August 25, 1969, BArch B102–152195; Braun, September 19, 1969; Limbourg to the Foreign Office, August 7, 1969; Limbourg, August 26, 1969, PA AA B63–435.
58. Braun, September 19, 1969, PA AA B63–435.
59. Lantzke to von Dohnanyi, September 30, 1969, BArch B102–152195.
60. Ibid; Klarenaar to the West German Embassy in Paris, October 7, 1969, PA AA B63–435.
61. *gwf*, November 21, 1969, p. 1316f.; Blomeyer, October 28, 1969, PA AA B63–435; Göpner, October 31, 1969, BArch B102–152195.

62. Plesser, November 28, 1969, BArch B102–152195.
63. For details on the Soviet-Italian contract, see Ambassador Steg to the Foreign Office, December 23, 1969, BArch B102–240343; Lantzke to Arndt, January 8, 1970, BArch B102–152196; and Spravka-kharakteristika na gruppu ENI (undated), RGAE 458–1–2679.
64. Schiller, February 1, 1970, BArch B102–152196.
65. A German translation of Osipov's speech is found in BArch B102–152196.
66. Schiller to Brandt, February 4, 1970, BArch B136–7686; Rudolph 2004, p. 294; Stent 1981, p. 169.
67. Lantzke, January 29, 1970, BArch B102–152196.
68. Mertens (Referat IV A 5) to Plesser, December 16, 1969, BArch B102–152196; Plesser, January 29, 1970, BArch B102–152196.
69. Schöllhorn to Franz, May 12, 1970, BArch B102–240343.
70. Gerke, January 27, 1971, BArch B102–208639.
71. Rummer, January 25, 1971; Lantzke to Schiller, March 24, 1971; BArch B102–240343. Bartels and Salinger from BEB met with Baranovsky and Michailov from the Soviet side in Helsinki.
72. Rummer, December 21, 1970, BArch B102–240343.
73. Weiss to Ehmke and Bahr, June 14, 1971, BArch B136–7667; Plesser, September 24, 1973, BArch B102–257471.
74. "France signs gas deal with Soviets," *Oil and Gas Journal*, August 14, 1972. France had signed an agreement in principle with the Soviet Union in July 1971 and the deal is now officially confirmed.
75. ÖMV, Sitzung des Aufsichtsrates, June 23, 1971, OeStA ÖIAG-Archiv, Box 137; "Austria to import more Soviet gas," *Oil and Gas Journal*, July 19, 1971.
76. Skachkov (chairman of the state committee for foreign economic relations of the Soviet Council of Ministers) to Dymshits (deputy chairman of the Soviet Council of Ministers), April 28, 1973, RGAE 458–1–3387; "Russian gas market expands to Finland," *Oil and Gas Journal*, December 20, 1971; "Finland gets pipe for gas line," *Oil and Gas Journal*, October 23, 1972; "Neste Oy übernimmt in Finnland den Gastransport," *gwf*, September 1971, p. 457; "Sweden may buy Soviet gas," *Financial Times*, March 31, 1970; "Williams Bros. studying line from Finland to Sweden," *Oil and Gas Journal*, May 7, 1973; "Erdgas für Schweden," *gwf*, February 1973, p. 99.
77. "Sowjet-Gas auch für die Schweiz?," *gwf*, October 1970, pp. 595 and 617; *Financial Times*, April 2, 1971; *gwf*, January 1972, p. 43; "Back on speaking terms," *Financial Times*, October 18, 1972.
78. "First major LNG-import bid advances," *Oil and Gas Journal*, June 21, 1971; "Soviet gas may warm the US" and "US confirms Soviet LNG-import talks," *Oil and Gas Journal*, December 6, 1971. Distrigas wanted to import around 6 bcm per year.
79. "Signing near for Far East Soviet LNG project," *Oil and Gas Journal*, November 13, 1972; "Treasury boss Connally backs Soviet oil, gas imports," *Oil and Gas Journal*, April 24, 1972; "More Algerian and Russian LNG is proposed in US," *Oil and Gas Journal*, August 21, 1972; "Russian LNG coming to US? Probably," *Oil and Gas Journal*, May 29, 1972.
80. "Talks advance on US supply of Soviet LNG," *Oil and Gas Journal*, July 9, 1973.

8 Constructing the Export Infrastructure

1. Kortunov to Baibakov, June 7, 1966, RGAE 458–1–114; Bokserman to Yudin (deputy head of Gosplan's division for Oil and Gas Industry), February 5, 1968, RGAE 458–1–972.

2. *Gazovaya promyshlennost,* July 1969, p. 44; *Izvestiya* quoted in *Oil and Gas Journal,* September 7, 1970, p. 53.
3. Kortunov to Baibakov, June 7, 1966, RGAE 458-1-114.
4. Smirnov to Isaev (deputy chairman of Gosplan) and Vasilenko (deputy chairman of the Board of Stroibank), January 7, 1970, RGAE 458-1-2045; Kortunov to Council of Ministers, April 7, 1970, RGAE 458-1-2051; Smirnov to Novikov (chairman of Gosstroi), January 13, 1970, RGAE 458-1-2045. 200,000 tons of 1,420 mm pipes were to be delivered during the second half of 1970.
5. Kortunov to the Council of Ministers, April 7, 1970, RGAE 458-1-2051; cf. *Gazovaya promyshlennost,* January 1971; Dertsakyan and Altshul (Giprospetsgaz), "Osnovnye proektnye resheniya severnoi sistemy magistralnykh gazoprovodov," *Gazovaya promyshlennost,* November 1969, p. 13f.
6. Protokol no. 13 tekhnicheskogo soveshchaniya u Ministra gazovoi promyshlennosti SSSR, December 24, 1969, RGAE 458-1-1482; Sorokin to Misnik, February 23, 1970; Mingazprom to the Council of Ministers, February 26, 1970, RGAE 458-1-2047; Kortunov to the Council of Ministers, April 7, 1970, RGAE 458-1-2051.
7. Orudzhev to Kosygin, September 17, 1974, RGAE 458-1-3728.
8. Dinkov to Gosplan, January 11, 1971, RGAE 458-1-2526; Chevardov (deputy head of the Transport Division), May 21, 1970, RGAE 458-1-2053.
9. Brezhnev 1971, pp. 13, 34, and 157; Kortunov, "Gazovaya promyshlennost SSSR v 1969 godu," *Gazovaya promyshlennost,* January 1969, p. 1; Predlozhenie v Gosplan SSSR ob osnovnykh napravleniyakh gazovoi promyshlennosti na 1971–1975 gg., May 13, 1968, RGAE 458-1-2547; Kortunov, "Uspekhi gazovoi promyshlennosti i perspektivy ee razvitiya," *Gazovaya promyshlennost,* March 1971, p. 3; see also report from the Party Congress in *Gazovaya promyshlennost,* May 1971, p. 1.
10. Kortunov to Baibakov, May 21, 1971, RGAE 458-1-2533. A second, parallel pipeline would start to be built in 1974 with the help of the newly contracted German pipes, see Kortunov to the Council of Ministers, June 15, 1971, RGAE 458-1-2536.
11. Kortunov to Council of Ministers, June 15 and 18, 1971, RGAE 458-1-2536; cf. Orudzhev to Council of Ministers, December 29, 1973.
12. *Gazovaya promyshlennost,* January 1974, p. 7.
13. Kortunov to the Council of Ministers, May 8, 1970, RGAE 458-1-2052.
14. Kortunov to the Council of Ministers, March 29, 1971, RGAE 458-1-2530.
15. Dinkov to Medvedkov, January 20, 1971, RGAE 458-1-2526.
16. Ibid.
17. Sidorenko to Dymshits, June 8, 1973, RGAE 458-1-3387.
18. Ibid.; Dinkov to Medvedkov, January 20, 1971, RGAE 458-1-2526; Dinkov to Galonskii, February 22, 1971, RGAE 458-1-2528; Doklad o razvitii gazovoi promyshlennosti na 1971–1975 gg., August 1970, RGAE 458-1-2548.
19. Kortunov to the Council of Ministers, August 9, 1971, RGAE 458-1-2539; Smirnov to Lalayants, October 13, 1971, RGAE 458-1-2542.
20. Sidorenko to Dymshits, June 8, 1973, RGAE 458-1-3387.
21. The Italian delay is further discussed in chapter 9.
22. Kiselev (chairman of the Belarusian Council of Ministers) to Kortunov, July 9, 1971; Kortunov to Kiselev, August 10, 1971, RGAE 458-1-2539.
23. Doklad o razvitii gazovoi promyshlennosti na 1971–1975 gg., August 1970, RGAE 458-1-2548; Kortunov, "Zadachi gazovoi promyshlennosti v 1972 godu," *Gazovaya promyshlennost,* February 1972, p. 1.
24. Orudzhev to Dymshits, February 22, 1973, RGAE 458-1-3397; Kurchenkov (deputy head of Glavtransgaz), February 14, 1973, RGAE 458-1-3384; Protokol soveshchaniya u tov. Dymshitsa, March 2, 1973, RGAE 458-1-3387.
25. Runov and Sedykh 1999.
26. Orudzhev to Dymshits, February 22, 1973, RGAE 458-1-3397.

27. Sidorenko to Dymshits, March 16, 1973, RGAE 458-1-3397; "Ekonomicheskoe sotrudnichestvo SSSR s sotsialisticheskimi stranami v oblasti gazovoi promyshlennosti," *Gazovaya promyshlennost,* February 1975; ÖMV, Bericht für den Aufsichtsrat über das 1. Halbjahr 1973, OeStA ÖIAG-Archiv, Box 138; Entwicklung von Stadtgasaufkommen, Erdgasförderung, Erdgasimport u. Gasverbrauch 1960–1976, BArch DG12–691; Fenske to Enkelmann, March 16, 1973, BArch DG 12–289. Corresponding figures were 1.3 bcm for 1970, 1.6 bcm for 1971, and 1.9 bcm in 1972.
28. Orudzhev to the Council of Ministers, May 3, 1973, RGAE 458-1-3397.
29. Burmistrov to Shelepin and Volkov, June 1, 1973, TsDAVO Ukrainy 2–13–7344.
30. Sidorenko to Dymshits, June 8, 1973, RGAE 458-1-3387; Burmistrov to Dymshits, June 30, 1973, RGAE 458-1-3387.
31. Sidorenko to Dymshits, June 8, 1973, RGAE 458-1-3387.
32. Ibid.; Batalin to the Council of Ministers, July 18, 1973; Kuzmin (deputy minister of foreign trade) to Dymshits, October 11, 1973, RGAE 458-1-3387.
33. Batalin (first deputy minister of construction of oil and gas facilities) to the Council of Ministers, July 18, 1973, RGAE 458-1-3387; ÖMV, Bericht für den Aufsichtsrat 1973, p. 26. OeStA ÖIAG-Archiv, Box 138; Orudzhev to the Council of Ministers, July 18, 1973, RGAE 458-1-3387.
34. Orudzhev to the Council of Ministers, July 18, 1973, RGAE 458-1-3387.
35. Protokol no. 26 soveshchaniya u Zamestitelya Predsedatelya SM SSSR, Predsedatelya Gossnaba tov. Dymshits, V. E., July 27, 1973, TsDAVO Ukrainy 337–15–436.

9 Trusting the Enemy: Importing Soviet Gas in Practice

1. Kortunov to Efremov, March 10, 1970, RGAE 458-1-2048; "Soviet-Czech gas pipe deal," *Financial Times,* December 29, 1970; Bundesstelle für Aussenhandelsinformation, February 13, 1971. The GDR's transit agreement was signed on July 2, 1971, see *gwf,* September 1971, p. 457.
2. Loens, September 30, 1969, PA AA B41–84; Plesser to Lantzke, October 2 and 9, 1969, BArch B102–152195.
3. Wedekind, July 14, 1969, BArch B102–152194; Kortunov to Lesechko, January 29, 1970, RGAE 458-1-2046; cf. *Oil and Gas Journal,* June 10, 1974, p. 38.
4. ÖMV, Sitzung des Aufsichtsrates, September 29, 1971, OeStA ÖIAG-Archiv, Box 138; "Italian pipeline for Soviet gas runs into more problems," *Financial Times,* June 15, 1972; "Erdgasleitungsvertrag ÖMV-ENI perfekt," *gwf,* September 1971, p. 457.
5. ÖMV, Sitzung des Aufsichtsrates, September 29, 1971, OeStA ÖIAG-Archiv, Box 138; "Projektierung und Bau der Trans-Austria-Gasleitung," *ÖMV-Zeitschrift* 2/1974, p. 2.
6. Ibid.; ÖMV, Bericht für den Aufsichtsrat über das 1. Halbjahr 1971, OeStA ÖIAG-Archiv, Box 138.
7. "Czech pipeline hits difficulties," *Financial Times,* March 4, 1971; "Zusatzvertrag der Ruhrgas mit den Russen über 4 Mrd cbm Erdgas abgeschlossen," *Telegraf,* July 16, 1971.
8. ÖMV, Bericht für den Aufsichtsrat über die ersten 3 Quartale 1972; ÖMV, Bericht für den Aufsichtsrat über das 1. Quartal 1973; ÖMV, Bericht für den Aufsichtsrat über die ersten 3 Quartale 1973; ÖMV, Bericht für den Aufsichtstsrat über das Jahr 1973; OeStA ÖIAG-Archiv, Box 138; "Projektierung und Bau der Trans-Austria-Gasleitung," *ÖMV-Zeitschrift* 2/1974, p. 2.
9. Otto Schedl, "Erdgas für Bayern—Energieverbund zwischen West und Ost, in Gasverwendung," *Zeitschrift für neuzeitliche Energieversorgung,* vol. 21, no. 8, 1970; "Ausbau der Gasversorgung in Bayern," *BLD,* November 9, 1970.
10. *Handelsblatt,* April 22 and July 7, 1970.

11. Geilenkeuser (member of Ruhrgas' management board), August 13, 1970, BayHStA MWi, file 27208.
12. "Verzögerung beim Bau der Russengas-Leitung?" *Nürnberger Nachrichten*, November 2, 1971; "Unsicherheit über die sowjetische Erdgaspolitik gegenüber Westeuropa," *Neue Zürcher Zeitung*, November 4, 1971.
13. Heitzer, "Pipelinebaue der UdSSR," September 14, 1971, BayHStA MWi, file 27208.
14. Heitzer to Ruhrgas, October 14, 1971, BayHStA MWi, file 27208.
15. Liesen to Bayer, November 19, 1971, BayHStA MWi, file 27208.
16. Presuhn and Kolb (members of Bayerngas' management board) to Schiller, March 3, 1971, BArch B102–277764.
17. Plesser, Bericht über die Einfuhr sowjetischen Erdgases, late 1970 (undated), BArch B102–208639.
18. Presuhn and Kolb (Bayerngas management board) to Schiller, March 3, 1971, BArch B102–277764.
19. Liesen to Bayer, November 19, 1971, BayHStA MWi, file 27208; Heitzer to Lantzke, May 25, 1971, BArch B102–277764.
20. Rummer to Lantzke, June 18, 1971, BArch B102–277764.
21. Rummer, August 27, 1971, BArch B102–77764.
22. Eder to the German Embassy in Algiers, March 8, 1972, BArch B102–152183; Jaumann, December 19, 1972, BayHStA StK, file 18790.
23. Jaumann to the Bavarian Parliament, December 21, 1972; Liesen to Bayer, November 19, 1971, BayHStA MWi, file 27208.
24. Jaumann to the Bavarian Parliament, December 21, 1972, BayHStA MWi, file 27208.
25. Jaumann, October 1, 1973, BayHStA StK, file 18790; Friderichs, October 1, BArch B102–257471.
26. Kalchenko (deputy chairman of the Ukrainian Council of Ministers) to Kosygin, September 21, 1973, RGAE 458–1–3384.
27. Ibid.
28. Kuzmin (deputy minister of foreign trade) to Dymshits, October 11, 1973, RGAE 458–1–3387; Ob obespechanii vvoda v deistvie moshnostei v IV kv. 1973 g. na Ukrainskoi sisteme gazoprovodov, obespechayushchikh postavku gaza na eksport (protocol from a meeting hosted by Gas Minister Orudzhev), November 4, 1973, TsDAVO Ukrainy 337-15–436.
29. Masherov och Kisilev to Kosygin, October 26, 1973; Neporozhnii (minister of energy and electrification) to Dymshits, October 30, 1973, RGAE 458–1–3384.
30. Osipanko (first deputy minister of chemical industry) to the Council of Ministers, November 22, 1973; Kostandov (minister of chemical industry) to the Council of Ministers, December 3, 1973; Sidorenko to the Council of Ministers, December 25, 1973; Kazanets (minister of ferrous metallurgy), December 11 and 17, 1973, RGAE 458–1–3384.
31. Titarenko to Dymshits, December 6, 1973, RGAE 458–1–3384.
32. To the general secretary of the central committee of the Communist Party, comrade L. I. Brezhnev, December 11, 1973, signed by 50 residents, TsDAVO Ukrainy, 337–15–440.
33. Smirnov to Dymshits, December 12, 1973, RGAE 458–1–3387; The plenum discussion is referred to in a later letter from Orudzhev and Shcherbina to the Central Committee, January 22, 1975, RGAE 458–1–3973.
34. Lyashko to the Soviet Council of Ministers, January 12, 1974, TsDAVO Ukrainy 2–13–8301.
35. Dymshits to Lyashko, January 12, 1974; Burmistrov to Tumanov (Ukrgazprom) and Plshchenko (Minkomchoz), TsDAVO Ukrainy 2–13–8301.

36. Rozenko to Dymshits, February 15, 1974, TsDAVO Ukrainy 2–13–8301; Orudzhev, "Zadachi razvitiya gazovoi promyshlennosti v 1974 g. v svete reshenii dekabrskogo (1973 g.) Pleniuma TsK KPSS," *Gazovaya promyshlennost*, February 1974, p. 3.
37. The disturbances in the start-up phase were discussed by the Foreign Office and the Federal Ministry of Economy in early November 1973, see BArch B102–257471.
38. See, for example, "Finnish-Soviet gas line open," *Financial Times*, January 11, 1974.
39. Sidorenko to Dymshits, June 8, 1973, RGAE 458–1–3387; Orudzhev and Shcherbina to the Council of Ministers, January 22, 1975, RGAE 458–1–3963.
40. "Feierliche Inbetriebnahme der Trans-Austria-Gasleitung am 20. Mai 1974 in Baumgarten und Schloss Eckartsau," *ÖMV-Zeitschrift* 2/1974; Lyashko to Kosygin, June 18, 1974, TsDAVO Ukrainy 2–13–8301.
41. Orudzhev and Shcherbina to the Council of Ministers, January 22, 1975, RGAE 458–1–3963; "Soviets' foreign oil, gas sales hit $7 billion," *Oil and Gas Journal*, April 28, 1975. All in all the Soviet Union exported 14.1 bcm of Soviet gas during 1974, of which 39 percent was for Western Europe.
42. Orudzhev to the Council of Ministers, October 23, 1974, and November 20, 1974, RGAE 458–1–3728; Orudzhev to the Council of Ministers, February 26, 1975, RGAE 458–1–3963. The unfortunate site was at 150 km distance from the Soviet border.
43. Jaumann to the Bavarian Parliament, February 19, 1975, BayHStA StK, file 18791.
44. Sackmann (state secretary at the Bavarian ministry of economy) to the Bavarian Parliament, May 16, 1974 and February 19, 1975, BayHStA StK, file 18791; Koch (Social Democratic Member of the Bavarian Parliament) to the Bavarian Parliament, February 24, 1975, BayHStA StK, file 18790; Bavarian Parliament, Stenographischer Bericht, 8. Wahlperiode, 11. Sitzung, February 26, 1975.
45. Jaumann to the Bavarian Parliament, April 8, 1975, July 29, 1977, and November 17, 1978, StK 18790. By the late 1970s the share of Soviet natural gas in Bavaria's total gas supply already amounted to nearly three-fourths.
46. Referat III C 4, Erdgaspreise für Haushaltskunden, Preisstand 1.7.1976 und 1.7.1977, BArch B102–208644; "Im Norden und in Bayern ist Erdgas am billigsten," *Die Welt*, October 17, 1978.
47. Jaumann, May 8, 1980; Sackmann to the Bavarian Parliament, October 26, 1977, BayHStA StK, file 18790; Jaumann to the Bavarian Parliament, February 19, 1975, BayHStA StK, file 18791; Bavarian Parliament, Stenographischer Bericht, 8. Wahlperiode, 11. Sitzung, February 26, 1975; Dahl 1997, p. 78.

10 Scale Up or Phase Out?

1. "Thoughts on the seventies," *Oil and Gas Journal*, January 5, 1970.
2. Ibid.; "High oil prices turn Europe towards gas," December 6, 1971.
3. "Soviet gas exports 'could ease energy crisis in Europe,'" *Financial Times*, November 15, 1973; *Financial Times*, November 22, 1971. UNECE was a strong supporter of this idea at the pan-European level.
4. Lumpe, July 3, 1972, BArch B102–240343.
5. Ibid.
6. Weiss to Abteilungsleiter IV, January 15, 1973; Wedekind to Lantzke, August 27, 1969, BArch B102–152195, citing press reports. See also Chapter 7.
7. *ÖMV-Zeitschrift* 2/1975, p. 2.
8. "Dutch gas: one bubble not to burst," *Financial Times*, February 15, 1974.
9. "Algerian LNG cutoff not tied to Arab embargo," *Oil and Gas Journal*, November 26, 1973; "Algeria may renegotiate LNG pacts," *Oil and Gas Journal*, December 17,

1973; ÖMV, Sitzung des Aufsichtsrates, December 16, 1974, OeStA ÖIAG-Archiv, Box 138.
10. Lumpe, April 22, 1974; Referat III C 4 and V B 1, October 4, 1974, BArch B102-240343; ÖMV Management Board to the Working Committee of the Supervisory Board, December 16, 1974, OeStA ÖIAG-Archiv, Box 138.
11. "French, Soviets sign 20-year gas deal," *Financial Times*, December 7, 1974.
12. "Soviet LNG project at critical point," *Oil and Gas Journal*, June 24, 1974.
13. "Ways seen to ease winter gas shortage," *Oil and Gas Journal*, July 7, 1975; "Sharing the gas shortage sidesteps the basic problem," *Oil and Gas Journal*, September 22, 1975; "Time seen wrong for Soviet-US natural gas deal," *Oil and Gas Journal*, February 9, 1976.
14. "Bulgaria getting first Soviet gas," *Oil and Gas Journal*, August 26, 1974; *Oil and Gas Journal*, June 10, 1974.
15. O gazifikatsii naselennykh punktov USSR, raspolozhennykh vdol trassy gazoprovoda Orenburg-Zapadnaya gosgranitsa SSSR, March 9, 1976, TsDAVO Ukrainy 2-14-386; Mikhail B. Korchemkin, "Russia's huge Gazprom struggles to adjust to new realities," *Oil and Gas Journal*, October 18, 1993. The Soyuz was equipped with worn-out imported compressors from Nuovo Pignone and Cooper-Bessemer (1,580 MW of total installed capacity), which as of 1993 were reported to be consuming no less than 25 percent of the gas transported.
16. ÖMV, Bericht für den Aufsichtsrat über das Jahr, 1975, OeStA ÖIAG-Archiv, Box 138; Stern 1980, p. 79.
17. "Gas vom Persischen Golf für die Sowjetunion: Der Schah öffnet Pipeline," *Handelsblatt*, October 26, 1970.
18. Mösges, July 15, 1974, BArch B102-238920. The document is a translation of a Canadian report. In 1973, Iran exported oil for $7.5 billion but natural gas for only $70 million.
19. Plesser to Kruse (Foreign Office), March 18, 1974, BArch B102-238920.
20. Plesser and Lantzke to Friderichs, January 23, 1974, BArch B102-238920.
21. Ibid.; Referat III B 3, June 3, 1973; Plesser and Lantzke to Friderichs, January 23, 1974; Ruhrgas AG, Vorstand, to Friderichs, March 10, 1975, BArch B102-238920.
22. Plesser and Lantzke to Friderichs, January 23, 1974, BArch B102-238920; ÖMV, Sitzung des Aufsichtsrates, March 14, 1974; ÖMV Management Board to the Working Committee of the Supervisory Board, November 20, 1975; ÖMV, Sitzung des Aufsichtsrates, October 14, 1974, OeStA ÖIAG-Archiv, Box 138.
23. Plesser and Lantzke to Friderichs, January 23 and March 4, 1974, BArch B102-238920.
24. Gespräch Staatssekretär Dr. Rohwedder—Stellvertreter Aussenhandelsminister Ossipow, April 22, 1974; Ambassador von Lilienfeld (Tehran) to the Foreign Office and the Ministry of Economy, May 6, 1974; Schoeller (Tehran) to the Foreign Office, August 30, 1971, BArch B102-238920.
25. Wieck (Tehran) to the Foreign Office and the Ministry of Economy, July 3 and 6, 1974; Henze to Foreign Office, July 2, 1974; Plesser, July 31, 1974, BArch B102-238920; "Iran and Soviets deadlocked on gas prices," *Oil and Gas Journal*, July 8, 1974.
26. Wieck (Tehran) to the Foreign Office and the Ministry of Economy, July 12, July 18, and August 18, 1974, BArch B102-238920. However, the Soviet embassy in Tehran declared that the Soviet Union regarded the incident as a true accident and not as an intentional disruption.
27. Wieck (Tehran) to the Foreign Office and the Ministry of Economy, August 20, 1974, and January 13 and 16, 1975; Abteilung III, November 27, 1974, BArch B102-238920.

28. Wieck (Tehran) to the Foreign Office and the Ministry of Economy, February 6, 1975; Ruhrgas AG (Management Board) to Friderichs, March 10, 1975; Balser (Moscow) to the Foreign Office, the Ministry of Economy, and the German Embassy in Tehran, April 10, 1975, BArch B102–238920.
29. Balser (Moscow) to the Foreign Office, the Ministry of Economy, and the German Embassy in Tehran, April 10, 1975, BArch B102–238920; ÖMV, Management Board to the Working Committee of the Supervisory Board, November 20, 1975, OeStA ÖIAG-Archiv, Box 138; Stern 1980, p. 79.
30. Balser (Moscow) to the Foreign Office, the Ministry of Economy, and the German Embassy in Tehran, April 10, 1975; Ruhrgas Management Board to Minister Friderichs, June 30, 1975, BArch B102–238920; ÖMV, Bericht für den Aufsichtsrat über das Jahr 1975, OeStA ÖIAG-Archiv, Box 138.
31. *Echo*, Economic Survey, no. 32 (April 1975).
32. ÖMV, Management Board to the Working Committee of the Supervisory Board, November 20, 1975, OeStA ÖIAG-Archiv, Box 138; Minister Friderichs to the German Parliament, June 18, 1976, BArch B102–208642.
33. ÖMV, Sitzung des Aufsichtsrates, December 16, 1976, OeStA AdR ÖIAG-Archiv, Box 138.
34. "High costs plague Russian Arctic gas," *Oil and Gas Journal*, June 20, 1977.
35. Gustafson 1989.
36. Ibid., p. 30; "High costs plague Russian Arctic gas," *Oil and Gas Journal*, June 20, 1977.
37. See, for example, "Soviets pushing projects to boost gas production," *Oil and Gas Journal*, May 22, 1978.
38. The pipeline from Siberia's giant Medvezhye field to Ukhta and Torzhok, which had been started to be constructed already in July 1970, but whose routing had been changed in 1971, could eventually be taken into operation in 1976. See, for example, *Gazovaya promyshlennost*, March 1977, p. 5. Its completion was coordinated with the extension of the Northern Lights system to Belarus and Ukraine, where the main line was drawn in the direction of Dolina and Uzhgorod, and with the expansion of the Czechoslovak transit route. For the expansion of the Czech route, see *Oil and Gas Journal*, October 18, 1976, and September 26, 1977. In June 1977, the Belarus-Uzhgorod section, with a capacity of 16 bcm, was activated, enabling Siberian gas to be pumped across the border to Central and Western Europe, see *Oil and Gas Journal*, June 13, 1977. For the first time, Western Europe was now directly connected with Siberia's immense gas riches.
39. "Soviets award $165 million gas-pipeline project," *Oil and Gas Journal*, December 20, 1976; "Soviets look to natural gas for energy production gains," *Oil and Gas Journal*, March 30, 1981. The capacity of the compressors was 18,200 hp (14 MW). GE's gas turbine technology was used on the Ukhta-Torzhok section and Rolls-Royce's aeroengines on the Tyumen-Ukhta section. See, for example, *Financial Times*, October 30, 1981.
40. Gustafson 1985, p. 5.
41. Engelmann, March 23, 1978, BArch B102–257471.
42. Ibid.
43. Liesen to Lambsdorff, April 24, 1978; Engelmann, March 23, 1978, BArch B102–257471.
44. Ibid.
45. Rohwedder to Genscher, May 3, 1978; Pfletschinger to Engelmann, May 5, 1978; Pfletschinger, May 6, 1978, BArch B102–257471.
46. Referat V B 1, May 1, 1978, BArch B102–257471.
47. As noted by ÖMV. See "Das Geschäftsjahr 1978," *ÖMV-Zeitschrift* 3/1979, p. 2.
48. Ritzel to the Foreign Office, June 8, 1979, BArch B102–313208; "Iran's oil exports hit by strikes," *Financial Times*, October 31, 1978; "Gas flow to Caucasus disrupted,"

Financial Times, January 24, 1979; "Iran planning to increase gas prices to the Russians," *Financial Times*, March 13, 1979; "Iran on verge of killing big IGAT II gas line to USSR," *Oil and Gas Journal*, June 18, 1979.

49. "Gas flow to Caucasus disrupted," *Financial Times*, January 24, 1979; "USSR building two new natural gas pipelines," *Oil and Gas Journal*, August 4, 1980; "Soviets revive trans-Caspian line plans," *Oil and Gas Journal*, October 13, 1980.
50. "Soviets make new offer to supply Sweden with gas," *Oil and Gas Journal*, May 30, 1977; "Soviets press construction of 56 in. gas pipelines," *Oil and Gas Journal*, June 14, 1982.
51. Gustafson 1989, p. 29 and Chapter 3; "Soviets push for construction of strategic Yamal gas line," *Oil and Gas Journal*, November 2, 1981.
52. Gustafson 1985, p. 8.
53. "Soviets push for construction of strategic Yamal gas line," *Oil and Gas Journal*, November 2, 1981.
54. "Soviets trying to convince W. Europe of gas line need," *Oil and Gas Journal*, March 30, 1981.
55. *New York Times*, July 29, 1981.
56. Brezhnev cited in "Soviets look to natural gas for energy production gains," *Oil and Gas Journal*, March 30, 1981.
57. Estrada et al. 1988, p. 106.
58. ÖMV, for example, was quite happy to see that its sales decreased from 3.9 bcm in 1979 to 3.8 bcm in 1980. "Aus dem Geschäftsbericht der ÖMV," *ÖMV-Zeitschrift* 3/1981, p. 4.
59. "West German group signs deal for Soviet gas," *Oil and Gas Journal*, November 30, 1981.
60. "LNG price parity with oil clouds future of European gas market," *Oil and Gas Journal*, April 19, 1982.
61. "Holland, Belgium mull smaller Soviet gas buys," *Oil and Gas Journal*, February 8, 1982. Eventually the Soviets decided not to join the Algerian demand for 100 percent crude oil parity. The price formula agreed on with the Western gas companies implied 20 percent coupling to crude oil, 40 percent to gas oil, and 40 percent to heavy fuel oil. See, for example, *Financial Times*, April 5, 1983.
62. "France agrees to buy gas from western Siberia," *Oil and Gas Journal*, February 1, 1982.
63. Gustafson 1985, p. 32f.
64. Ibid., p. 31f.; "Holland, Belgium mull smaller Soviet gas buys," *Oil and Gas Journal*, February 8, 1982.
65. "Austria may trim planned Soviet gas buys," February 15, 1982; "1,5 Milliarden Kubikmeter Erdgas aus der Sowjetunion," *ÖMV-Zeitschrift* 3/1982, p. 60; "Belgium, Spain delay commitment for Soviet gas," *Oil and Gas Journal*, August 2, 1982. Austrian gas demand fell by 5.5 percent in 1981.
66. "Reagan extends limits on trade with USSR," *Oil and Gas Journal*, June 28, 1982.
67. Ibid.
68. "UK opposes US sanctions against Soviets," *Oil and Gas Journal*, July 12, 1982; "Belgium, Spain delay commitment for Soviet gas," *Oil and Gas Journal*, August 2, 1982.
69. "Soviets, Germans agree on gas line credit," *Oil and Gas Journal*, July 19, 1982; "Controversy building over US sanctions against Yamal pipeline," *Oil and Gas Journal*, August 16, 1982; "Soviets press construction of 56 in. gas pipelines," *Oil and Gas Journal*, June 14, 1982.
70. "EEC: US embargo against USSR is illegal," *Oil and Gas Journal*, August 23, 1982; "Dutch court rules against US Yamal sanctions," *Oil and Gas Journal*, September 27, 1982.

71. "US scaling back sanctions against Yamal line," *Oil and Gas Journal*, September 13, 1982; "Two West German firms added to US blacklist," *Oil and Gas Journal*, October 11, 1982; "End to Yamal pipeline embargo eyed," *Oil and Gas Journal*, November 8, 1982; "US embargo of Yamal gas line damages West more than Soviets," *Oil and Gas Journal*, October 18, 1982; "US halts sanctions against Yamal line," *Oil and Gas Journal*, November 22, 1982.
72. *Oil and Gas Journal*, November 2, 1981, p. 61.
73. "Norway: Troll gas can't stop Yamal line," *Oil and Gas Journal*, July 5, 1982; "End to Yamal pipeline embargo eyed," *Oil and Gas Journal*, November 8, 1982.
74. *Oil and Gas Journal*, November 2, 1981, p. 61.
75. From 1977, Dutch gas reserves actually started to decline. "Dutch gas reserves end 17-year climb," *Oil and Gas Journal*, April 4, 1977.
76. "Norway: Troll gas can't stop Yamal line," *Oil and Gas Journal*, July 5, 1982.
77. "Security of supply" and "Soviet line gets boost from Norwegian strikes," *Oil and Gas Journal*, November 9, 1981.
78. "EEC: US embargo against USSR is illegal," *Oil and Gas Journal*, August 23, 1982; "Dutch court rules against US Yamal sanctions," *Oil and Gas Journal*, September 27, 1982.
79. Hormats quoted in *Oil and Gas Journal*, November 2, 1981, p. 61.
80. "Dutch court rules against US Yamal sanctions," *Oil and Gas Journal*, September 27, 1982; "End to Yamal pipeline embargo eyed," *Oil and Gas Journal*, November 8, 1982; "Soviet supply a hot issue," *Financial Times*, December 13, 1982.
81. Sackmann to the Bavarian Parliament, October 26, 1977, BayHStA StK, file 18790; "Forum Semmering November 1975," *ÖMV-Zeitschrift* 4/1975, p. 3f.; ÖMV, Sitzung des Aufsichtsrates, December 21, 1976, OeStA ÖIAG-Archiv, Box 138.
82. Sackmann to the Bavarian Parliament, October 26, 1977, BayHStA StK, file 18790.
83. "ÖMV und Erdgas," *ÖMV-Zeitschrift* 1/1983, p. 4, and 3/1989, p. 46.

11 From Soviet to Russian Natural Gas

1. "Soviet gas industry to surpass most 1985 targets," *Oil and Gas Journal*, January 28, 1985.
2. Gustafson 1989, p. 139; "Haste brings problems for Siberian gas pipeline," *Financial Times*, January 12, 1984; "Explosion 'has cut Soviet gas by 20 percent,'" *Financial Times*, July 10, 1989.
3. Gustafson 1989, p. 159; "Haste over gas in Siberia risks prospects," *Financial Times*, June 7, 1982, referring to a critical article published in Pravda.
4. "Yamal pipelaying nearly complete," *Oil and Gas Journal*, April 25, 1983; "Haste brings problems for Siberian gas pipeline," *Financial Times*, January 12, 1984.
5. "The Russians hold the ring," *Financial Times*, March 2, 1984.
6. "Soviet Union seen dominating European energy market," *Oil and Gas Journal*, May 13, 1985, referring to a study by George Washington University's International energy institute; Gustafson 1989, p. 34; "Soviet pipeline offer for Turkey," *Financial Times*, February 3, 1983; "Greek-Soviet cooperation agreement signed," *Financial Times*, February 23, 1983; "Soviets aim for top capacity of Yamal line," *Oil and Gas Journal*, March 26, 1984; "Supply of imported gas 'should be increased,'" *Financial Times*, June 11, 1984; "Putting something in the pipeline," *Financial Times*, June 26, 1986; "British gas faces clash on imports," *Financial Times*, December 29, 1986. As of spring 1984, the Soviets claimed that "experts on economics in the Spanish government are fully convinced that purchases of Siberian gas meet the country's interests." In the British debate, the possibility of Soviet gas imports to Britain was at this stage mainly pointed at by Jonathan Stern.

7. "Turkey seeks bids for gas pipeline," *Financial Times*, February 28, 1986; "Turkey in talks on Soviet gas," *Financial Times*, January 23, 1987; "International briefs," *Oil and Gas Journal*, March 23, 1987; "Greece in deal to buy Soviet natural gas," *Financial Times*, October 8, 1987; "Athens and Moscow sign $2.3bn gas contract," *Financial Times*, July 27, 1988.
8. Helga Steeg, "Gas' role in Europe tied to competitive pricing," *Oil and Gas Journal*, October 7, 1985.
9. "Systems go on gas," *Financial Times*, June 15, 1984; "US backing may not be enough," *Financial Times*, May 23, 1983; "Europe warned against second Soviet gas line," *Financial Times*, May 11, 1983; "Turkey and Iran agree to study oil and gas pipelines," *Financial Times*, January 23, 1985.
10. "Soviet Union seen dominating European energy market," *Oil and Gas Journal*, May 13, 1985; "Lockout halts Norwegian North Sea production" and "Norway's labor dispute," *Oil and Gas Journal*, April 14, 1986.
11. "USSR to pace hefty world gas supply, demand growth to 2000," *Oil and Gas Journal*, September 29, 1986.
12. "Gas target: Sweden," *Oil and Gas Journal*, February 22, 1988; "Norway, Soviet Union set sights on gas markets in Scandinavia," *Oil and Gas Journal*, June 13, 1988; "Sibiriengas på väg till Sverige," *Dagens industri*, October 23, 1989.
13. "Nytt utspel från Sovjet om gasen," *Dagens industri*, March 22, 1988; "Parkinson cautious over imports of Soviet gas," *Financial Times*, May 25, 1989; "Pricing policies defended by British Gas chairman," *Financial Times*, November 9, 1989.
14. "USSR strives to hike gas exports," *Oil and Gas Journal*, June 20, 1988; "A new era in gas," *Financial Times*, April 19, 1990.
15. "Soviets push for construction of strategic Yamal gas line," *Oil and Gas Journal*, November 2, 1981; "Soviet oil exports drop in volume and value," *Oil and Gas Journal*, April 30, 1990; "Soviet oil, gas industry languishes," *Oil and Gas Journal*, May 15, 1989; "Europe's cross-border gas trade on the rise," *Oil and Gas Journal*, July 8, 1991; "USSR revenues from 1988 natural gas exports fall," *Oil and Gas Journal*, August 14, 1989. The absolute figure for 1988 was 5.84 billion rubles, down from 7.7 billion in 1985. After 1988 the revenues started increasing again, though at a modest pace, to 6.13 billion rubles in 1989 and 6.5 billion in 1990. See *Oil and Gas Journal*, August 14, 1989, p. 21.
16. "Gazprom moving towards new era in Russian natural gas industry," *Oil and Gas Journal*, September 7, 1992.
17. "Energileveranserna till Litauen nedskurna," *TT*, April 19, 1990; "Moscow closes Lithuanian oil pipeline" and "Lithuania looks for more than words of support from the West," *Financial Times*, April 19, 1990; "Sovjet: Ingen totalblockad mot Litauen," *TT*, April 22, 1990.
18. "Fuel shortages intensify in former Soviet Union," *Oil and Gas Journal*, December 30, 1991; "OGJ newsletter," *Oil and Gas Journal*, July 6, 1992; Mikhail B. Korchemkin, "Russia's huge Gazprom struggles to adjust to new realities," *Oil and Gas Journal*, October 18, 1993.
19. "OGJ newsletter," *Oil and Gas Journal*, July 20, 1992.
20. "East Germany forging ties with foreign firms," *Oil and Gas Journal*, May 14, 1990; "OGJ newsletter," *Oil and Gas Journal*, October 21, 1991.
21. "East Germany faced with gas crisis," *Financial Times*, December 11, 1991; "Eastern Germany gas supply cutoff looms," *Oil and Gas Journal*, December 9, 1991; "Disruption threat averted in German gas supply row," February 21, 1992.
22. "Hungary petroleum privatization limited by economic concerns," *Oil and Gas Journal*, July 4, 1994; "Eastern Europe moving to solve energy problems," *Oil and Gas Journal*, July 23, 1990; "OGJ newsletter," *Oil and Gas Journal*, November 23, 1992.

23. "OGJ newsletter," *Oil and Gas Journal*, February 10, February 17, and March 16, 1992; "CIS members seek oil and gas supplies from Middle East," *Oil and Gas Journal*, March 2, 1992; Stern 2005, p. 85.
24. "OGJ newsletter," *Oil and Gas Journal*, January 20, 1992.
25. "Gasstopp även till Litauen," *TT*, June 28, 1993.
26. Mikhail B. Korchemkin, "Russia's huge Gazprom struggles to adjust to new realities," *Oil and Gas Journal*, October 18, 1993; "Ryssland stoppade gas till Estland," *TT*, June 25, 1993.
27. "Gasstopp även till Litauen," *TT*, June 28, 1993; "Ryssland återupptar gasleveranserna till Estland," *TT*, June 29, 1993; "OGJ newsletter," *Oil and Gas Journal*, July 12, 1993.
28. "Gazprom blames Ukraine for disruption in Russian gas flow to Europe," *Oil and Gas Journal*, October 26, 1992; "OGJ newsletter," *Oil and Gas Journal*, November 2, 1992.
29. "OGJ newsletter," *Oil and Gas Journal*, November 23, 1992.
30. "OGJ newsletter," *Oil and Gas Journal*, March 14, 1994; "Italy promoting bigger role for natural gas in its energy mix," *Oil and Gas Journal*, November 7, 1994.
31. "Weak oil prices seen hindrance to pace of increase in gas use," *Oil and Gas Journal*, June 27, 1994.
32. "Gaz de France sets sight on worldwide gas operations," *Oil and Gas Journal*, November 30, 1992; Stern 2005, pp. 91 and 112. A turning point came only in June 2002, when German chancellor Gerhard Schröder signed an agreement with the presidents of Russia and Ukraine.
33. Stern 2005, pp. 113 and 142.
34. "EC seeks Soviet energy charter," *Financial Times*, October 19, 1990; Stern 2005, p. 138. 2005, pp. 113 and 142.
35. Ibid., pp. 131ff.
36. See table in Stern 2005, p. 110.
37. Victor and Victor 2006, p. 149; Porozumienie między Rządem Rzeczypospolitej Polskiej a Rządem Federacji Rosyjskiej o budowanie systemu gazociągów dla tranzytu gazu rosyjskiego przez terytorium Rzeczypospolitej Polskiej i dostawach gazu rosyjskiego do Rzeczypospolitej Polskiej, August 25, 1993.
38. EuroPol Gaz, official website, www.europolgaz.com.pl, accessed on May 8, 2008.
39. Ibid.; Victor and Victor 2006, p. 154.
40. Stern 2005, p. 89.
41. Ibid., pp. 90–92.
42. Ibid., p. 98.
43. "UK to resume natural gas imports," *Oil and Gas Journal*, February 17, 1992; "Gazprom and Neste plan North European gas pipeline," *Oil and Gas Journal*, February 10, 1997; "Industry briefs," *Oil and Gas Journal*, July 14, 1997; "OGJ newsletter," *Oil and Gas Journal*, July 26, 1999.
44. "OGJ newsletter," *Oil and Gas Journal*, July 26, 1999; Högselius 2005.
45. "Gazprom seeks to broaden global energy role," *Oil and Gas Journal*, July 7, 2005; "UK's North Sea gas infrastructure must compete with LNG," *Oil and Gas Journal*, August 25, 2003. Gazprom sold 2 bcm to Britain in 2003 and 4 bcm in 2004.
46. "North European Gas Pipeline agreement signed," *Oil and Gas Journal*, September 26, 2005.
47. Nord Stream, "Gasunie and Gazprom agree on Nord Stream terms," November 6, 2007 (press release).
48. The Russian government in November 2007 announced that it had dropped the idea of constructing a second pipeline along the Belarusian-Polish transit route. See, for example, *RIA Novosti*, "Russia drops second leg of gas pipeline via Belarus," November 1, 2007; Fredholm 2006, p. 4ff.

12 Conclusion

1. IEA 2004, p. 318.
2. Burgbacher cited in "Energieprobleme in Europa—Möglichkeiten ihrer Lösung," *gwf,* December 22, 1967.

Bibliography

Archival Sources

Russian State Archive of the Economy, Moscow (RGAE)

Ministry of Gas Industry (f. 458, 1965–1975)
Ministry of Oil Industry (f. 8627, 1939–1957)
Permanent representation of the USSR at COMECON (f. 302, 1964–1987)
Production Committee for Gas Industry (f. 279, 1956–1965)

Central State Archive of the Highest Organs of Government and Administration of Ukraine, Kiev (TsDAVO Ukrainy)

Ukrainian Council of Ministers (f. 2, 1945–1978)
Ukrainian Gosplan (f. 337, 1961–1973)

Austrian State Archives, Vienna (OeStA)

Foreign Office (II-Pol)
Österreichische Mineralöl-Verwaltung (ÖMV) (ÖIAG-Archiv)
Vereinigte Österreichische Eisen- und Stahlwerke (VÖEST) (ÖIAG-Archiv)

Federal German Archives, Koblenz (BArch)

Chancellor's Office (B 136, 1955–1979)
Ministry of Economy (B 102, 1955–1979)
Nachlass Karl Schiller (N 1229)
Nachlass Ulf Lantzke (N 1360)

Federal German Archives, Berlin (BArch)

Ministry of Coal and Energy (DG 12)
Sozialistische Einheitspartei Deutschlands (SED) (DY 30)

Political Archive of the German Foreign Office, Berlin (PA AA)

File assembled during editing of Akten zur Auswärtigen Politik (B 150)
Minister's office (B 1)
Soviet Union (B 41)
West-East Trade (B 63)

Bavarian State Archives, Munich (BayHStA)

Ministry of Economy and Transportation (MWi)
Nachlass Otto Schedl (NL Schedl)
State Chancellery (StK)

Archives of the United Nations Office at Geneva (UNOG Archives)

ECE Working Party for Gas (GX.11/13)

Printed Primary Sources

Edited Source Collections

Akten zur auswärtigen Politik der Bundesrepublik Deutschland 1967, ed. Hans-Peter Schwarz, 3 volumes, Munich: Oldenbourg, 1998.
Akten zur auswärtigen Politik der Bundesrepublik Deutschland 1969, ed. Hans-Peter Schwarz, 3 volumes, Munich: Oldenbourg, 2000.

Newspapers and Magazines

Dagens industri
Economist
Financial Times
Frankfurter Allgemeine Zeitung
Frankfurter Rundschau
Handelsblatt
Industriekurier
Izvestiya
Neue Zeitung
Neue Zürcher Zeitung
Neues Deutschland
New York Times
Pravda
Die Presse
Der Spiegel
Stuttgarter Zeitung
Süddeutsche Zeitung
Wall Street Journal
Die Welt
Die Zeit

Specialist Journals

gwf (Gas- und Wasserfach)
Gazovaya promyshlennost
Oil and Gas Journal

Secondary Literature

Badenoch, A., and Fickers, A. 2010 (eds.), *Materializing Europe: Transnational Infrastructures and the Project of Europe*, Basingstoke and New York: Palgrave Macmillan.

Birkenfeld, W. 1964, *Der synthetische Treibstoff 1933–1945: Ein Beitrag zur nationalsozialistischen Wirtschafts- und Rüstungspolitik*, Göttingen: Musterschmidt-Verlag.

Bokserman, Yu. I. 1958, *Razvitie gazovoi promyshlennosti SSSR*, Moscow: Gosudarstvennoe nauchno-tekhnicheskoe izdatelstvo neftyanoi i gorno-toplivnoi literatury.

Brezhnev, L. 1966, "Report of the Central Committee of the Communist Party," in *23rd Congress of the Communist Party of the Soviet Union, Moscow, from March 29 to April 8, 1966*, Moscow: Novosti Press Agency.

———. 1971, "Report of the Central Committee of the Communist Party," in *24th Congress of the Communist Party of the Soviet Union, March 30–April 9, 1971*, Moscow: Novosti Press Agency.

Chadwick, M., Long, D., and Nissanke, N. 1987, *Soviet Oil Exports: Trade Adjustments, Refining Constraints and Market Behaviour*, Oxford: Oxford Institute of Energy Studies.

Dahl, K. 1997, *Erdgas in Deutschland: Entwicklung und Bedeutung unter Berücksichtigung der Versorgungssicherheit und des energiepolitischen Ordnungsrahmens sowie des Umweltschutzes*, PhD thesis, Clausthal: Technische Universität.

Davis, J. 1984, *Blue Gold: The Political Economy of Natural Gas*, London: Allen & Unwin.

Derzhavnyi komitet naftovoi, gazovoi ta naftopererobnoi promyslovosti Ukrainy 1997, *Nafta i gaz Ukrainy*, Kiev: Naukova Dumka.

Estrada J., Bergesen, H. O., Moe, A., and Sydnes, A. K. 1988, *Natural Gas in Europe: Markets, Organisation and Politics*, London: Pinter.

Fernandez, R. 2009, "Russian gas exports have potential to grow through 2020," *Energy Policy*, 37, 10.

Fredholm, M. 2006, *Gazprom in Crisis*, Conflict Studies Research Center, Russian Series 06/48, Defence Academy of the United Kingdom.

Gheorghe, A. V., Masera, M., Vries, L. de, Weijnen, M., and Kröger, W. 2007, "Critical infrastructures: The need for international risk governance," *International Journal of Critical Infrastructures*, 3, 1/2, 3–19

Goldthau, A. 2008, "Rhetoric versus reality: Russian threats to European energy supply," *Energy Policy*, 36, 686–692.

Gupta, E. 2008, "Oil vulnerability index of oil-importing countries," *Energy Policy*, 36, 3, 1195–1211.

Gustafson, T. 1985, *Soviet Negotiating Strategy: The East-West Gas Pipeline Deal, 1980–1984*, Santa Monica, CA: Rand.

———. 1989, *Crisis amid Plenty. The Politics of Soviet Energy under Brezhnev and Gorbachev*, Princeton: Princeton University Press.

Högselius, P. 2005, *Die deutsch-deutsche Geschichte des Kernkraftwerkes Greifswald: Atomenergie zwischen Ost und West*, Berlin: Berliner Wissenschafts-Verlag.

Holmberg, R. 2008, *Survival of the Unfit: Path-Dependence and the Estonian Oil Shale Industry*, Linköping: Department of Technology and Social Change.

Hughes, T. P. 1983, *Networks of Power: Electrification in the Western World 1880–1930*, Baltimore: Johns Hopkins University Press.

IEA 2004, *Security of Gas Supply in Open Markets. LNG and Power at a Turning Point*, Paris: International Energy Agency.

———. 2011, *World Energy Outlook 2011*, Paris: International Energy Agency.

Kaiser, K. 1968, *German Foreign Policy in Transition: Bonn between East and West*, Oxford: Oxford University Press.

Kosygin, A. 1966, "Report on the directives for the five-year economic development plan of the USSR for 1966–1970," in *23rd Congress of the Communist Party of the Soviet Union, Moscow, from March 29 to April 8, 1966*, Moscow: Novosti Press Agency.

Laurien, H. 1974, *Stand und Entwicklung des westeuropäischen Erdgas-Verbundnetzes, Vortrag gehalten am 24. Oktober 1974 in Saarbrücken*, Köln: Deutsche Verkehrswissenschaftliche Gesellschaft.

Luhmann, N. 1995, *Social Systems*, Stanford: Stanford University Press (German original published in 1984).

Mawdsley, E., and White, S. 2000, *The Soviet Elite from Lenin to Gorbachev: The Central Committee and its Members, 1917–1991*, Oxford: Oxford University Press.

Misa, T., and Schot, J. 2005, "Inventing Europe: Technology and the hidden integration of Europe," *History and Technology*, 21, 1, 1–19.

OÖ Ferngas, 2007, *OÖ Ferngas 1957–2007: 50 Jahre Erdgasinfrastuktur in Oberösterreich*, Linz: OÖ Ferngas.

Rambousek, H. 1977, *Die ÖMV-Aktiengesellschaft: Entstehung und Entwicklung eines nationalen Unternehmens der Mineralölindustrie*, Vienna: Wirtschaftsuniversität Wien (PhD thesis).

Remington, R. A. 1971, *The Warsaw Pact: Case Studies in Communist Conflict Resolution*, Cambridge, MA: MIT Press.

Rudolph, C. 2004, *Wirtschaftsdiplomatie im Kalten Krieg. Die Ostpolitik der westdeutschen Großindustrie, 1945–1991*, Frankfurt am Main and New York: Campus Verlag.

Runov, V. A., and Sedykh, A. D. 1999, *Aleksei Kortunov*, Moscow: Izdatelstvo Moskovskoe voenno-istoricheskoe obshchestvo.

Semjonow, J. 1973, *Erdöl aus dem Osten: die Geschichte der Erdöl- und Erdgasindustrie in der Sowjetunion*, Düsseldorf: Econ.

Slavkina, M. V. 2002, *Triumf i tragediya: Razvitie neftegazovogo kompleksa SSSR v 1960–1980-e gody*, Moscow: Nauka.

———. 2005, "Istoriya prinyatiya resheniya o promyshlennom osvoenii Zapadnoi Sibiri," in Borodkin, L. I. (ed.), *Ekonomicheskaya istoriya. Obozrenie*. Moscow: MGU, pp. 146–162.

Söderbergh, B. 2010, *Production from Giant Gas Fields in Norway and Russia and Subsequent Implications for European Energy Security*, Uppsala: Uppsala University.

Solanko, L., and Sutela, P. 2009, "Too much or too little Russian gas for Europe?" *Eurasian Geography and Economics*, 50, 58–74.

Stent, A. 1981, *From Embargo to Ostpolitik: The Political Economy of West German-Soviet Relations, 1955–1980*, Cambridge: Cambridge University Press.

Stern, J. 1980, *Soviet Natural Gas Development to 1990: The Implications for the CMEA and the West*, Lexington, MA: Lexington Books.

———. 2005, *The Future of Russian Gas and Gazprom*, Oxford: Oxford University Press.

Thue, L. 1995, "Electricity rules: The formation and development of the Nordic electricity regimes," in Kaijser, A., and Hedin, M. (eds.), *Nordic Energy Systems: Historical Perspectives and Current Issues*, Canton, MA: Science History Publications.

Umbach, F. 2009, "Global Energy Security and the implications for the EU," *Energy Policy*, 38, 3, 1229–1240.

Van der Vleuten, E., and Kaijser, A. 2006 (eds.), *Networking Europe: Transnational Infrastructures and the Shaping of Europe*, Sagamore Beach: Science History Publications.

Victor, N. and Victor, D. 2006, "Bypassing Ukraine: Exporting Russian gas to Poland and Germany," in Victor, D., Jaffe, A., and Hayes, M. (eds.), *Natural Gas and Geopolitics. From 1970 to 2040*, Cambridge: Cambridge University Press.

Von Dannenberg, J. 2007, *The Foundations of Ostpolitik: The Making of the Moscow Treaty between West Germany and the USSR*, Oxford and New York: Oxford University Press.

Webb, T., and Barnett, N. 2006, "Gas: Russia's secret agenda. Energy supply is a 'political weapon,'" *Independent*, January 8.

Yergin, D. 1991, *The Prize: The Epic Quest for Oil, Money and Power*, New York: Simon & Schuster.

Index

Abdessalam, Belaid, 73, 170
Adenauer, Konrad, 76, 85
Aderklaa, 45
Adriatic Sea, 48, 70, 72–3, 172
AEG-Kanis, 189–90
Aerohydrodynamics Research Institute (Russia), 16
Afghanistan, 40, 184, 220
AGA, see American Gas Association
Algeria, 28–9, 34–9, 48–50, 52, 55, 57–8, 62, 70–3, 77, 87, 90, 116, 125, 132, 158–9, 169–70, 177, 187–8, 190–2, 202, 207, 217–18, 225
Algerian LNG, 48, 71–2, 77, 169–70, 187
Algiers, 73, 87
Allardt, Helmut, 121
Alps, 39, 70, 154
Alsthom-Atlantique, 190
American Gas Association (AGA), 27
von Amerongen, Otto Wolff, 126, 129
Angola, 171
anticorrosion technology, 24, 39
Arab countries, 57–8, 168–70, 219, 228–9
Arab oil embargo (1973–1974), 10, 167–70, 219, 228–9
Arctic system-building, 138–43
Armenia, 21–2, 205, 207
Arndt, Klaus Dieter, 107–8, 110
Astara, 172, 174
Auersthal, 46, 50
Augsburg, 71, 157
Austria, 3, 9–10, 36–40, 45–66, 70–1, 73–9, 82, 86–92, 102, 106, 108, 114–15, 120–1, 128–9, 132–3, 148, 151–4, 158, 162–3, 169–72, 174–7, 188, 193–5, 203, 206–7, 209, 218–19, 222, 224–5, 227–8, 230–1
  and border opening with Hungary, 203
  and gas transit enablement, 151–4
  and interest in Yamal gas, 188
  and Iranian deal, 177
  and Trans-Austria pipeline, 171, 193, 207, 209, 231
Austria Ferngas, 48–50, 55, 59, 62, 64–5, 72, 158, 230
Austrian Control Bank, 63
Austrian-Soviet negotiations, 38–40, 45, 58–66, 73–80, 86, 89–103, 106–9, 153, 203, 222
Austrian State Treaty, 46–7, 54
Azerbaijan, 21, 40, 147, 172, 174, 205–7
Azovstal metallurgical plant, 16

Baden-Württemberg, 69, 71–2, 82–3, 87, 109–10, 119, 156–7, 163
Bahr, Egon, 74, 76, 79–88, 106–8, 110, 118–20, 130, 174, 218, 226
Baibakov, Nikolai, 13, 15–17, 33, 41–2, 178
Baltic Sea, 133, 190, 202, 205–6, 212, 214–15, 216, 220, 232
Baltic states, 21, 41, 96–102, 133, 136, 138–9, 144, 163, 180, 205–6, 208, 214–16, 223
Baranovsky, Yuri, 63
BASF, 212, 215
  see also Wintershall
Bashkiria, 17, 198
Bauer, Ludwig, 52, 56, 94, 132, 163, 177, 183
Baumgarten, 91–4, 154, 163, 206
Bavaria, 67–88, 109–19, 126, 154–9, 162–3, 165–6, 182, 193, 224, 226–7, 230
  and Yamal pipeline, 182
Bavarian Ministry of Economy, 73, 82, 110, 116, 155, 159, 166

Bavarian Radio, 155
Bayerische Ferngas AG, *see* Bayerngas
Bayerngas (Bayerische Ferngas AG), 71–2, 77, 82, 110–12, 116, 119, 154–9, 162–5, 173, 182, 193, 230
Belarus, 9, 10, 21, 41, 95–102, 136, 138–9, 144, 146–7, 160, 208–9, 213–15, 220, 223, 228, 232
Belarusian-Polish link, 213–15
Belgium, 3, 34–5, 60, 72, 131–3, 158, 170–3, 175, 180, 182, 199, 218–19
Berlin Wall, 3, 5, 7, 10, 29, 33, 109, 203, 217, 220
van Beveren, Jos, 56, 81–2
Black Sea, 21, 209
Bock, Fritz, 55
Bogomyakov, G. P., 178
Bokserman, Yuli, 24–5, 100, 137
Bonn, 58, 74, 78, 86, 106, 108, 112, 115, 123, 126, 157
Bosnia-Hercegovina, 172
Brandt, Willy, 4, 10, 74, 76, 78, 80, 82–4, 86–8, 105–8, 118–21, 135, 174–5, 218, 226
Bratislava, 35, 50, 90, 210
Bratstvo (Brotherhood) pipeline, 35, 39, 50, 59, 66, 90, 127, 135, 144, 149, 151, 159–60, 162, 164, 171, 212
Braudel, Fernand, 234–5
Brezhnev, Leonid, 9, 34, 37, 40–1, 67–8, 87, 142, 161, 170, 174, 182–5
Britain, *see* United Kingdom
Brown & Root, 133
de Bruijne, Dirk, 123
Brunet, Jean-Claude, 126
Bryansk, 21
Budapest appeal, 105–6, 108

Cameroon, 185
Cape of Good Hope, 175
carbon dioxide, 2, 167, 201
Carinthia, 154
Carter, Jimmy, 185
Caspian Sea, 21, 31
Caucasus, 21, 145, 183
Central Asia, 22–4, 26, 28, 31–4, 40, 42, 95, 136, 139, 144–5, 209

Central Committee of the Soviet Communist Party, 17, 139–40, 147, 161, 164
Central European Pipeline (CEL), 70
Chapelle, Jean, 127
Chelyabinsk, 23–4
Chernobyl disaster, 201, 225
China, 2, 75, 190
Christian Democratic Union (Germany), 74–5, 84, 106, 120
Christian Social Union (Bavaria), 74
CIS (Commonwealth of Independent States), 204, 206, 210, 224, 232
Clark, 91
CoCom, 24, 26
Cold War, 2–6, 26, 74, 184, 197, 203–4, 210, 212, 217–22, 224, 229, 232–4
Cologne, 80, 108, 112
COMECON, 35, 171
compressor embargo, 188–90
compressor technology, 4, 8, 23, 25–7, 42, 46, 91–3, 97, 99–100, 128, 142–50, 152, 154, 164–5, 171–2, 179, 188–90, 198–9, 231
conditioning facility, 157, 231
Congress of the Soviet Communist Party
 1956, 15, 21
 1961, 19
 1966, 9, 41
 1971, 142
 1981, 185
Connole, William R., 27–8, 35
de Corval, Gérard, 28
countertrade, 6, 24, 36, 42, 55, 61, 72, 78, 81–2, 90, 131, 139, 142, 150, 182, 184, 230, 231
Couture, Jean, 126
Creusot-Loire, 189
Cuban missile crisis, 29, 33, 217
Czechoslovakia, 9, 26, 35–6, 47–50, 55, 59, 61, 72, 89–94, 99, 105–6, 108, 134, 148, 151–4, 159–61, 164, 184, 203, 207, 218–19, 227, 231–2
 compressor technology, 26
 Soviet invasion of, 9, 105–6, 108, 134, 184, 218
 transit pipeline construction, 152

d'Estaing, Valéry Giscard, 170
Danube River, 77, 90, 231
Dashava gas field (in western Ukraine), 13–14, 20–1, 24–5, 96, 98–102, 146
Davignon, Viscount Etienne, 193
détente, 8, 35–6, 52, 74, 83–4, 105–6, 109, 120, 150, 168
Dikanka, 98, 148
Dinkov, Vasily, 101
Distrigas (USA), 133
Distrigaz (Belgium), 158, 172, 175, 180, 188
DIW, *see* German Institute of Economic Research (Deutsches Institut für Wirtschaftsforschung)
Dnepropetrovsk, 20–1
von Dohnanyi, Klaus, 107, 109, 111–13, 115–16, 118–20, 122–3, 126
Donets basin (Donbass), 20
Dresser, 189–90
Drogobych, 161
Druzhba (Friendship) oil pipeline, 35, 152
dumping of gas, 7, 221
Düsseldorf, 81–2
Dymshits, Venyamin, 147–9, 161–2

E.ON group, 215
East Germany, 9, 74, 85, 135, 161, 171, 205–6, 234
East Prussia, 21
East-West industrial exchange program, 27
Eastern bloc, 46, 67, 203
EC, *see* European Communities
EC Commission, 117
EEC, *see* European Economic Community
Efremovka-Kiev pipeline, 98–101
Ekofisk gas field, 192
El Paso Natural Gas, 133, 175
Emden, 192
Emmel, Egon, 109
Energy Charter, 211
energy weapon, 1–2, 7, 11, 37, 191, 220–4
Engine of Revolution, 25

ENI, 36, 38–9, 42–3, 51–3, 55, 57–8, 60–2, 65, 70–2, 105, 126–32, 146, 153, 163, 171–3, 176, 180, 187–8, 209–10, 225, 227, 229–30
environmental characteristics of natural gas, 14, 47, 167, 178, 195, 198, 205, 219, 225
Erhard, Ludwig, 74, 76, 85
Essen, 69, 115, 129, 174
Esso, 28, 69, 71, 73, 76, 122–6, 131, 169, 225
Estonia, 21, 148, 205, 207–8, 216
EU-Russia Summit, 211
EURATOM, 67
Eurogasco, 45
European Coal and Steel Community, 67, 107
European Communities (EC), 3, 107, 117, 122, 127–8, 190–3, 211
European Economic Community (EEC), 9, 51–4, 67, 168, 191, 222, 225–6, 231–2
Europol, 213, 216
Exhibition of Achievements of the National Economy (VDNKh), 26
export infrastructure, 10, 120, 125, 135–50, 160, 178, 180, 182–3, 191, 198, 204, 207, 223, 227, 231–3

Federal Geological Survey (Bundesanstalt für Bodenforschung), 124
Federal Republic of Germany, *see* Germany
Finland, 3, 9–10, 57, 67, 89, 131–3, 135, 138, 147, 151, 163, 166, 169, 200, 202, 212, 214, 218, 219, 222, 224, 226–8, 234
Finsider, 24, 52, 55
five-year plan, 23, 142, 152, 207
Ford, Gerald, 171
Foreign Trade Bank of the USSR, 63
France, 3, 34, 39–40, 55, 58, 61, 65–6, 68, 71, 86–7, 125–30, 132, 151, 169–70, 173–4, 176, 180–3, 187–9, 193–5, 210, 219, 226, 231
Frankfurt, 81, 110

French-Algerian conflict, 34
Friderichs, Hans, 159, 177, 181
Friendship of the Peoples system, 21, 205
Fukushima disaster, 2, 225
Fulda, 165
Funcke, Friedrich, 123

Galician gas fields, 13–14, 20–1, 23, 36, 95–102, 135, 138, 144–5, 161, 216
Gas-Union, 71, 158, 173
Gasunie, 180, 188, 191, 193, 215
 see also NAM Gas Export
Gasversorgung Süddeutschland (GVS), 71, 156–8, 173
Gaz de France (GdF), 34, 39, 55, 57, 65, 128, 132, 158, 173–4, 176, 180, 187–8, 195, 201, 210, 224–5, 230
Gazli gas field (in Uzbekistan), 23
Gazocean, 39, 133
Gazovaya promyshlennost (journal), 18, 25, 32, 137
Gazprom, 204, 206, 208–15, 223
GdF, see Gaz de France
Gebersdorf, 165
Geilenkeuser, Hans, 155
General Electric, 179, 189
Genova, 70
Georgia, 21–2, 205
German Industrial Trade Fair, 108–9
German Institute of Economic Research (Deutsches Institut für Wirtschaftsforschung) (DIW), 107, 110
German-Soviet negotiations, 105–34
Germany (Federal Republic of), 2–3, 9, 53, 57–8, 67–8, 74, 78, 82, 84, 87, 106, 108–9, 120, 122, 135–9, 146, 149–51, 161, 166, 180, 189, 212, 219, 221–2, 226, 228
Giprospetsgaz, 40–1
Girotti, Raffaele, 163
Glasgow, 190
Glavgaz SSSR, 15, 17, 19–22, 24–7, 31–6, 38–40
Gomułka, Władysław, 78
Gorbachev, Mikhail, 205
Gorky (Nizhny Novgorod), 25
Górzyca, 213

Gosplan (Soviet State Planning Commission), 13, 15, 23, 33–4, 38–42, 95, 97, 99, 141, 143–4, 160–2, 178, 207, 233
Gossnab (Soviet State Committee for Material-Technical Supply), 148, 161–2, 207
grand coalition (in Austria), 54
grand coalition (in Germany), 74, 120–1
Greece, 3, 199–200, 212, 234
Greifswald, 215
Gromyko, Andrei, 119, 130
GVS, see Gasversorgung Süddeutschland

H-Gas, see high-calorific gas
Haferkamp, Wilhelm, 107
The Hague, 117
Haig, Alexander, 185
Hallstein Doctrine, 75, 78
Hamburg, 57, 81, 107, 190, 227
Handelsblatt, 155
Hassi R'Mel gas field (in Algeria), 28
Heitzer, Hans, 73, 115, 155–6
Herbst, Axel, 126
Hessen, 69, 165
high-calorific gas (H-Gas), 114, 157, 165–6
Hitler, Adolf, 13–14, 51, 68
Hoesch, 121
Hungary, 39, 43, 52, 54–5, 65, 68, 153, 171–2, 203, 206
Hveding, Vidkun, 192
hydrogen sulfide, 39

Iberian peninsula, 72, 200
IGAT-1 pipeline, 172, 175, 182, 206
IGAT-2 pipeline, 172–7, 182, 206
IGU, see International Gas Union
Ingolstadt, 70–1, 165
intentional disruptions, 8, 10, 181, 192, 204–9, 211, 220, 223, 232
International Energy Agency (IEA), 2, 200
International Gas Union (IGU), 26–7, 57, 81–2, 108, 229
Iran, 10, 38, 40–2, 172–7, 180–3, 185–6, 191, 194, 202, 206–7, 219–20
Iranian national gas company (NIGC), 172, 174–5

Iranian Revolution, 182, 185, 206, 219–20
Iraq, 57
isolation, politics of, 67–9
Israel, 57–8, 170, 190
Italy, 3, 38–40, 42–3, 49–50, 52–8, 60–6, 70–1, 77–8, 105, 125–33, 153–4
and Austrian gas contract, 60–6
and Austrian gas price negotiations, 60–2
and Bavarian plans, 70–1, 77–8
and Mingazprom, 42
and Six-Days War, 55–8
and Willy Brandt, 125–33
Izvestiya, 137

Japan, 2, 39, 40, 57, 89, 133, 172, 218
Jaumann, Anton, 158–9, 165–6
John Brown Engineering, 189, 198
Johnson, Lyndon, 84, 86
Jonava, 161

Kaliningrad (Königsberg), 21, 214
Kalush, 97
Kamenets-Podolsk, 97
Kangan gas field (in Iran), 172, 175
Karadag gas field (in Azerbaijan), 21–2
Kaun, Heinrich, 82–3, 108
Kazan, 20
Kekkonen, Urho, 163
Kharkov, 21, 162
Khrushchev, Nikita, 15, 18, 24–5, 27, 29, 32–5, 41, 197
Kiesinger, Kurt Georg, 74, 84, 86–7, 106, 118, 120–1, 127
Kiev, 13–14, 20–1, 24–5, 96, 98–101, 148, 160, 162, 209, 213–14
Kissinger, Henry, 171
Klaipeda, 21
Klaus, Josef, 54
Koller, Herbert, 51–2, 65
Komi ASSR, 138–47, 163, 176, 179, 183
Königsberg, *see* Kaliningrad
Koper, 48, 72
Kortunov, Alexei, 15–28, 31, 33–4, 36, 38, 40–2, 56, 81, 92, 97, 99–100, 138–9, 142–9, 153, 178, 222–3

and Arctic system-building, 139, 142–3
death of, and Ukrainian crisis, 143–7
defining export strategy, 31, 33–4, 36, 38, 40–2
early life of, 16–17
and rise of the Soviet gas industry, 15–28
and the Soviet-Austrian gas trade, 92, 97, 99–100
Kosovo, 203
Kosygin, Alexei, 34, 40–2, 62, 119, 142, 147, 160, 162, 164
Kozyrev, Andrei, 208
Kratzmüller, Emil, 123
Kröning, Rudolf, 80–1
Kursk, 21
Kuwait, 57
Kuybyshev (Samara), 20
Kuzmin, Mikhail, 160

L-Gas, *see* low-calorific gas
Lambsdorff, Otto Graf, 181–2
Lanc, Erwin, 163
Landshut, 71, 165
Lantzke, Ulf, 107, 109, 111–12, 118, 125, 127–30, 173
large technical system (LTS), 5–7, 195, 220, 223, 229–32
Latvia, 9, 21, 95–7, 99, 138, 146–7, 205–6, 208, 228, 232–3
Lenin Prize, 26
Leningrad, 20, 25–6, 41, 100, 108, 132, 138–9, 145, 163, 189, 198
see also St. Petersburg
Libya, 34–5, 38, 55, 57, 60, 62, 115–16, 169, 224, 232
Liepaja, 21
Liesen, Klaus, 131, 156, 181–3
Linz, 47–8, 51, 53, 74, 77–8
liquefied natural gas (LNG), 34, 39, 48–9, 58, 71–2, 77, 116, 125, 132–3, 158, 169–73, 175–6, 185, 187, 192, 200, 205
liquid petroleum gas (LPG), 39–40
Lithuania, 9, 21, 95–6, 99, 138, 146–7, 161, 204–5, 208, 228, 232
Livorno, 190

LNG, *see* liquefied natural gas
low-calorific gas (L-Gas), 114, 157, 165
LPG, *see* liquid petroleum gas
LTS, *see* large technical system
Lukesch, Rudolf, 50–4, 56–8, 65, 74, 78–9
Lvov, 95–6, 160–2
Lwówek, 213
Lyashko, Alexander, 162, 164
Mannesmann, 24, 52, 55–6, 65, 79–82, 84, 86, 100, 121, 140, 143, 189

Mannesmann-Röhrenwerke, 121
Manshulo, Andrei, 109
Mao Zedong, 75
Marseille, 58, 70
Marshall Plan, 37, 47
Masherov, Piotr, 160
Matzen gas field and storage facility, 91, 94, 102, 169
Medvedev, Dmitry, 223–4
Medvezhye gas field (in Siberia), 140–1, 143, 179
Medvezhye-Nadym pipeline, 141
Metalimex, 50, 59, 93
Middle East, 57, 70, 200
Mikoyan, Anastas, 34
Minenergo (Soviet Ministry of Energy), 33, 149
Mingazprom (Soviet Ministry of Gas Industry), 40, 101, 108, 125, 135–50, 155, 160–1, 166, 171–2, 177–9, 183–4, 191, 197–8, 203–4, 208, 223, 228–30, 233
  export strategy of, 135–8
Ministry for Construction of Oil and Gas Facilities, *see* Minneftegazstroi
Ministry of Chemical Industry (Soviet Union), 102, 161
Ministry of Economy and Finance (France), 127
Ministry of Education and Science (Germany), 120
Ministry of Ferrous Metallurgy (Soviet Union), 24, 161
Ministry of Food Industry (Soviet Union), 102

Ministry of Foreign Trade (Soviet Union), 24, 36–7, 39, 42, 53, 56, 63, 99, 109, 111, 131, 133, 175, 178, 184, 227
Ministry of Gas Industry, *see* Mingazprom
Ministry of Geology (Soviet Union), 32, 34
Ministry of Heavy Machine Building (Soviet Union), 146, 149
Ministry of Industry (France), 126
Ministry of Interior (Soviet Union), 149
Ministry of Petrochemical Machine Building (Soviet Union), 149
Minneftegazstroi (Soviet Ministry for Construction of Oil and Gas Facilities), 147–9, 160–1, 177–8, 183, 197
Minsk, 21, 96, 99–102, 140, 146
Mitterand, François, 187
Mitterer, Leo, 152–3
Mobil Oil, 70
Moldova, 21, 205, 220, 223
Molotov-Ribbentrop Pact, 13, 216
Mommsen, Ernst Wolf, 79–81
Monfalcone, 158, 172
Moscow, 14, 20–1, 23, 25–6, 33, 55–6, 61, 109, 112, 116, 119, 126, 130, 131, 133, 145, 149, 160, 162, 175, 181–4, 214
Munich, 70–1, 73, 82, 87, 111, 118, 155–6, 159, 165

Nadym-Ukhta-Torzhok route, 140
Naftogaz, 208, 210, 213
Nagorno-Karabakh, 205
NAM Gas Export, 55, 60, 69, 71, 80
  see also Gasunie
National Iranian Oil Company (NIOC), 182
NATO, *see* North Atlantic Treaty Organization
Nazih, Hassan, 182
Nazism, 9, 13, 45–6, 51, 68–9, 74, 95, 106
Neef, Fritz, 80
NEGP (North European Gas Pipeline), *see* Nord Stream
neoliberalism, 202
Neporozhnii, Piotr, 160
Neste (Finland), 132, 214

Netherlands, 3, 10, 28, 34–8, 48–50, 55, 60, 62, 64–5, 69, 71, 80, 87, 89, 107, 112, 124–5, 128, 155–6, 168–70, 180–2, 188, 191, 193, 211, 215, 217–19, 225
Nevsky machine-building factory (in Leningrad), 25–6, 189, 198
NIGC, *see* Iranian national gas company
Nigeria, 185, 200
NIOC, *see* National Iranian Oil Company
NIOGAS, 47, 48, 65
Nixon, Richard, 118, 124, 132–3, 171
Nord Stream pipeline, 214–17, 232
North Africa, 28, 57, 71–2
see also Algeria; Libya
North Atlantic Treaty Organization (NATO), 3–4, 33, 36, 39, 52, 67–8, 74–5, 130, 190, 222, 229
pipe embargo (1962), 36, 39, 52, 74–5
North German coal, 68–72, 74, 76, 78
North Ossetia, 205
North Sea gas, 2, 28, 62, 107, 157, 165–6, 201–2
North Star project, 133, 171
Northern Lights system (Siyanie severa), 138–43, 179, 183
Norway, 2, 168, 191–2, 201–2, 206–7, 212, 214, 224, 231–2, 234
Nuovo Pignone, 189–90, 199
Nuremberg (Nürnberg), 157, 159, 165–6
Nürnberg-Würzburg pipeline, 165–6

Oberlaa, 45
Oberösterreichische Ferngas (OÖ Ferngas), 47, 48, 50, 65
Occidental Petroleum, 133
October Revolution, 14, 25, 145
Odessa, 21
Odvarka, Josef, 160
ÖIAG, 51
Oil and Gas journal, 4, 20, 49, 98, 127, 136, 139, 147, 176, 217
oil crisis (1973–1974), 10, 167–70, 219, 228–9
see also Arab oil embargo
oil crisis (1979), 185–7, 220
Oil Ministry (Soviet Union), 15, 147

ÖMV (Austrian Mineral Oil Administration), 38–9, 101–2, 115, 120, 132, 148, 152–4, 162–3, 169, 171–3, 176–7, 180, 183, 188, 193–5, 206, 209–10, 224–5, 227–30
versus Austria Ferngas, 48–50
and Austrian-Soviet contract, 63–6
and Bavaria, 70
and Iranian deal, 177, 180, 183
and price negotiations, 58–63
and Rudolf Lukesch, 51–4
and Six-Days War, 55–8
and Yamal pipeline, 188
and Yugoslavia, 171–2
OÖ Ferngas, *see* Oberösterreichische Ferngas
Orange Revolution (in Ukraine), 214–15
Order of Lenin, 17
Orenburg, 171, 176, 180
Organization of Petroleum Exporting Countries (OPEC), 1, 219
Ortoli, François-Xavier, 127
Orudzhev, Sabit, 147–9, 159–60, 162, 164, 178, 181
Osipov, Nikolai, 56, 58, 111–13, 115–18, 121, 129, 154, 163, 174, 181, 224
Ostpolitik (German Eastern policy), 4, 10, 78, 83–4, 88, 106, 118, 120, 122, 226
Ostrogozhsk, 145

Panhandle-Hugoton gas field (USA), 28
Parkinson, Cecil, 202
Patolichev, Nikolai, 108–10, 112–13, 116, 129, 181–2
People's Party (ÖVP, in Austria), 54
Perle, Richard, 192
Persian Gulf, 175
Petrol gas company (Yugoslavia), 171, 195
pipe industry, 4, 6, 17, 23–6, 33, 36, 42–3, 51–2, 62, 75, 97, 109, 113, 139, 143, 159, 184
and Arctic system-building, 139
rise and stagnation of, 6, 8, 23–6
Plesser, Norbert, 77, 86, 110, 116–19, 123–4, 126, 130, 173–4

Podgorny, Nikolai, 34, 53–4, 73
Poland, 13–14, 21, 78, 82, 95, 100, 135, 138, 171, 187–8, 206, 212–16, 220
Politik der Bewegung (Policy of Movement), 68, 84
Poltava, 98
Pompidou, Georges, 127
Poznań, 213
Prague Spring, 91, 184
Pravda, 15
price negotiations
  Austria-Soviet Union, 58–63
  Germany-Soviet Union, 112–18
Pritchard (American company), 91
Prodi, Romano, 211
Putin, Vladimir, 203, 213, 223–4

Qatar, 200

Reagan, Ronald, 10, 184–5, 188–90, 192–3, 220
Riga, 21, 96, 99–100, 146, 206
Rolls-Royce, 179
Romania, 39, 52, 68, 78, 171, 199
Rosenheim, 77
Rovno, 97, 101, 161
Ruben, Vitalii, 96–7, 99, 146–7
Ruhrgas, 48, 69, 71, 76, 107, 111–24, 126–31, 152, 154–9, 162–3, 165, 168, 172–6, 180–3, 185–6, 188, 199, 201, 203, 205–6, 209–10, 212, 214–15, 224–5, 228–30
Runge, Hans Carsten, 123
Rusk, Dean, 84
Ruská, 90
Ryabenko, Alexander, 41

Saar Ferngas, 71, 158, 173
Sackmann, Franz, 86
Saharan gas, 28, 34–5, 48, 58, 60, 72, 158, 170, 187, 192–3, 200
  see also Algerian LNG
St. Petersburg, 215
  see also Leningrad
Salzgitter Ferngas, 121, 174
Samara, see Kuybyshev
Saratov, 13–14, 20, 25
Saudi Arabia, 57

Schedl, Otto, 69–77, 79, 81–2, 86–8, 110–12, 114–16, 118, 159, 165, 226–7, 230
Scheel, Walter, 120
Schelberger, Herbert, 112–19, 121–4, 129, 131, 159
Schiller, Karl, 74, 76, 78–81, 84, 86, 107, 109–10, 118, 120, 127, 129
Schlieker, Willy, 76–7
Schloss Hernstein-Berndorf, 58, 60
Schmidt, Helmut, 175, 191
Schröder, Gerhard, 68
Schwarz, Hans-Otto, 82
Scotland, 189–90
Sedin, Ivan, 13–15
Serbia, 172
Shcherbina, Boris, 34, 137, 178
Shebelinka gas field (in eastern Ukraine), 21, 98–9, 144–9, 159, 161, 163
Shell, 28, 45, 69, 71, 73, 76, 122–6, 131, 163, 169, 225
Siberia, 2, 8, 10, 31–43, 55–8, 67, 95, 99, 117, 121, 133–47, 150, 155–6, 163, 167, 171, 176–9, 183–92, 197–9, 203, 212, 217, 233–4
  and Arctic system-building, 138–43
  and Austrian-Soviet contract, 55–6
  and Brezhnev, 185–6
  and Galician gas, 95
  and Finland, 163
  and Iran, 176, 183
  and Ukranian crisis, 143–7
  and United States, 133, 171, 186, 190–2
  and Yamal pipeline, 188, 190–2
Sidorenko, Alexander, 34, 148–9
Six-Days War, 55–8
Sleipner gas field (off Norway), 201
Slochteren gas field (in the Netherlands), 28, 34, 48, 81
Slovakia, 203, 213
  see also Czechoslovakia
SMV (Soviet Mineral Oil Administration), 46–8
Social Democratic Party (SPÖ, in Austria), 54, 74, 84, 107, 120, 165, 226
socialism, 9, 14, 19, 35, 39, 68, 171–2, 187

Societé Commerciale du Methane Saharien (COMES), 28
Solidarnośc, 187
   see also Lech Walesa
Sonatrach, 34, 55, 72–3, 87–8, 90, 170, 187, 191, 225
Sopex, 172, 175
Sorokin, Alexei, 26, 31, 56–8, 81–6, 108, 111–12, 115, 117, 129
Sorsa, Kalevi, 163
Soviet-Austrian Commission for Economic Cooperation, 132
Soviet Council of Ministers, 24, 34, 41–2, 90, 95, 100–1, 138, 140, 143, 147–9, 153, 160, 182
Soviet-Finnish agreement (1971), 132
Soviet Ministry of Foreign Trade, 56, 109, 131, 133
Soviet Union, collapse of, 3, 5, 7, 10, 202–4, 207–8, 214, 220, 223, 232
Soyuz (Union) pipeline, 171, 177, 180
Soyuznefteexport, 63–5, 116, 132, 227
Spain, 35, 49, 71, 131–3, 171, 195, 200–1, 219
Sputnik satellite, 17
Stalin, Josef, 13–15
Stance, Maruice, 133
Staribacher, Josef, 163, 188
State Committee for Material-Technical Supply, *see* Gossnab
State Planning Commission, *see* Gosplan
Statfjord gas field (off Norway), 192
Stavropol, 20, 26
Stavropol-Moscow-Leningrad pipeline, 26
Steeg, Helga, 200
Steirische Ferngas, 47, 48, 50, 65
Stockholm Environmental Conference, 167
Strauss, Franz-Josef, 84, 86
Streibl, Max, 182
Styria, 47, 154, 172
   see also Steirische Ferngas
Südostdeutsche Ferngas AG, 45
Suez canal, 57
sulfur dioxide, 167
supply disruptions, 7, 21, 37, 85, 94, 117, 120, 156, 163, 166, 169, 181, 190–5, 206, 208, 221–4, 227, 232–3

surplus for export, 26–9, 35
Sverdlovsk, 23–4
Sweden, 3, 9–10, 64, 67, 89, 131–3, 171, 199, 201–2, 206, 214, 216, 218–19, 222, 226
Swissgas, 158
Switzerland, 3, 34–5, 39, 69, 72, 126–7, 131–3, 153, 158, 171, 199, 219
synthetic oil production, 69
system-building, 5–10, 14, 20–4, 27–8, 36, 43, 89–103, 136–50, 155, 197, 220, 225, 229–32, 234–5
   Arctic, 138–43
   coalitions, 5
   and Lenin, 14
   Ukrainian, 144, 147, 150, 155
   Western, 27–8

TAL, *see* Trans-Alpine Pipeline
Tallesbrunn gas field and storage facility (in Austria), 169
Tarvisio, 61, 154
Tbilisi, 22
TDOs, *see* Temporary Denial Orders
Technical Works of Stuttgart, 82
Temporary Denial Orders (TDOs), 190
TEN, *see* Trans-European Networks
Tenneco, 133
Ternopol, 98, 144
Texas Eastern Transmission, 133
Thatcher, Margaret, 189, 200
Three Mile Island, 185
Thyssen, 52–3, 55–6, 65, 79, 81–2, 86, 121
Thyssengas, 48, 69, 80–1, 130, 174
Titarenko, Alexei, 161
Tolloy, Giusto, 58
Tončić-Sorinj, Lujo, 56
Trans-Alpine Pipeline (TAL), 70
Trans-Austria pipeline, 171, 193, 207, 209, 231
Trans-Canada pipeline, 23
Trans-European Networks (TEN), 213
Trans-European Pipeline, 5, 39, 55–6, 58–9, 61, 65–6, 77, 80, 86, 105, 126, 129
Trans-Mediterranean pipeline, 187, 192–3, 202

Trans-Saharan pipeline, 200
transit through Czechoslovakia and Austria, 151–4
Trefgarne, Lord, 202
Trieste, 43, 70, 177
Troll gas field (off Norway), 201, 214
Tsarapkin, Semyon, 106, 108, 115
Tuimazinsk, 17
Turkey, 3, 172–3, 199–200, 209, 212, 234
Turkmenistan, 183, 209
Tyrol, 77
Tyumen, 31–4, 38, 40–1, 136–9, 142–3, 178–9, 197

Ufa, 20
Ukraine, 1, 5, 9–10, 13, 20–1, 36, 41, 50, 90, 95–103, 135–6, 138–40, 143–50, 159–64, 171, 178, 197, 204–5, 207–15, 217, 220, 232
  and Austrian-Soviet contract, 90, 95–103
  and Kortunov's death, 143–7
  and regional gas crisis (1969), 101
  as victim, 97–101
Ukrainian Council of Ministers, 95, 100–1, 148–9, 160
Ukhta-Torzhok pipeline, 138–9, 143
Ulbricht, Walter, 78, 83
Ulm, 157
UNECE, see United Nations Economic Committee for Europe
United Austrian Iron and Steel Works, see VÖEST
United Kingdom, 34–5, 37, 57, 62, 71, 115, 189, 192, 200–2, 214–15, 218
United Nations, and Afghanistan, 184
United Nations Economic Committee for Europe (UNECE), 26–8, 230
United States, 6, 18–19, 24–7, 37, 45, 84, 91–2, 131–4, 155, 171, 184–8, 191
  compressor technology, 25–7
  and opposition to Yamal pipeline, 184–8
  and pipe manufacturing, 24
  and political relations with the Soviet Union, 133
Urals, 22–4, 26, 32–3, 38, 40, 136, 179
Urengoi gas field (in Siberia), 133, 140, 179, 185, 198–9, 225, 231
Urengoi-Uzhgorod export pipeline, 225, 231
  see also Yamal pipeline
den Uyl, Joop, 169
Uzbekistan, 22, 23, 40, 145
Uzhgorod, 90, 100–1, 144, 146–9, 159, 161, 163, 171, 183, 198, 225, 231

Vacuum Oil, 45
Valdai-Latvia pipeline, 146–7
VDNKh, see Exhibition of Achievements of the National Economy
VGW (Verband der deutschen Gas- und Wasserwerke), 81–2
Vienna, 45, 58, 60, 63, 113–14, 153
Vienna Public Works (Wiener Stadtwerke), 47–8, 65
Vilnius, 21, 96, 146, 204
VNG, 205–6
VÖEST (United Austrian Iron and Steel Works), 51–6, 58–61, 65, 74, 78, 81, 99
Vojvodina, 172
Volchkov, Stanislav, 108, 112, 114, 119
Volga, 20–1, 31
Vonhoff, Hendrik, 191, 193
Vuktylskoe gas field (in Komi ASSR), 138, 143
Vuktylskoe-Ukhta-Torzhok pipeline, 143
Vyakhirev, Rem, 209
Vyborg, 215

Waidhaus, 114, 159
Waldheim, Kurt, 62
Walesa, Lech, 187, 212
Wandel durch Annäherung (change through rapproachment), 74
Warsaw Pact, 9, 78, 91–2, 105, 218
Wedekind, Gerhard, 110, 112, 123
Wehner, Herbert, 86
Weise, Jürgen, 117
Wiener Elektrizitätswerke, 45
Wiener Stadtwerke, see Vienna Public Works
Wilhelmshaven, 170
Wingas, 206, 212–13

Wintershall, 205–6, 211–12, 215
  see also BASF
Wodak, Walter, 61
Woratz, Gerhard, 78, 107
World War II, 13, 28–9, 69
Würzburg, 157, 165–6

Yamal pipeline, 179–95, 197–200, 217–19, 222, 229
  and compressor embargo, 188–90
  envisaging, 179–83
  and Europe's contested vulnerability, 190–5
  and United States opposition, 184–8

Yamal-Nenets national region, 137, 180, 198
Yamburg gas field (in Siberia), 180, 198
Yeltsin, Boris, 212, 223
Yerevan, 22
Yugoslavia, 39, 48–9, 52, 54–5, 65, 72–3, 87, 153, 158, 169, 171–2, 190, 193, 203, 231, 234
Yushchenko, Viktor, 215

Zeebrugge, 201
Zwerndorf gas field (in Austria and Czechoslovakia), 47, 50, 59, 93

Printed and bound in Great Britain by
CPI Antony Rowe, Chippenham and Eastbourne